MULTIMEDIA NETWORKS

MULTIMEDIA NETWORKS

PROTOCOLS, DESIGN, AND APPLICATIONS

Hans W. Barz

Gregory A. Bassett

Library of Congress Cataloging-in-Publication Data applied for.

ISBN: 9781119090137

A catalogue record for this book is available from the British Library.

Set in 10/12pt, TimesLTStd by SPi Global, Chennai, India.

1 2016

Contents

Preface

It is obvious to almost any consumer with a PC, Smartphone, radio, TV, or even a "normal" telephone that the methods of communicating, listening to, or viewing content have changed radically over the past years. Napster, iPhone, Skype, Netflix, Android, Internet Radio, YouTube, VoIP, streaming programs from TV networks (the list is almost endless) have all profoundly changed the way we talk, listen, and watch.

What sets multimedia delivery somewhat apart from other technology topics is its unusually high level of complexity. Different user platforms, different network media, different types of content, and of course, many different vendors and standards all make this topic a challenging one.

The motivation for this book is based on the many new multimedia technologies that have appeared or changed in the past 15 years, the immense number of new standards, and the lack of overarching books on the subject. At the time of writing, there is no book available that covers the range of topics you find here, which are needed to cope with multimedia usage from a network point of view. It goes without saying that there are many excellent books on the individual subjects covered in this book. In fact, we cite many of them as references. While they cover specific topics in detail, this book provides the "big picture" and ties together the various principles for the entire multimedia space.

Another driver for this book is the authors' involvement in deploying multimedia applications and networks in large multinational companies. These deployments and operational involvement provided insight into what is important and what can safely be ignored. Many new multimedia technologies went live during our working time in networks, and the reader will profit from the combined practical experience and technological background of the authors.

This book focuses primarily on principles but sometimes dives into the details to show how the principles really work in practical situations. As a prerequisite for reading the book, you need basic knowledge of networks, but no additional literature is required. Since this book is aimed to give an overview in a rather active field, we provide the reader with extensive references – over 600 recent literature links – to dive deeper. In addition, the detailed index and the abbreviation list will help you easily find the treatment of a specific subject.

Acknowledgments

First and foremost, we must thank our spouses and families. The writing of a reference book is a long and difficult task. Without their patience and support over many months, including during holidays, we could not have completed this book.

We also thank Professor Bernhard Plattner from ETH Zürich (Switzerland) for his support of a lecture by Hans Barz from 2010 to 2014 on this subject at the ETH.

Finally, and equally important, we thank our former colleagues from the networking group at Hoffmann-La Roche in Basel, Switzerland (PGIN), for working together with us to create and implement many of the multimedia solutions described here in the book. Our special thanks to Andre Rudin for his expert review of our material, Andy Fringeli for world-class operational insight, and Brian O'Connor for all things related to voice. There are other members of the PGIN team too numerous to mention, but they know who they are.

About the Authors

Hans W. Barz

Hans Barz received a Master's Degree in Economical Engineering from the Technical University in Karlsruhe, Germany (Prof. H. Maurer) and a PhD in Computer Science from University Bonn, Germany (Prof. H. Zima) with a theme on the power of synchronization mechanisms. He worked for over 25 years with the IT staff at the headquarters of Ciba-Geigy, Novartis, Syngenta, and Hoffmann-La Roche in Basel. For many years, he held the position of Global Head of Architecture. At that time, he was responsible for major rollouts of new technologies (TCP/IP, X.400/X.500 backbone, E-Business-Infrastructure), technology trials (DECnet Phase V, OSI, PKI), and infrastructure consolidations in mergers and split-offs. In Hoffmann-La Roche, he was, for many years, Global Head of Network Services with engineering and operation responsibility for networks, telephony, and multimedia services in more than 80 countries.

He has lectured on telecommunication subjects at the University Freiburg (Germany), University Basel (Switzerland), and ETH Zürich (Switzerland). He did publish a number of articles on parallel processing and a book on telecommunication in 1994 (Hanser Verlag, in German).

Gregory A. Bassett

Greg Bassett received a Bachelor's Degree in Computer Engineering from Syracuse University, in Syracuse, NY, USA. The first 20 years of his career was spent working for Digital Equipment Corporation is a wide variety of roles including instructor (including teaching the first DECnet course offered), software engineer, technical leader for networking and personal computers, and program management.

For the next 18 years, he worked for the pharmaceutical companies of Syntex in Palo Alto, CA, USA, and Hoffmann-La Roche in Basel, Switzerland. He spent 8 years as a network engineer. For the remaining 10 years, he was an Enterprise Architect responsible for networking including data, fixed and mobile voice, video, and call center. He earned the Cisco Certified Internetworking Engineer (CCIE) in Routing and Switching certification in 2002.

During his years as an Enterprise Architect for Roche, he worked closely with Hans Barz to develop networking strategies and implementation plans.

Abbreviations

This list contains more than 1500 abbreviations. Many abbreviations are in bold. Those abbreviations in bold are handled in this book and mentioned in the index. When searching abbreviations in the index, it may be helpful first to expand the abbreviation in this list and then look it up in the index.

3DTV:	3-Dimensional TV
3GPP:	3rd-Generation Partnership Project
AAA:	Authentication, Authorization, and Accounting
AAC:	Advanced Audio Coding
AAL:	ATM Adaption Layer
AAM:	Audio Activity Metric
AAR:	AA Request
ABNF:	Augmented Backus – Naur Form
ABR:	Available Bit Rate
AC:	Alternating Current
ACA:	Accounting Answer
ACD:	Automatic Call Distribution
ACE:	Advanced Coding Efficiency
ACELP:	Algebraic Code Excited Linear Prediction
ACF:	Admission ConFirm
ACK:	ACKnowledge
ACM:	Address Complete Message
ACR:	Automatic Call Return/Accounting Request
ACS:	Auto Configuration Server
AD:	Active Directory (Microsoft)
A/D:	Analog-to-Digital
ADC:	Analog-to-Digital Converter
ADPCM:	Adaptive Differential Pulse Code Modulation
ADSL:	Asymmetric Digital Subscriber Line
ADTS:	Audio Data Transport Stream
ADU:	Application Data Unit
AEC:	Adaptive Error Concealment/Acoustic Echo Canceller
AES:	Audio Engineering Society, Advanced Encryption Standard
AF:	Adaption Field/Assured Forwarding

AFC:	Automatic Frequency Control
AFX:	MPEG-4 Animation Framework eXtension
AGCF:	Access Gateway Control Function
A-GNSS:	Assisted Global Navigation Satellite System
AGW:	Access GateWay
AHS:	Adaptive HTTP Streaming
AHT:	Average Holding Time
AIT:	Application Information Table
AKA:	Authentication and Key Agreement
ALC:	Asynchronous Layered Coding
ALF:	Application Layer Framing/Adaptive Loop Filtering
AL-FEC:	Application Layer-FEC
ALG:	Application Layer Gateway
ALI:	Automated Location Identifier
ALL:	Application Layer Logic
ALM:	Application Layer Multicast
ALS:	Audio Lossless Coding
ALTO:	Application Layer Traffic Optimization
AM:	Amplitude Modulation
AMF:	Action Message Format
A-MGF:	Access Media Gateway Function
AMI:	Alternate Mark Inversion
AMP:	Adaptive Media Play-out/Access Network Provider
AMR:	Adaptive Multirate
ANDSF:	Access Network Discovery and Selection
ANI:	Automatic Number Identification
ANSI:	American National Standard Institute
AOC:	Advice Of Charge
AOI:	Area Of Interest
AOR:	Address Of Record
AP:	Authentication Proxy
APC:	Adaptive Predictive Coding
APE:	Application Protocol Entities
API:	Application Programming Interface
APN:	Access Point Name
APT:	Actual Presentation Timestamp
AQM:	Active Queue Management
ARED:	Adaptive Random Early Detection
ARF:	Access Relay Function
ARIB:	Association of Radio Industries and Business
ARJ:	Admission Reject
ARP:	Allocation and Retention Priority
ARQ:	Automated Repeat-reQuest/Admission Request
ARTS:	Advanced Real Time Simple profile
AS:	Application Server/Application-Specific maximum bandwidth/Application Sharing

ASA:	Abort Session Answer
ASF:	Advanced Streaming Format
AS-FE:	Application Support Functional Entity
ASI:	Asynchronous Serial Interface
ASM:	Any Source Multicast
ASN.1:	Abstract Syntax Notation 1
ASO:	Arbitrarily Slice Ordering
ASP:	Application Service Provider/Application Server Processes
ASR:	Abort Session Request/Automatic Speech Recognition
ATCF:	Access Transfer Control Function
ATGW:	Access Transfer GateWay
ATIS:	Alliance for Telecommunication Industry Solutions
ATM:	Asynchronous Transfer Mode
ATS:	Adaptive Transport Stream
ATSC:	Advanced Television Systems Committee
AT&T:	American Telephone & Telegraph company
AU:	Access Unit
AUC:	Authentication Center
AUCX:	AUdit Connection
AUEP:	Audit EndPoint
AUID:	Application Usage Identification
AUR:	Address of Record
AUTN:	AUTHentificatioN Token
AutoIP:	Automatic IP addresses
AV:	Audio-Visual/Audio Video
A/V:	Audio/Video
AVC:	Advanced/Audio Video Codec
AVI:	Audio Video Interleave
AVO:	Audiovisual Object
AVP:	Audio Video Profile/Audio Visual Profile
AVPF:	Audio Video Profile Feedback
AVS:	Audio Video coding Standard
AVT:	Audio Video Transport
B2BUA:	Back-to-Back User Agent
BAP:	Body Animation Parameters
BAT:	Bouquet Association Table
BBC:	British Broadcasting Service
BBF:	BroadBand Forum
BC:	BroadCast
BCF:	Bandwidth Confirm Message
BCG:	Broadband Content Guide
BCH:	Bose-Chaudhuri – Hocquenghem code
BCP:	Best Current Practice
BD:	Blue-ray Disk
BDP:	Body Definition Parameters
BE:	Best Effort

BER:	Basic Encoding Rules
BFCP:	Binary Floor Control Protocol
BGCF:	Breakout Gateway Control Function
BGD:	Broadband Gateway Device
BGMP:	Border Gateway Multicast Protocol
BGP:	Border Gateway Protocol
BHCA:	Busy Hour Call Attempts
BHT:	Busy Hour Traffic
BICC:	Bearer Independent Call Control
BIFS:	Binary Format for Scenes
BiM:	Binary MPEG Format for XML
BIOP:	Broadcast Inter ORB Protocol
BISDN:	Broadband ISDN
BLF:	Busy Lamp Field
BLP:	Bitmask Lost Packet
BMA:	Boundary Matching Algorithm
BMFF:	Base Media File Format
BMP:	BitMaP
BNF:	Backus-Naur Form
BNG:	Broadband Network Gateway
bpp:	bits per pixel
BRA:	Basic Rate Access
BRAS:	Broadband Remote Access Server
BRI:	Basic Rate Interface
BRJ:	Bandwidth Reject Message
BRQ:	Bandwidth Request
BSAC:	Bit-Sliced Arithmetic Coding
BSP:	Broadband Service Provider
BSR:	BootStrap Router
BSS:	Base Station System
BT:	British Telecom
BTS:	Base Transceiver Station
CA:	Conditional Access/Civic Address
CABAC:	Content-Adaptive Binary Arithmetic Coding
CAC:	Connection/Call Admission Control
CAF:	Control of Agent Features
CAN:	Connectivity Access Network
CAP:	Competitive Access Provider
CAPI:	Common ISDN Application Programming Interface
CARP:	Cache Array Routing Protocol
CAS:	Channel Associated Signaling/Conditional Access System
CAT:	Conditional Access Table
CAT:	Customized Alerting Tone
CATV:	Community Access Television/Community Antenna Television/CAble TeleVision
CAVLC:	Content-Adaptive Variable Length Coding

CB:	Coding Block/Communication Barring
C-BGF:	Core Border Gateway Function
CBMS:	Convergence of Broadcast and Mobile Services
CBR:	Constant Bit Rate
CBS:	Committed Burst Size
CBT:	Center-Based Tree/Core-Based Tree
CCBS:	Call Completion to Busy Subscriber
CCBS:	Completion of Communication to Busy Subscribers
CCF:	Common File Format/Charging Collection Function
CCI:	Congestion Control Identifier
CCIR:	Consultative Committee International Radio (now ITU-R)
CCITT:	Consultative Committee International Telecommunication Union (now ITU-T)
CCM:	Codex Control Messages
CCMP:	Centralized Conference Manipulation Control
CCN:	Content-Centric Networking
CCNR:	Completion of Communication on No Reply
CCS:	Common Channel Signaling
CCSA:	China Communication Standards Association
CCXML:	CallControl XML
CD:	Communication Deflection/Committee Draft
CDB:	Coded Data Buffer
CDF:	Content Delivery Function/Charging Data Function
CD&LCF:	Content Distribution and Location Control Function
CDMA:	Code Division Multiple Access
CDN:	Content Delivery Network
CDNI:	Content Delivery Network Interconnection
CDP:	Cisco Discovery Protocol
CDR:	Call Detail Record/Charging Data Record/Committed Data Rate/Common Data Representation
CDS:	Content Download Service/Content Directory Service
CD&SF:	Content Delivery and Storage Function
CDV:	Cell Delay Variation
CE:	Congestion Experience/Core Experiment
CEC:	Commission of the European Communities
CEI:	Commission Electrotechnique Internationale (in English IEC)
CELP:	Code(-book) Excited Linear Prediction
CEN:	Comité Européen de NormaliSation
CENC:	Content ENCoding/Common ENCryption
CENELEC:	Comite Europeen de NormaliSation Electrotechnique
CEPT:	European Conference of Postal and Telecommunications Administrations
CER:	Confidential Encoding Rules
CFB:	Communication Forwarding Busy
CFF:	Common File Format
CFNL:	Communication Forwarding on Not Logged in
CFNR:	Communication Forwarding No Reply

CFU:	Communication Forwarding Unconditional
CGF:	Charging Gateway Function
CGI:	Cell Global Identification/Common Gateway Interface
CGS:	Coarse Granularity Scalability
C/I:	Carrier to Interference ratio
CI:	Common Interface/Content Identifier
CIC:	Circuit Identification Code
CID:	Context Identifier
CIF:	Common Image Format/Common Intermediate Format/Common Interface Format
CII:	Content Identification and other Information
CIPA:	Camera & Imaging Products Association
CIPID:	Contact Information for the Presence Information Data
CIR:	Cyclic Intra Refresh
CK:	Ciphering Key
CLC:	Channel CLose
CLEC:	Competitive Local Exchange Carrier
CLI:	Call Line Identification
CLIP:	Call Line Identification Presentation
CLUE:	ControLling mUltiple streams for tElepresence
CLUT:	Color LookUp Table
CM:	Connection Manager/Content Marker
CMD:	Carousel Multicast Download
CMS:	Content Management System
CMTS:	Cable Modem Termination System
CN:	Core Network
CNAME:	CaNonical Name
CNCGF:	Customer Network Gateway Configuration Function
CND:	Customer Network Device
CNG:	Comfort Noise Generation/Customer Network Gateway
CO:	Central Office
CoD:	Content on Demand
CONF:	CONFerence calling
CORBA:	Common Object Request Broker Architecture
CoS:	Class of Service
CP:	Content Protection/Customer Premises/Control Point
CPB:	Coded Picture Buffer
CPC:	Continuous Packet Connectivity
CPCM:	Content Protection and Copy Management
CPDT:	Cascaded Pixel-Domain Transcoders
CPE:	Customer Premise Equipment
CPF:	Content Provider Function
CPG:	Call ProGress
CPI:	Common Interface Plus
CPL:	Call Processing Language
CPN:	Customer Premise Network

CPVR:	Client Personal Video Recorder
CQ:	Custom Queuing
CR:	Carriage Return
CR:	Content Recommendation
CR LF:	Carriage Return Line Feed
CRC:	Cyclic Redundancy Check
CRID:	Content Reference IDentifier
CR-LSP:	Constraint-Based LSP
CRM:	Customer Relationship Management
CRS:	Customized Ringing Signal/Content Recommendation Service
CRT:	Carriage ReTurn
CRTP:	Compressed RTP
CS:	Circuit Switched/Communication Session/Companion Screen
CSA:	Companion Screen Application, Common Scrambling Algorithm
CSCF:	Call State Control Function
CS-DON:	Cross Sessions DON
CSF:	Content Security Function
CSFB:	Circuit Switch FallBack
CS-MGW:	Circuit-Switched Mobile GateWay
CSMO:	Circuit Switched Mobile Originating
CSP:	Content Service Provider
CSQ:	Contact Service Queues
CSRC:	Contributing SouRCe
CSS:	Companion Screen and Supplementary streams/Cascaded Style Sheet
CSV:	Capture Scene View
CT:	Coding Tree/Composition Time
CTB:	Coding Tree Block
CTF:	Charging Trigger Function
CTI:	Computer Telephony Integration
CTU:	Coding Tree Unit
CU:	Coding Unit
CUG:	Closed User Group
CU-RTC-Web:	Customizable, Ubiquitous Real Time Communication over the Web
CVBS:	Color-Video Blanking System
CVT:	Commercial Video Player Format
CW:	Communication Waiting
CWMP:	CPE WAN Management Protocol
CWT:	Continuous Wavelet Transform
D2D:	Device-to-Device
DA:	Directory Agent
DAAP:	Digest Access Authentication/Digital Audio Access Protocol
DAB:	Digital Audio Broadcasting
DAC:	Digital-to-Analog Converter
DAI:	DMIF Application Interface
DASH:	Dynamic Adaptive Streaming over HTTP
DAVIC:	Digital Audio VIsual Council

dB:	DeciBel
DBC:	Dictionary-Based Coding
DBE:	Data path Border Element
dBOV:	dB OVerload
DC:	Direct Current
DCEP:	Data Channel Establishment Protocol
DCF:	Disengage ConFirm/DRM Content Format
DCH:	Dedicated CHannel
DCIA:	Distributed Computing Industry Association
DCP:	Device Control Protocol
DCT:	Discrete Cosine Transform
DDB:	Downloadable Data Block
DDC:	Device Discovery and Control
DDD:	Device Description Document
DDL:	Description Definition Language
DECE:	Digital Entertainment Consortium Ecosystem
DECT:	Digital Enhanced Cordless Telecommunications
DER:	Distinguished Encoding Rules
DF:	Digital Fountains/Do not Fragment
DFCA:	Dynamic Frequency and Channel Allocation
DFT:	Delay Factor/Discrete Fourier Transform
DFXP:	Distribution Format eXchange Profile
DHCP:	Dynamic Host Configuration Protocol
DHT:	Distributed Hash Table
DIAL:	DIscovery And Launch
DIBR:	Depth Image-Based Rendering
DID:	Direct Inward Dialing
DIDL:	Digital Item Declaration Language
DiffServ:	Differential Services
DII:	Downloadable Indication Message
DL:	DownLink
DLCX:	DeLete Connection
DLNA:	Digital Living Network Alliance
DLSR:	Delay Last Sender Report
DM:	Dense Mode
DMAP:	Digital Media Access Protocol
DMC:	Digital Media Controller
DMIF:	Delivery Multimedia Integration Framework
DMP:	Digital Media Player
DMPr:	Digital Media Printer
DMR:	Digital Media Renderer
DMS:	Digital Media Server
DND:	Do Not Disturb
DNG:	Delivery Network Gateway
DNI:	Digital Number Identification/DMIF Network Interface
DNS:	Domain Name Service

DNS-SD:	DNS Service Discovery
DOCSIS:	Data Over Cable Service Interface Specifications
DON:	Decoding Order Number
DOP:	Default Outbound Proxy
DoS:	Denial of Service
DP:	Data Partitioning/Discovery Point
DPAP:	Digital Photo Access Protocol
DPB:	Decoded Picture Buffer
DPCCH:	Dedicated Physical Control CHannel
DPCM:	Differential Pulse Code Modulation
DPDCH:	Dedicated Physical Data CHannel
DR:	Designated Router
DRB:	Data Radio Bearer
DRC:	Dynamic Resolution Conversion
DRJ:	Disengage ReJect
DRM:	Digital Rights Management/Digital Radio Mondiale
DRQ:	Disengage ReQuest
DRVCC:	Dual Radio Voice Call Continuity
DS:	Differentiated Services/Digital Signal/DMIF Signaling
DSCP:	Differentiated Services Code Points
DSG:	DOCSIS Set-Top box Gateway
DSI:	Downloadable Server Initiate
DSL:	Digital Subscriber Line
DSLAM:	Digital Subscriber Line Access Multiplexer
DSM-CC:	Digital Storage Media Command and Control
DSP:	Digital Signal Processor
DSPD:	Digital Still Storage Devices
DSS:	Digital Subscriber Signaling
DST:	Discrete Sine Transform
DTA:	Digital Terminal Adapters
DTAP:	Direct Transfer Application Part
DTCP:	Digital Transmission Content Protection
DTD:	Document Type Declaration
DTH:	Direct To Home
DTLS:	Datagram Transport Layer Security
DTLS-SRTP:	Datagram Transport Layer Security – Secure RTP
DTM:	Dual Transfer Mode
DTMF:	Dual-Tone MultiFrequency
DTN:	Delay Tolerant Networks
DTS:	Decoding Time Stamp
DTT:	Digital Terrestrial Television
DTV:	Digital TeleVision
DTVB:	Digital Television Video Broadcasting
DTX:	Discontinuous Transmission
DVB:	Digital Video Broadcasting
DVB-C:	Digital Video Broadcasting over Cable

DVB-H:	Digital Video Broadcasting for Handhelds
DVB-RC:	Digital Video Broadcasting Return Channel
DVB-S:	Digital Video Broadcasting over Satellite
DVB-SH:	Digital Video Broadcasting – Satellite services to Handhelds
DVBSTP:	Digital Video Broadcast SD&S Transport Protocol
DVB-T:	Digital Video Broadcasting Terrestrial
DVC:	Distributed Video Coding
DVD:	Digital Versatile Disk
DVI:	Digital Visual Interface
DVMRP:	Distance Vector Multicast Routing Protocol
DVR:	Digital Video Recorder, Digital Video Resolution
DWDM:	Dense Wavelength Division Multiplexing
DWT:	Discrete Wavelet Transform
EATF:	Emergency Access Transfer Function
EAV:	End-of-Active Video
EB:	Encoding Block
EBCOT:	Embedded Block Coding with Optimized Truncation
EBS:	Excess Burst Size
EBU:	European Broadcasting Union (UER)
EC:	Echo Cancellation
ECF:	Event Charging Function/Elementary Control Function
ECM:	Entitlement Control Message/Evolved packet system Connection Management
ECMA:	European Computer Manufacturers Association
ECN:	Explicit Congestion Notification
ECRTP:	Enhanced CRTP
E-CSCF:	Emergency CSCF
ECT:	Explicit Communication Transfer
E-DCH:	Enhanced Dedicated Channel
EDD:	Error Detection Delay
EDF:	Event Charging Function
EDGE:	Enhanced Data rates for Global Evolution
EDTV:	Enhanced-Definition TeleVision
EF:	Expedited Forwarding/Elementary Function
EFC:	Explicit Forward Congestion
EFCI:	Explicit Forward Congestion notification Indication
EFF:	Elementary Forwarding Function
EFR:	Enhanced Full Rate
eHRPD:	Evolved High Rate Packet Data
EIT:	Event Information Table
EKT:	Encrypted Key Transport
ELIN:	Emergency Location Identification Number
eMBMS:	evolved MBMS
EMM:	Entitlement Management Message /Evolved packet system Mobility Management
EMMA:	Extensible MultiModal Annotation

eMSS:	Element Management System
eMSS:	Enhanced Mobile Switching center Server
EN:	European Norm
eNB:	Evolved Node B
ENUM:	E.164 NUmber Mapping
EP:	Error Protection
EPA:	Event Publication Agent
EPC:	Enhanced Packet Core
EPCF:	EndPoint ConFiguration
ePDG:	Evolved Packet Data Gateway
EPG:	Electronic Program Guide
EPON:	Ethernet PON
EPS:	Enhanced Packet System
EPT:	Earliest Presentation Time
ER:	Explicit Route/Error Resilient
EREC:	Error Resilient Entropy Encoding
ERO:	European Radiocommunication Office (part of CEPT)
ES:	Encoding Symbol/Elementary Stream
ESC:	Event State Compositor
ESCR:	Elementary Stream Clock Reference
ESG:	Electronic Service Guide
ESI:	Encoding Symbol Identifier
E-SMLC:	Evolved Serving Mobile Location Center
ESP:	Enhanced Service Provider
EtE:	End-to-End
ETM:	Excess Traffic Marker
ETS:	European Telecommunication Standard
ETSI:	European Telecommunication Standards Institute
E-UTRAN:	Evolved Universal Terrestrial Radio Access
EV-DO:	EVolution Data Only
EVDO:	EVolution Data Optimized
EXIF:	EXchangeable Image File
EZW:	Embedded Zero-tree Wavelength
FA:	Flexible Alerting
FAP:	Facial Animation Parameters
FB:	Functional Block/FeedBack
FBA:	Facial and Body Animation
FC:	Floor Chair
FCC:	Fast Channel Change/Federal Communication Commission
FCD:	Final Committee Draft
FCI:	Feedback Control Information
FCS:	Floor Control Server/Frame Check Sequence
FDIS:	Final Draft International Standard
FDM:	Frequency-Division Multiple
FDMA:	Frequency-Division Multiple Access
FDP:	Facial Definition Parameters

FDT:	File Delivery Table
FE:	Functional Entity/Fast Ethernet
FEC:	Forward Error Correction/Forwarding Equivalence Class
FECC:	Far-End Camera Control
FF:	Forward Feedback/Fast Forward/Feed Forward
FFS:	For Further Study
FGI:	Feature Group Indicator
FGS:	Fine Grained Scalability
FIR:	Finite Impulse Response/Full Intra Request
FLUTE:	File deLivery over UnidirecTional sErvice
FLV:	FLash Video
FM:	Frequency Modulation
FMFM:	Find Me Follow Me
FMO:	Flexible Macroblock Ordering
FMS:	Flash Media Server
FMT:	Feedback Message Type
FoIP:	Fax over IP
FoR:	Follow on Request
FP:	Floor Participant
FPS:	Frames Per Second
FQDN:	Fully Qualified Domain Name
FR:	Full Rate
FSK:	Frequency Shift Keying
FSM:	Finite State Machine
FTI:	File Transfer Information
FTP:	File Transfer Protocol
FTTB:	Fiber To The Building
FTTC:	Fiber To The Cabinet/Fiber To The Curb
FTTH:	Fiber To The Home
FTTN:	Fiber To The Node
FTTP:	Fiber To The Premises
FU:	Fragmentation Unit
FUS:	Firmware Update System
FUSS:	FUS Stub
FVV:	Free Viewpoint Video
GAA:	Generic Authentication Architecture
GAN:	Generic Access Network
GARP:	Generic Attribute Registration Protocol
GBR:	Guaranteed Bit Rate
GBS:	Generic data Broadcasting & Service information protocols
GCC:	Generic Conference Control
GCF:	Gatekeeper Confirm
GDR:	Gradual Decoder Refresh
GEM:	Globally Executable MHP Support
GENA:	General Event Notification Architecture
GERAN:	Global system for mobile communication/Edge Radio Access Network

GGSN:	Gateway GPRS Support Node
GHRD:	Generalized Hypothetical Reference Decoder
GIF:	Graphics Interchange Format
GLBP:	Gateway Load Balancing Protocol
GMC:	Global Motion Compensation
GMLC:	Gateway Mobile Location Center
GMSC:	Gateway Mobile Switching Center
GMSK:	Gaussian Minimum Shift Key
GMT:	Greenwich Mean Time
GOB:	Group of Blocks
GOP:	Group of Pictures
GOS:	Grade Of Service
GOV:	Group of Video Objects
GPON:	Gigabit PON
GPRS:	General Packet Radio Service
GPS:	Global Positioning System
GRE:	Generic Routing Encapsulation
GRJ:	Gatekeeper Reject
GRQ:	Gatekeeper Discovery
GRUU:	Globally Routable UA URI's
GSE:	Generic Stream Encapsulation
GSI:	Global Standards Initiative
GSM:	Global System for Mobile communication
GSMA:	GSM Association
GSQ:	Grade Of Service
GSTN:	General Switched Telephone Network
GTP:	Global Title
GTP:	GPRS Tunneling Protocol
GVRP:	GARP VLAN Registration Protocol
GWF:	Gateway Function
HbbTV:	Hybrid Broadcast Broadband TV
HC:	Header Compression
HCR:	Huffman Codeword Reordering
HD:	High Definition
HDLC:	High-Level Data Link Control
HDMI:	High-Definition Multimedia Interface
HDS:	HTTP Dynamic Streaming
HDTV:	High Definition TV
HE:	HeadEnd
HEC:	Header Extension Code
HEL:	Header Extension Length
HET:	Header Extension Type
HEVC:	High Efficiency Video Coding
HFC:	Hybrid Fiber Coax
HG:	Home Gateway
HGI:	Home Gateway Initiative

HHI:	Heinrich Hertz Institute
HID:	Home Infrastructures Devices
HILN:	Harmonic Individual Line and Noise
HLR:	Home Location Register
HLS:	HTTP Live Streaming
HN:	Home Network
HNAP:	Home Network Administration Protocol
HND:	Home Network Device
HN-DP:	HN Discovery Point
HNED:	Home Network End Device
HNID:	Home Network Infrastructure Device
HNN:	Home Network Node
HNS:	Home Network Segment
HN-SP:	HN Streaming Point
HNT:	Hosted NAT Traversal
HPEL:	Half PixEL
HRD:	Hypothetical Reference Decoder
HRPD:	High Rate Data Packet
HS:	Hypertext Streaming
HSPA+:	evolved High Speed Packet Access
HSPA:	High Speed Packet Access
HSPDA:	High SPeed Downlink Access
HSRP:	Host Standby Router Protocol
HSS:	Home Subscriber Server
HTC:	Headend Time Clock
HTCP:	Hyper Text Caching Protocol
HTML:	Hyper Text Markup Language
HTML5:	HTML 5
HTTP:	Hyper Text Transfer Protocol
HTTPMU:	HTTP Multicast over UDP
HTTPS:	HTTP Secure
HTTPU:	HTTP Unicast over UDP
HVC:	High-Performance Video Coding
HVXC:	Harmonic Vector eXcitation Coding
I3A:	International Imaging Industry Association
IAB:	Internet Architecture Board
IACK:	InfoRequest Ack
IAD:	Integrated Access Device
IAM:	Initial Address Message
IANA:	Internet Assigned Numbers Authority
IAX:	Inter-Asterisk Exchange Protocol
IBCF:	Interconnection Border Control Function
ICANN:	Internet Corporation for Assigned Names and Numbers
ICE:	Interactive Connectivity Establishment
ICID:	IMS Charging IDentifier
ICM:	Incoming Call Management

ICMP:	Internet Control Message Protocol
ICP:	Internet Caching Protocol
ICS:	Internet protocol multimedia subsystem Centralized Services
I-CSCF:	Interrogating Call Session Control Function
ICSI:	ICS Identification
ICT:	Information and Communication Technology
ID:	Identifier
IDCT:	Inverse Discrete Cosine Transform
IDFT:	Inverse Discrete Fourier Transform
IDL:	Interface Definition Language
IDMS:	Inter-Destination Media Synchronization
IDR:	Integrated Decoder Receiver/Instantaneous Decoding Refresh
IE:	Information Element
IEC:	International Electrotechnical Commission
IEEE:	Institute of Electrical and Electronics Engineers
I-ENUM:	Infrastructure ENUM
IESG:	Internet Engineering Steering Group
IETF:	Internet Engineering Task Force
IF:	Intermediate Frequency
IGD:	Internet Gateway Device protocol
IGMP:	Internet Group Membership Protocol
IGRP:	Internet Gateway Routing Protocol
IGRS:	Intelligent Grouping and Resource Sharing
IIF:	IPTV Interoperability Forum
IIOP:	Internet Inter-ORB Protocol
IIP:	Internet Imaging Protocol
IIS:	Internet Information Services
ILBC:	Internet Low Bit Rate Codec
ILEC:	Incumbent Local Exchange Carrier
IM:	Instant Messaging
IMA:	IMA
IMEI:	International Mobile Equipment Identity
IMG:	Internet Media Guides
IMPI:	IMS Private User Identity
IMPS:	Instant Messaging and Presence Services
IMPU:	Internet protocol Multimedia subsystem Public User identity
IMS:	Internet Protocol Multimedia Subsystem/3G IP Multimedia System
IMS-GW:	IMS Media GateWay
IMSI:	International Mobile Subscriber Identifier
IM-SSF:	Internet protocol Multimedia subsystem Service Switching Function
IMTC:	International Multimedia Telecommunication Consortium
IN:	Intelligent Network
INAK:	Info request AcK
INAP:	Intelligent Network Application Part
I-NNI:	Interconnect NNI
IOI:	Inter-Operator Identifier

IOR:	Interoperable Object Identifier
IP:	Internet Protocol/Intellectual Property
IPDC:	Internet Protocol DataCast
IPDV:	Inter Packet Delay Variation
IPHC:	IP Header Compression
IPI:	IP Infrastructure/IP Interface
IPMP:	Intellectual Property Management and Protection
IP-PBX:	IP Private Branch Exchange
IPPM:	IP Performance Metrics
IPR:	Intellectual Property Rights
IPSEC:	IP protocol SECurity
IPT:	IP Telephony
IPTC:	International Press Telecommunication Council
IPTV:	Internet Protocol Television
IPv4:	IP Version 4
IPv6:	IP Version 6
IPX:	Internetwork Packet Exchange
IRC:	Internet Relay Chat
IRD:	Integrated Receiver/Decoder
IRQ:	Information ReQuest
IRR:	InfoRmation Response
IRSG:	Internet Research Steering Group
IRT:	"Institut für Rundfunktechnik"
IRTF:	Internet Research Task Force
IS:	International Standard
ISC:	Internet System Consortium/IMS Service Control
ISDN:	Integrated Services Digital Network
ISI:	InterSymbol Interference
ISIM:	IP multimedia Service Identity Module
ISMA:	Internet Streaming Media Alliance
ISN:	Initial Sequence Number
ISO:	International Standard Organization
ISOBMFF:	ISO Base Media File Format
ISP:	Internet Service Provider
ISR:	Idle mode Signal Reduction
ISUP:	ISDN User Part
ITF:	Internet Protocol Television Terminal Functions
ITOT:	ISO transport services on top of the TCP
ITU:	International Telecommunication Unit (UIT, former CCITT)
ITU-R:	ITU for Radio
ITU-T:	ITU for Telecommunication
IUA:	ISDN Q.921-User Adaption Layer
IVR:	Interactive Voice Response
IW:	Interworking
JAIN:	Java APIs for Integrated Networks
JBIG:	Joint Bi-level Image processing Group

JFIF:	JPEG File Interchange Format
JMF:	Java Media Framework
JMVM:	Joint Multiview Video Model
JPEG:	Joint Photographic Experts Group
JPEG-LS:	JPEG LossLess
JS:	JavaScript
JSEP:	JavaScript Session Establishment Protocol
JSON:	JavaScript Object Notation
JSVM:	Joint Scalable Video Model
JTC:	Joint Technical Committee
JVT:	Joint Video Team (ITU, ISO)
KBR:	Key Based Routing
KLT:	Karhunen–Loéve Transform
KMM:	Key Management Message
KMS:	Key Management Message
KSM:	Key Stream Message
KTA:	Key Technology Area
LAI:	Location Area Identity
LAN:	Local Area Network
LAPD:	Link Access Procedure D-Channel
LAR:	Log Area Ratio
LASeR:	Lightweight Application Scene Representation
LATM:	Low-overhead MPEG-4 Audio Transport Multiplex
LC:	Low Complexity
LCD:	Liquid Crystal Display
LCF:	Location ConFirm
LCR:	Least Cost Routing
LCS:	Location Control Services
LCT:	Layered Coding Transport
LD:	Long Distance/Low Delay
LDAP:	Lightweight Directory Access Protocol
LDP:	Label Distribution Protocol
LDPC:	Low-Density Parity Checking Code
LDTV:	Low Definition TeleVision
LEC:	Local Exchange Carrier
LER:	Label Edge Router
LF:	Line Feed
LFE:	Low Frequency Enhancement
LIA:	Location-Info-Answer
LIME:	Lightweight Interactive Multimedia framework for IPTV services
LIR:	Location-Info-Request
LLC:	Logical Link Control
LLDP:	Link Layer Discovery Protocol
LLDP-MED:	Link Layer Discovery Protocol Media Endpoint Discovery
LLMNR:	Link-Local Multicast Name Resolution
LMB:	Live Media Broadcast

LMDS:	Local Multipoint Distribution Service
LOAS:	Low Overhead Audio Stream
LOD:	Level Of Detail
LOF:	Loss of Frame
LPC:	Linear Predictive Coding
LPCM:	Linear Pulse Code Modulation
LPT:	Latest Presentation Time
LRF:	Location Retrieval Function
LRJ:	Location ReJect
LRQ:	Location ReQuest
LSB:	Least Significant Bit
LSF:	Line Spectral Frequencies
LSP:	Label Switched Path/Line Spectral Pairs
LSR:	Last Sender Report/Label Switching Router
LSSU:	Link Status Signal Unit
LTC:	Linear Time Code
LTE:	Long-Term Evolution
LTP:	Long-Term Prediction
LTRF:	Long-Term Reference Frame
LTRP:	Long-Term Reference Picture
LUF:	Look-Up Function
LVDS:	Low Voltage Differential Signaling
LZ77:	Lempel-Ziv 1977
LZW:	Lempel-Ziv-Welch
M2PA:	MTP2 Peer-to-Peer Adaptation layer
M2UA:	MTP2 User Adaption layer
M3UA:	MTP3 User Adaption layer
MAA:	Multimedia-Auth-Answer
MAC:	Move, Add, and Change/Medium Access Control/Multiplexed Analog Component
MAD:	Minimal Absolute Difference
MANE:	Media Aware Network Element
MAP:	Multicast Adaption Protocol/Mobile Application Part
MAR:	Multimedia-Auth-Request
MAWG:	Media Annotation Working Group
MB:	Macro Block
MBAFF:	Macro Block Adaptive Frame/Field
MBAmap:	MacroBlock Allocation map
MBFT:	Multipoint Binary File Transfer
MBM:	Motion Boundary Marker
MBMS:	Multimedia Broadcast Multicast Service
MBR:	Maximum Bit Rate
MBwTM:	Media Broadcast with Trick Mode
MC:	MultiCast/Motion CompenStation/Multipoint Controller/Media Capture
MCC:	Mobile Country Code
MCCF:	Media Control Channel Framework

McCPF:	Multicast Control Point Functional block
MCF:	Message Confirm/Media Control Function
MCID:	Malicious Communication Identification
MCN:	Media Channel Negotiation
MCP:	Motion Compensated Prediction
McRF:	Multicast Replication Functional block
MCS:	Multipoint Communication Service
MCSAP:	Multipoint Communication Service Access Point
MCTF:	Motion Compensated Temporal Filtering
MCU:	Multipoint Control Unit
MDA:	Multi-Dimensional Adaption
MDB:	Multiprotocol Decapsulation Buffer
MDC:	Multiple Description Coding
MDCT:	Modified Discrete Cosine transform
MDCX:	MoDify Connection
MDF:	Media Delivery Function
MDI:	Media Delivery Index
M-DMC:	Mobile Digital Media Controller
M-DMD:	Mobile Digital Media Downloader
M-DMP:	Mobile Digital Media Player
M-DMS:	Mobile Digital Media Server
M-DMU:	Mobile Digital Media Uploader
mDNS:	Multicast DNS
MDP:	Media Presentation Description
ME:	Mobile Equipment
MEDIACTRL:	MEDIA ConTRoL
MEGACO:	MEdia GAteway COntrol protocol
MELP:	Mixed Excitation Linear Prediction
MF:	Media Function/MultiFrequency
MFC:	MultiFrequency Compelled
MFN:	MultiFrequency Network
MFWG:	Media Fragment Working Group
MG:	Media Gateway
MGC:	Media Gateway Control
MGCF:	Media Gateway Control Function
MGCP:	Media Gateway Control Protocol
MGS:	Medium Granularity Scalability
MGW:	Media GateWay/Mobile GateWay
MHD:	Mobile Handheld Device
MHP:	Multimedia Home Platform
MHT:	Mean Holding Time
MI:	Material Information
MIB:	Management Information Base
MIDI:	Musical Instrument Digital Interface
MIME:	Multipurpose Internet Mail Extension
MIU:	Media Interoperability Unit

MLD:	Multicast Listener Discovery/MultiLoop Decoding
MLQ:	Maximum Likelihood Quantization
MLR:	Media Loss Rate
MMCO:	Memory Management Control Operation
MME:	Mobile Management Entity
MMRP:	Multicast Multiple Registration Protocol
MMS:	Microsoft Media Server/Multimedia Messaging Service
MMT:	Media Transport Standard
MMTel:	MultiMedia TELephony communication service
MMUSIC:	Multiparty MUtimedia SessIon Control
MNC:	Mobile Network Code
M-NCF:	Mobile Network Connectivity Function
MoCA:	Multimedia over Cable
MoFRR:	Multicast only Fast ReRoute
MOH:	Music On Hold
MO-LR:	Mobile Originating Location Request
MOS:	Mean Opinion Scouring
MOSPF:	Multicast Extension to OSPF
MoU:	Minutes of Use
MP:	Multipoint Processor
MP3:	MPEG audio Player 3
MP4:	MPEG-4 file format
MPC-MLQ:	Multipulse LPC with Maximum Likelihood Quantization
MPD:	Media Presentation Description/Multiprotocol Decapsulation
MPE:	MultiProtocol Encapsulation
MPE:	MultiPulse Exited
MPEG:	Moving Pictures Expert Group
MPEG-TS:	Moving Pictures Expert Group-Transport Stream
MPLS:	Multiprotocol Label Switching
MPQM:	Moving Pictures Quality Metric
MPTCP:	Multipath TCP
MPTS:	Multiple Program Transport Streams
MPV:	MPEG-2 transPort Video
MR:	Multiresolution/Media Renderer
MRB:	Media Resource Broker
MRCP:	Media Renderer Control Point/Media Resource Control Protocol
MRF:	Multimedia Resource Function
MRFC:	Multimedia Resource Function Controller
MRFP:	Media Resource Function Processor
MRIB:	Multicast Routing Information Base
MRS:	Material Resolution Server/Media Resource Server
MS:	Media Server/Multiple Server
MSAP:	MAC Service Access Points
MSAS:	Media Synchronization Application Server
MSB:	Most Significant Bit
MSC:	Mobile Switching Center

MSCML:	Media Server Control Markup Language
MSCP:	Media Server Control Point
MSD:	Master–Slave Determination
MSE:	Mean Square Error
MSI:	Media–Service Interface (Cisco)
MSIN:	Mobile Subscriber Identification
MSINDN:	Mobile Subscriber ISDN
MSML:	Media Server Markup Language
MSP:	Media Service Provider
MSRN:	Mobile Station Roaming Number
MSRP:	Message Session Relay Protocol
MSU:	Message Signal Unit
MTA:	Media Terminal Adapter
MTAP:	Multiple-Time Aggregation Packet
MTBF:	Mean Time Between Failure
MTC:	Midi Time Codes
MT-LR:	Mobile Termination Location Request
MTP:	Media Transfer Protocol
MTP1:	Media Transfer Protocol 1
MTP2:	Media Transfer Protocol 2
MTP3:	Media Transfer Protocol 3
MTRF:	Mobile Terminating Roaming Forwarding
MTRR:	Mobile Terminating Roaming Retry
MTS:	MPEG-2 Transport Stream
MTU:	Maximum Transfer Unit
MUX:	Multiplex/Multiplexer
MV:	Motion Vector
MVC:	Multiview Video Coding
MVRP:	Multiple VLAN Registration Protocol
MWI:	Message Waiting Indication
MXM:	MPEG Extensible Middleware
NAB:	National Association of Broadcasters
NACC:	Network Assisted Cell Change
NACF:	Network Attachment Configuration Function
NACK:	Negative ACKnowledgment
NADA:	Network Assisted Dynamic Adaption
NAI:	Network Access Identifier
NAK:	No AcKnowledge
NAL:	Network Abstraction Layer
NALU:	Network Abstraction Layer Unit
NAPT:	Network Address and Port Translation
NAPTR:	Name Authority PoinTeR
NAS:	Network Access Stratum/Network Access Storage
NASS:	Network Attachment SubSystem
NAT:	Network Address Translation
NB:	NarrowBand/National Body

ND:	Network Device
NDS:	Network Domain Security/Novell Directory Server
NENA:	National Emergency Number Association
NGMN:	Next Generation Mobile Network Alliance
NGN:	Next-Generation Network
NGOV:	Next Gen Open Video
NGW:	Network Gateway
NIC:	Network Interface Card
NID:	Network ID
NIF:	Nodal Interface Function
N-ISDN:	Narrowband ISDN
NIST:	National Institute of Standards and Technology
NIT:	Network Information Table
NLSML:	Natural Language Semantic Markup Language
NMS:	Network Management System
NNI:	Network Network Interface
NNTP:	Network News Transport Protocol
NPP:	Noise Preprocessing
NPT:	Network Play Time/Normal Play Time
NPVR:	Network PVR
NR:	Noise Reduction
NRI:	NAL Reference ID
NRZ:	NonReturn to Zero
NSAP:	Network Service Access Point
NT:	Notification Type/Network Termination
NTFY:	NoTiFY
NTP:	Network Time Protocol
NTS:	Notification Subtype
NTSC:	National Television Systems Committee (FCC)
NTT:	Nippon Telephone and Telegraph company
OA&M:	Operations, Administration, and Maintenance
OAM&P:	Operations, Administration, Maintenance, and Provisioning
OBO:	Output Back-Off
OC:	Optical Carrier
OCAP:	OpenCable Application Platform
OCI:	Object Content Information
OCS:	Online Charging System
OD:	Object Descriptor
OFDM:	Orthogonal Frequency Division Multiplexing
OFDMA:	Orthogonal Frequency Division Multiplexing Access
OICD:	Open Internet Content Distribution
OIF:	Outbound Interface Table
OIP:	Originating Identification Presentation
OIPF:	Open IPTV
OIR:	Originating Identification Restriction
OLC:	Open Logical Channel

OLED:	Organic Light Emitting Diode display
OMA:	Open Mobile Alliance
OMG:	Object Management Group
OMT:	Object Modeling Technique
OMUX:	Output MUltipleXer
ONID:	Original Network ID
OPC:	Originating Point Code
ORB:	Object Request Broker
ORTC:	Object Real-Time Communication
OS:	Origin Server
OSA:	Open Service Architecture
OSI:	Open Systems Interconnection
OSN:	Original Sequence Number
OSPF:	Open Shortest Path First
OTDOA:	Observed Time Difference Of Arrival
OTI:	Object Transmission Information
OTT:	Over the Top
OUI:	Organizational Unique Identifier
OWD:	One-Way Delay
PA:	Presence Agent
PAFF:	Picture Adaptive Frame/Field
PAL:	Phase Alternating Line (ITU)
PAM:	Pulse Amplitude Modulation
PAT:	Program Association Table
PB:	Prediction Block
PBS:	Peak Burst Size
PBX:	Private Branch Exchange
PC:	Point Code/Personal Computer/Parental Control
PCC:	Policy and Charging Control
PCEF:	Policy Enforcement Function
PCH:	Paging Channel
PCM:	Pulse-Code Modulation
PCMA:	Pulse-Code Modulation A-Law
PCMU:	Pulse-Code Modulation U-Law
PCN:	PreCongestion Notification
PCP:	Port Control Protocol
PCR:	Program Clock Reference
PCRF:	Policy and Charging Control Function
P-CSCF:	Proxy Call Session Control Function
PD:	Phase Discriminator
PDCCH:	Physical Downlink Control Channel
PDCP:	Packet Data Convergence Protocol
PDD:	Post Dial Delay
PDN:	Packet Data Network
PDP:	Packet Data Protocol/Policy Decision Point/Plasma Display Panel
PDR:	Peak Data Rate

PDU:	Protocol Data Unit
PDV:	Packet (cell) Delay Variation
PE:	Protocol Engines
PEL:	PixEL
PEP:	Policy Enforcement Point
PER:	Packet Encoding Rules
PES:	Packetized Elementary Streams
PESQ:	Perceptual Evaluation of Speech Quality
PFC:	Previous Frame Concealment
PGM:	Pragmatic General Multicast
P-GW:	Packet data network GateWay
PHB:	Per-Hop Behavior
PHY:	PHYsical layer
PID:	Packet IDentifier
PIDF:	Presence Information Data Format
PIDF-LO:	PIDF-Location Object
PIE:	Proportional Integral controller Enhanced
PII:	Personally Identifiable Information
PIM:	Protocol-Independent Multicast
PIM-DM:	PIM Dense Mode
PIM-SM:	PIM Sparse Mode
PINT:	PSTN and INternet inTerworking
PiP:	Picture in Picture
PKI:	Public Key Infrastructure
PLC:	Packet Loss Concealment
PLI:	Picture Loss Indication
PLL:	Phase Locked Loop
PLMN:	Public Land Mobile Network
PLR:	Packet Loss Rate
PLS:	Pronunciation Lexicon Specification
PMP:	Port Mapping Protocol
PMT:	Program Map Table
PNA:	Progressive Networks Architecture/Presence Network Agent
PNG:	Portable Network Graphics
POC:	Picture Order Count
PoE:	Power over Ethernet
POIS:	Placement Opportunity Information Service
PON:	Passive Optical Networks
POPAI:	Point-Of-Purchase Advertising International
POTS:	Plain Old Telephony
PPA:	Push-Profile-Answer
PPP:	Point-to-Point Protocol
PPR:	Push-Profile-Request
PPV:	Pay Per View
PQ:	Priority Queue
PRACK:	Provisional Responses ACK

PRC:	Primary Reference Clock
PRI:	Primary Rate Interface
PRN:	Provide Roaming Number
PS:	Program Stream/Packet Switcher/Presence Server
PS HO:	Packet Switched Handover
PSAP:	Public Safety Answering Point
PSC:	Parameter Set Concept/Personalized Service Composition
PSD:	Power Spectrum Density
PSI:	Program-Specific Information
PSI/SI:	Program-Specific Information/Service Information
PSK:	Phase Shift Keying
PSNR:	Peak-Signal-to-Noise Ratio
PSQM:	Perceptual Speech Quality Measurement
PSS:	Packet-switched Streaming Service
PSTN:	Public-Switched Telephone Network
PT:	Payload Type
PtM:	Point to Multipoint
PTP:	Picture Transfer Protocol/Precision Time Protocol
PtP:	Point to Point
PTR:	PoinTeR
PTS:	Presentation Time Stamp
PTT:	Postal Telephone and Telegraph
PUA:	Presence User Agent
PUC:	Public Utilities Commission
PVC:	Permanent Virtual Circuit
PVR:	Personal Video Recorder
PVRG:	Portable Video Research Group
QAM:	Quadrature Amplitude Modulation
QAPD:	Quality Adaptive Progressive Download
QCIF:	QoS Class Identifier/Quarter CIF
QoE:	Quality of Experience
QoS:	Quality of Service
QP:	Quantification Parameter
QPEL:	Quarter PixEL
QPSK:	Quadrature Phase-Shift Keying
QQIC:	Querier's Query Interval Code
QQIC:	Query's Query Interval Code
QRC:	Query Response Channel
QRV:	Query's Robustness Variable
QSIG:	Q SIGnaling
Q-SIP:	QoS aware SIP
QTFF:	Quick Time File Format
QVGA:	Quarter Video Graphics Array
RAC:	Resource Available Confirm Message
RACF:	Resource and Admission Control Function
RACS:	Resource Admission Control System

RADA: Remote Access Discovery Agent
RAI: Resource Available Indicate
RAMS: unicast-based Rapid Acquisition of Multicast RTP Sessions
RAN: Radio Access Network
RAP: Rate Adoption Protocol/Random Access Point
RAR: Reauthorization Request
RAS: Registration, Admission, and Status
RATA: Remote Access Transport Agent
RBOC: Regional Bell Operating Company
RC: Rendering Control/Reception report Count
RCDO: Reduced Complexity Decoding Operation
RCEF: Resource Control Enforcement Function
RCF: Registration Confirm
RCS: Rendering Control Service
RDB: RTP Decapsulation Buffer
RDF: Routing Determination Function
RDO: Rate Distortion Optimization
RED: Random Early Delete
REL: RELease
RET: RE-Transmission
RF: Rate Distortion/Radio Frequency
RFC: Request For Comments
RFI: Request For Information
RFP: Request For Proposal
RG: Residential Gateway
RGB: Red, Green, Blue color model
RIP: Routing Information Protocol
RIR: Random Intra Refresh
RLC: Radio Link Control/ReLease Complete/Run Length enCoding
RLE: Run Length Encoding
RLS: Resource Location Server
RM: Remote Management
RMCAT: RTP Media Congestion Avoidance Techniques
RMRG: Reliable Multicast Research Group
RMS: Remote Management System
RNA: Ring No Answer
RNC: Radio Network Controller
R-NNI: Roaming NNI
ROADM: Reconfigurable optical add-drop multiplexer
ROAP: RTCWEB Offer Answer Protocol
ROHC: RObust Header Compression
ROPE: Recursive Optimal Per-pixel Estimate
RP: Rendezvous Point
RPE: Regular Pulse Excited
RPF: Reverse Path Forwarding
RPID: Rich Presence Information Data

RPLM:	Reference Picture List Modification/Reference Picture List Reordering
RPMR:	Reference Picture Marking Repetition
RPS:	Reference Picture Set
RPSI:	Reference Picture Selection Indication
RPT:	Rendezvous Point Tree
RQNT:	NoTification ReQuest
RR:	Receiver Report
RRC:	Radio Resource Control
RRJ:	Registration ReJect
RRQ:	Registration ReQuest
RS:	bandwidth modifier for RTCP Sender reports/Recording Session
RS-DVR:	Remote Service – Digital Video Recorder
Rseq:	Response Sequence
RSI:	Receiver Summary Information
RSIP:	ReStart In Progress
RSVP:	Resource reSerVation Protocol
RTA:	Registration-Termination-Answer/Receive-Terminate-Acknowledgment
RTC:	Real Time Communication
RTCP:	Real-time Transport Control Protocol
RTMP:	Real Time Messaging Protocol Chunk Stream
RTP:	Real-time Transport Protocol
RTR:	Registration-Termination-Request
RTSP:	Real-Time Streaming Protocol
RTT:	Round-Trip Time
RUI:	Remote User Interface
RVC:	Reconfigurable Video Coding
RVLC:	Reversible Variable Length Coding
SA:	Service Agent/Security Association
SAA:	Server-Assignment-Answer
SABR:	Smooth Adaptive Bit RatE
SAD:	Sum of Absolute Differences/Service Action Data
SADS:	Service and Application Discovery and Selection
SAE:	System Architecture Evolution
SAF:	Simple Aggregation Format
SAH:	Service Access History
SALT:	Speech Application Language Tags
SAO:	Sample Adaptive Offset
SAOL:	Structured Audio Orchestra Language
SAP:	Service Access Point/Session Announcement Protocol
SAR:	Server-Assignment-Request
SAV:	Start of Active Video
SAVP:	Secure Audio Video Profile
SB:	Source Block
SBC:	Subband Coding/Session Border Controllers
SBE:	Signaling Border Element
SBL:	Source Block Length

SBN:	Source Block Number
SbS:	Side by Side
SC:	Subcommittee/Synchronization Client
SCCP:	Skinny Client Control Protocol/Signal Connection Control Part
SC&DF:	Service Control and Delivery Function
SCF:	Service Control Function
SCIM:	Service Capability Interaction Manager
SCP:	Service and Content Protection/Service Control Point
SCPD:	Service Control Protocol Description
SCR:	System Clock Referenced
SCS:	Service Capability Server
S-CSCF:	Serving Call Session Control Function
SCTE:	Society of Cable Telecommunications Engineer
SCTP:	Stream Control Transmission Protocol
SCXML:	State Chart XML
SD:	Standard Definition/Service Discovery
SDAP:	Service Discovery Application Profile
SDCP:	Standardized Device Control Protocol
SDD:	Service Description Document
SDDT:	Simplified DCT-Domain Transcoders
SDES:	Source DEScription
SDF:	Service Data Flows/Service Discovery Function
SDH:	Synchronous Digital Hierarchy
SDI:	Serial Digital Interface
SDP:	Session Description Protocol, Service Delivery Platform
SD&S:	Service Discovery & Selection
SDT:	Signal Dependant Transform/Service Description Table
SDTV:	Standard-Definition TV
SDV:	Switched Digital Video
SECAM:	SEquentiel Couleur A Memoire
SEG:	Security Gateway
SEI:	Supplemental Enhancement Information
SER:	Symbol-Error rate
SF:	Service Function
SFM:	Source-Filtered Multicast
SFMT:	Sub-Feedback Message Type
SFN:	Single-Frequency Network
SFTP:	Simple FTP
SFU:	Selective Forwarding Unit
SG:	Signaling Gateway
SGCP:	Simple Gateway Control Protocol
SGSN:	Serving GPRS Support Node
SGW:	Security GateWay
S-GW:	Serving GateWay
SHE:	Super HeadEnd
SI:	Service Information/Switching I pictures/Still Image

SIB:	System Information Block
SID:	Silence Insertion Descriptor
SIF:	Signaling Information Field/Source Input Format
SIM:	Subscriber Identity Module
SIMPLE:	SIP for Instant Message Leveraging Extension
SIO:	Service Information Octet
SIP:	Session Initiation Protocol
SIS:	Systems for Interactive Services/Subscriber Information Service
SISR:	Semantic Interpretation for Speech Recognition
SIV:	Speaker Identification and Verification
SKMF:	Service Key Management elementary Function
SL:	Subscriber Location/Synchronization Layer
SLA:	Service Level Agreement
SLD:	Single Loop Decoding
SLEP:	Systematic Lossy Error Protection
SLF:	Subscription Location Function
SLI:	Slice Loss Indication
SLP:	Service Location Protocol
SLS:	Signaling Link Selection
SM:	Stream Monitor/Sparse Mode
SMATV:	Satellite Master Antenna TeleVision
SMD:	Scheduled Multicast Download
SMF:	Service Membership elementary Function
SMIL:	Synchronized Multimedia Integration Language
SMPT:	Society of Motion Picture and Television
SMPTE:	Society of Motion Picture and Television Engineers
SMS:	Short Message Service
SMSC:	Short Message Service Center
SMTP:	Simple Mail Transfer Protocol
SMTS:	Single MPEG Transport Stream
SNAP:	SubNetwork Attachment Point
SNHC:	Synthetic Natural Hybrid Coding
SNMP:	Simple Network Management Protocol
SNR:	Signal-to-Noise Ratio
SNTP:	Simple Network Time Protocol
SOAP:	Simple Object Access Protocol
SOHO:	Small Office Home Office
SONET:	Synchronous Optical NETwork
SP:	Service Protection/Service Provider/Switching Picture/Signaling Point
SPC:	Signaling Point Code
SPDY:	SPeeDY
SPEERMINT:	Session PEERing for Multimedia INTernet
SPI:	Security Parameter Index
SPIM:	SPam over Instant Messaging
SPIMAP:	Serial Peripheral Interface MAP
SPIT:	SPam over Internet Telephony

SPP:	Service Purchase and Protection
SPPP:	SPam over Presence Protocol
SPT:	Shortest Path Tree/Service Point Trigger
SPTS:	Single Program Transport Stream
SQNR:	Signal to Quantization Noise Ratio
SR:	Sender Report/Scheduled Recording/Speech Recorder
SRC:	Session Recording Client
SRGS:	Speech Recognition Grammar Expression
SRM:	System Renewability Message/Session Resource Manager
SRP:	Session Recording Protocol
SRS:	Scheduled Recording Service/Session Recording Server
SRTP:	Secure RTP
SRV:	SeRVice resource record
SRVCC:	Single Radio Voice Call Continuity
SS:	Single Server
SS7:	Signaling System 7
SSC:	Shared Service Control
SSD:	Sum of Squared Differences/Service Shared Data
SSDP:	Simple Service Discovery Protocol
SSE:	Server Sent Events
SSF:	Service Switching/Selection Function
SSID:	Service Set Identifier
SSL:	Secure Socket Layer
SSM:	Single Source Multicast
SSML:	Speech Synthesis Markup Language
SSN:	SubSystem Number
SSP:	Sip Service Providers/Service Switching Point
SSRC:	Synchronization Source
STAP:	Single-Time Aggregation Packet
STB:	Set-Top Box
STC:	System Time Clock
STP:	Signal Transfer Point
STSA:	Stepwise Temporal Sublayer Access
STUN:	Simple Traversal of User Datagram Protocol
SUA:	SCCP User Adaption layer
SUT:	System Under Test
SVC:	Scalable Video Codec/Scalable Video Coding
SVG:	Scalable Vector Graphics
SVLTE:	Simultaneous Voice LTE
TA:	Trunk Amplifier/Terminal Adapter
TaB:	Top and Bottom
TAI:	Targeted Advertisement Insertion
TALI:	Transport Adaptation Layer Interface
TAPI:	Telephone Application Programming Interface
TB:	Transform Block
TBCP:	Talk Burst Control Protocol

TCAP:	Transaction Capabilities Application Part
TCE:	Terminal Capabilities Exchange
TCP:	Transport Communication Protocol
TCS:	Terminal Capabilities Set
TDD:	Time Division Duplexing
TDM:	Time Division Multiplexing
TDT:	Time and Date Table
TE:	Terminal Equipment/Traffic Engineering/Technology Engines
TE:	Trigger Event
TEI:	Terminal Endpoint Identifiers
TEMI:	Timed External Media information
TEN:	Trigger Event Notification
T/F:	Time/Frequency Coder
TFA:	TransFer Allowed
TFP:	TransFer Prohibited
TFTP:	Trivial File Transfer Protocol
TGREP:	Telephony Gateway REgistration Protocol
TH:	Transport-stream Header
TIA:	Telecommunications Industry Association
TIAS:	Transport-Independent Application-Specific
TIP:	Telepresence Interoperability Protocol
TISPAN:	Telecoms and Internet converged Services and Protocols for Advanced Networks
TLD:	Top Level Domain
TLS:	Transport Layer Security
TLV:	Tag Length Value/Type Length Value
TM:	Technical Module
TMMBR:	Temporary Maximum Media Stream Rate
TMR:	Transport Medium Requirement
TN:	Telephone Number
TO:	Transport Object
TOI:	Transport Object Identifier
TOR:	The Onion Router
ToS:	Type of Service
TOT:	Time Offset Table
TP:	Transition Point
TPDU:	Transport Protocol Data Unit
TPF:	Transport Processing Function
TPKT:	ISO transport services on top of the TCP PacKeT
TPS:	Transmission Parameter Signaling
TRD:	Target Region Descriptor
TRIP:	Telephony Routing over IP
TS:	Transport Stream
TSA:	Temporal Sublayer Access
TSAP:	Transport layer Service Access Point
TSI:	Transport Session Identifier

TSID:	Transport Stream ID
TSN:	Transport Sequence Number
TSP:	Telephone Service Provider/Transport Stream Packet
T-STD:	(MPEG-2) Transport stream System Target Detector
TSTR:	Temporal-Spatial Trade-off Request
TSTV:	Times Shifted TV
TTA:	Telecommunication Technology Association
TTC:	Telecommunication Technology Committee
TTCN:	Testing and Test Control Notation
TTI:	Transmission Time Interval
TTL:	Time To Live
TTS:	Text To Speech
TTWG:	Timed Text Working Group
TU:	Transaction User
TURN:	Traversal Using Relays around NAT
TV:	TeleVision
TVA:	TV Anytime
TVC:	Telescopic Vector Composition
TVE:	TV Everywhere
TZ:	Time Zone
UA:	User Agent
UAA:	User-Authorization-Answer
UAC:	User Agent Client
UAR:	User-Authorization-Request
UAS:	User Agent Server
UC:	UniCast
UCF:	Unregistration ConFirm
UCM:	Unified Communication Manager
UD:	Unicast Download
UDA:	UPnP Device Architecture
UDN:	Uniform Device Name
UDP:	User Datagram Protocol
UE:	User Equipment
UEP:	Unequal Error Protection
UER:	Union European de Radiodiffusion (EBU)
UGC:	User Generated Content
UHD:	Ultra High Definition
UHDTV:	Ultra-High-Definition TV
UI:	User Interface
UICC:	Universal Integrated Circuit Card
UIT:	Union Internationale de Telecommunication (in English ITU)
UL:	UpLink
ULE:	Unidirectional Lightweight Encapsulation
UltraHD:	Ultra High Definition
UML:	Universal Modeling Language
UMTS:	Universal Mobile Telecommunications System

U–N:	User to Network
UNI:	User–Network Interface
UPC:	Universal Product Code
UPnP:	Universal Plug and Play
UPSF:	User Profile Service Function
URC:	Unregistration Confirm
URI:	Uniform Resource Identifier
URJ:	Unregistration Reject
URL:	Uniform Resource Locator
URN:	Uniform Resource Name
URQ:	Unregistration Request
USB:	Universal Serial Bus
USGSN:	UMTS Serving GPRS Support Node
USI:	Usage State Information
USIM:	Universal Subscriber Identity Module
USR:	User-to-UseR
UTC:	Universal Time, Coordinated
UTF:	Unicode Transformation Format/Universal character set Transformation Format
UTF-8:	UTF for 8 bit
UTRAN:	Universal Terrestrial Radio Access Network
U–U:	User-to-User
UUID:	Universal Unique IDentifiers
UV:	UltraViolet
VA:	Voice Admit
VAD:	Voice Activated Detector
VANC:	VoLGA Access Network Controller
VBO:	Video Back Office
VBR:	Variable Bit Rate
VBV:	Video Buffering Verifier
VC:	Video Codec
VC1:	Video Codec 1
VC9:	Video Codec 9
VCB:	Virtual Code Block
VCEG:	Video Coding Experts Group (ITU)
VCL:	Video Coding Layer
VCR:	Video Cassette Recorder
VGA:	Video Graphics Array
VGCF:	Voice Gateway Control Function
VHO:	Video Hub Office
VHS:	Video Home system
VLAN:	Virtual LAN
VLBV:	Very Low Bit Rate Video
VLC:	Variable-Length Coder
VLD:	Variable-Length Decoder
VLR:	Visitor Location Register

VM:	Verification Model
VO:	Video Objects
VoATM:	Voice over ATM
VoD:	Video-on-Demand
VoIP:	Voice over IP
VOL:	Video Objects Layer
VoLGA:	Voice over LTE via Generic Access
VoLTE:	Voice over LTE
VoMBB:	Voice over Mobile Broadband
VOP:	Video Objects Plane
VOS:	Virtual Origin Server
VP8:	Video comPression 8
VP9:	Video comPression 9
VPLS:	Virtual Private LAN Service
VPN:	Virtual Private Network
VPWS:	Virtual Private Wire Service
VQ:	Vector Quantization
VRML:	Virtual Reality Modeling Language
VRRP:	Virtual Router Redundancy Protocol
VRU:	Voice Response Unit
VSO:	Video Serving Office
VSS:	Virtual Switching System
VUI:	Video Usability Information
W3C:	World Wide Web Consortium
WAN:	Wide Area Network
WC:	Wall Clock
WCCP:	Web Cache Control Protocol/Web Cache Coordination Protocol
WCDMA:	Wideband Code Division Multiple Access
WD:	Working Draft
WDM:	Wavelength Division Multiplex
WebRTC:	Web Real Time Communication
WFQ:	Weighted Fair Queuing
WG:	Working Group
WGDTB:	Working Group on Digital Television Broadcasting
WiFi:	Wireless Fidelity
WLAN:	Wireless Local Area Network
WMA:	Windows Media Audio
WMDRM:	Windows Media Digital Rights Management
WMDRM-ND:	Windows Media Digital Rights Management for Network Devices
WMM:	WiFi MultiMedia
WMSP:	Windows Media HTTP Streaming Protocol
WMV:	Windows Media Video
WPP:	Wavefront Parallel Processing
WRED:	Weighted Random Early Detection
WRR:	Weighted Round Robin
WWW:	World Wide Web

XCAP:	XML Configuration Access Protocol
XCON:	centralized CONnferencing framework
XDM:	XML Document Management
XDMS:	XDM Server
XHTML:	eXtensible HTML
XML:	eXtensible Markup Language
XMP:	eXtensible Metadata Platform
XMPP:	eXtensible Messaging and Presence Protocol
XMT:	eXtensible MPEG-4 Textual Format
XOR:	eXclusive OR
XRS:	unknown message ReSponse
XSL:	XML Schema Document
YIQ:	Chroma, Saturation, Color (NTSC)
YUV:	Chroma, Hue, Tint (PAL)
z2z:	Zeroconf-to-Zeroconf
Zeroconf:	Zero Configuration Networking
ZRTP:	Zimmermann's RTP

1

Introduction

This book is aimed for students and practitioners with a basic knowledge in computer networks. One possible source on the basics of computer networks is the book by Tanenbaum and Wetherall [1].

We will provide the reader with all concepts of currently used Internet Protocol (IP) technologies to deliver multimedia efficiently to the user. The book does not treat any non-IP multimedia technologies – for example, TV/Audio transmission by satellite, cable, or other terrestrial methods. For examples of these technologies, see Ref. [2].

We provide enough detail to understand the operation and transmission of multimedia data over a packet network. However, it is not intended to provide sufficient detail for an implementation of a complete multimedia application. For that level of detail, we provide links to the relevant standards and other literature.

The transport of multimedia over the network requires timely and errorless transmission much more strictly than any other data. The criticality of transport has led to specialized protocols and to special treatment within multimedia applications (telephone, IP-TV, streaming, and others) to overcome network issues.

The percentage of multimedia traffic over the Internet is already substantial today and, based on 2014 estimations, will grow to 79% of all traffic by 2018. Therefore, the knowledge of protocols used and handling of the traffic needs to increase as well.

The next chapter will exemplify the typical requirements of multimedia applications, and we will show their protocols and implementation in later chapters.

Before we treat the underlying network functions in Chapter 4 and synchronization in Chapter 5, we will treat coding and compression of multimedia in Chapter 3. At first glance, the need for the treatment of coding and compression may not be obvious, since in computer networking, one typically assumes that the lower layers are independent of the application layers. However, experienced networkers know that even data-driven applications occasionally show unacceptably long response times over Wide Area Networks (WANs). In such situations, network and application engineers jointly analyze this in network labs, and after that, the application and/or network gets changed. Multimedia applications require an absolutely optimal interworking with the network since it cannot afford response time variations of a second as a data-driven application. This optimization requires knowledge

Multimedia Networks: Protocols, Design, and Applications, First Edition. Hans W. Barz and Gregory A. Bassett.
© 2016 John Wiley & Sons, Ltd. Published 2016 by John Wiley & Sons, Ltd.

of what options are possible on both sides. Consequently, we have to treat coding and compression for multimedia in Chapter 3.

While the next chapters handle technologies, we also have to understand how multimedia fit into existing network types, which standard organizations are creating the specifications, and how the market functions. These elements strongly influence the real use of multimedia.

1.1   Types of Networks

We will classify networks according to the following types: Internet, Telecommunication providers, Companies, Universities, and Home. It will become obvious that certain multimedia applications only work well in certain network types. Each network type is characterized by the following elements:

Administration: Organization of the administration
Redundancy/stability: Network is constructed to fulfill certain stability goals
Service quality: Availability of service classes guaranteeing traffic quality
Monitoring: Extent of network monitoring
Standards: Standards on which the network is based
Operator: Who operates the network.

1.1.1   Internet

The Internet has central bodies for standardization – Internet Engineering Task Force (IETF) – and addressing – Internet Corporation for Assigned Names and Numbers (ICANN). However, the Internet has no central administration that rules all interconnection and technical implementation details. The Internet offers a lot of redundancies, but there is no authority that plans this in detail. Service qualities do not exist and cannot be contracted. The network is monitored in parts but not as a whole.

The standards of the IETF govern all traffic and protocols. In some locations, additional protocols may work, but this is not true globally.

A plethora of telecommunication providers and institutions operate the Internet.

1.1.2   Telecommunication Provider Networks

Each telecommunication service provider (Telco) has a home territory where he owns cable infrastructure. Historically, each country has one major network provider who had once the monopoly for all telecommunication services within the country (incumbent). With the exception of the United States, the incumbents still own the majority of all cable infrastructure within their home country and typically all connections to the households. The particular situation in the United States slows the implementation of High-End TV as presented in Chapter 11.

Major Telcos contract network links or network capacity from other Telcos and build international cable infrastructure jointly in consortiums. The biggest Telcos of the world are American Telephone and Telegraph (AT&T, United States), Verizon (United States), British Telecom (BT, United Kingdom), Orange (France), German Telekom, Nippon Telegraph and Telephone Corporation (NTT, Japan), Telefonica (Spain), and Colt (United Kingdom).

Each Telco's network is centrally administered, and all redundancy is centrally planned. The network is monitored with the exception of links to the households. Service quality can also be provided at additional cost over contracted links in other countries. The Telcos offer services to their clients (companies, universities, end users, other providers) in different quality and throughput levels:

- Pure connectivity – that is, a link or part of a link with given throughput – based upon IEEE (International Electrical and Electronics Engineers) standards
- Data traffic based upon IETF standards
- Voice traffic based upon International Telecommunication Unit (ITU) standards.
- Television distribution based upon Digital Video Broadcast (DVB) or Advanced Television Systems Committee (ATSC) standards.

All major Telcos offer additional value-added services – for example, provide outsourcing services for companies.

The term Internet Service Provider (ISP) is used for those Telcos that provide Internet service. Mostly, all Telcos are also ISPs.

1.1.3 Company Networks

National and international companies own the networks in their premises – that is, data network equipment as well as telephone switches plus end-devices and cabling. One or multiple Telcos provide WAN connections; also tunnels over the Internet deliver the connectivity. The networks are centrally administered and monitored. All redundancy is well planned. Quality guarantees are possible if contracted from the Telcos plus internal management on LANs.

The networks use IETF, ITU, and diverse provider protocol standards. The connections to the Internet use firewalls plus Network Address Translation (NAT) and allow Virtual Private Networks (VPNs) connections to the company. Often, companies have dedicated links to other businesses to exchange well-controlled information.

All companies' networks tend to have a star-like topography with the headquarters at the center. This topology has consequences for some multimedia services – see Section 12.1.

1.1.4 University Networks

University networks are similar to company networks within one country. Of course, interconnections of university networks also exist. In contrast to business networks, interconnections to other parties are typically without controls. Since the main difference is security, we can see a university network as a company network, and we will not treat them specifically any further.

1.1.5 Home Networks

Home networks have a simple topology. Redundancy, quality, and monitoring generally do not exist in home networks. The quality of operation depends on the end customer and is usually of inferior quality. The network uses IETF and specific home network protocols; in Chapter 10,

Table 1.1   Characteristics of Network Types

	Internet	Telcos	Company	Home
Administration	Distributed	Central	Central	Central
Redundancy	Yes: not really planned	Yes: planned	Yes: planned	No
Quality	No	Possible	Possible	No
Monitoring	Partially	Yes	Yes	No
Standards	IETF	IETF and ITU	IETF & ITU and others	IETF plus Home protocols
Operator	Many	One	One	One, without knowhow

we will treat home network protocols. The diversity of end devices is significant in relation to the network size.

A home network is typically connected to the Internet via a Telco's DSL or cable router. In some cases, Telcos offer quality assurance in combination with additional services.

1.1.6   Overview

Table 1.1 summarizes the characteristics per type. In subsequent chapters, the reader will understand which multimedia application best operates in which network type. We summarize this in Section 12.5.

1.2   Standard Organizations

The ITU has investigated all committees in "Information and Communication Technology," and their report lists 227 organizations [3]. We will only mention the most important ones with respect to multimedia:

ITU: International Telecommunication Union – formerly Consultative Committee International Telecommunication Union (CCITT) – standardized all historical telephone standards. As well, numerous compression standards were created in the subgroup "Moving Picture Expert Group" (MPEG) and "Joint Photographic Experts Group" (JPEG). The MPEG group and the JPEG group are collaborations – Joint Telecommunication Committee (JTC) – between the ITU-Telecommunication (ITU-T) and ISO/IEC.
The International Standards Organization (ISO) and the International Electrical Committee (IEC) do not play a significant role in multimedia.
IETF: The Internet Engineering Taskforce (IETF) governs all Internet protocols; all publications are "Request for Comments" (RFC) and carry continuous numbers. An RFC has a status: Informational, Experimental, Proposed Draft, Draft Standard, Standard, and Historic; however, even Proposed Drafts and Draft Standards are often executed as being a Standard.

Some collaborations with other standard bodies exist. In Chapter 7, we will list a common standard with the ITU.

ETSI: European Telecommunication Standard Institute is the follow-up organization of Conférence Européenne des Administrations des Postes et des Télécommunications (CEPT). ETSI Standards are binding in the European Union but also used elsewhere. Asia and Latin America use, for example, ETSI mobile telephone standards.
ETSI publishes the standards of DVB as European Standards. One can compare ETSI with the Federal Communication Commission (FCC) in the United States.

DVB: Digital Video Broadcasting is a consortium of more than 270 companies, Telcos, and representatives from standard bodies. The consortium creates standards for the transmission of digital voice and video.

ATSC: Advanced Television Systems Committee is the US counterpart of the DVB.

W3C: The World Wide Web Consortium also defines multimedia standards when they are integrated with browsers.

DLNA: Digital Living Network Alliance is a consortium of more than 250 companies. DLNA set home network standards, which include Universal Plug and Play Forum (UPnP) protocols.

Others: Worthwhile to mention are audio standards by the Audio Engineering Society (AES), interoperability specifications by the Internet Multimedia Telecommunication Forum (IMTC), and home – Telco gateway specifications by the Home Gateway Initiative (HGI). Some others will be explained as and when they occur in the text.

1.3   Market

The market has many players, and we try to segment them, but since all of them try to enter other market segments, the classification might get blurred over time.

Computer software suppliers: Here, we see Microsoft, Google, and Apple delivering the so-called TV solutions (playing only movies in most countries), defining standards in contrast to standard committees, and offering streaming players. Microsoft, in addition, provides VoIP telephone/video server plus presence (Lync). Microsoft also produced software solutions for the operation of streaming service (Mediaroom) but sold this to Ericsson. Google with YouTube offers a public usable video streaming platform.
Their strategy has always been to maintain the user interface and hinder others to encroach on it.

Network hardware/software supplier: Most prominent here is Cisco, but there are, of course, also Nortel, Ericsson, Avaya, and Siemens. However, the most complete platform on IP multimedia products is Cisco with VoIP telephone server, high-end video conferencing, streaming platforms, cache engines, set-top boxes, digital signs, video surveillance, and media monitoring solutions. The other mentioned companies once produced non-VoIP telephone systems and grew from phones into this business. For videoconferencing, Polycom is also a mentionable provider.
Their strategy is to own the user interface within companies and to expand the network for better multimedia services.

Telcos and cable TV providers: There are a lot of them, and we cannot even try to provide a list but the biggest were mentioned before in Section 1.1.2. Apart from all connectivity services, they also offer telephone, TV distribution, and streaming services.

Their strategy is to maintain customers by bundling their services – for example, the so-called triple play (Internet connection, fixed and mobile telephone) or quadruple play by adding high-end Internet TV access. The bigger a provider is, the better bundles he can offer.

TV production studios: Again, there are a lot of them globally. In the past, they produced the TV program movies and contracted the transmission to Telcos and cable providers. Now, they themselves offer streaming platforms.

Streaming platforms: Of course, for movies, there are Netflix, Hulu, Amazon, and Google with YouTube. Also, the Telcos and cable TV providers have movie streaming platforms but often only accessible at their customers' homes. Some streaming providers started to produce movies similarly to the TV production studios or the movie industry.

Moreover, we also see many audio streaming platforms – for example, Spotify.

2

Requirements

In this chapter, we discuss typical requirements for a number of multimedia usage areas. An experienced reader probably knows certain but not all needs of the introduced areas. In the final section of this chapter, we list the requirements with a few criteria. At the very end of this book, we will revisit the requirements and discuss the characteristics of their implementation.

2.1 Telephony

Telephony is the oldest multimedia technology that even our grandparents (see Figure 2.1) used – commonly named Plain Old Telephony Service (POTS). The first patent for a phone was awarded to A. G. Bell in 1876, but the inventor was J. P. Reis in 1860. The first telephone company probably started in 1878 in the United States. Calls in other countries became available in the 1950s and typically cost $10 for 3 minutes – roughly $90 in today's dollars. International calls without operator assistance were first possible in 1970s. At the same time, Fax machines were broadly used with multinational companies. Many of the requirements grew over decades. The most extensive requirements today exist in company telephone systems and Telcos switches.

For fixed telephones, we expect fast setup of calls, absolutely reliable connections, no downtime, and no noise disturbance during calls. In case we call a remote site on the planet, we may allow less quality and less stability. Even when talking to a remote site, we assume that we can interact seamlessly. Typically, in companies, availability above 99,999% during working hours is expected – less than 5 minutes downtime per year. A typical computer system does not fulfill this requirement at all. Contracts of telephone providers for a household offer mostly much less – maybe 99%.

However, with mobile phones, the user is willing to accept much less: longer setup of calls, occasional drop of connections, and noise during calls.

Phones should also be available as applications on all types of computers; in such cases, they are often used with headsets. Some users ask for this functionality, but most users still prefer a separate device.

In the following, we list only the most important requirements for company fixed telephone systems – the so-called Private Branch Exchange (PBX). Due to decades of development, PBX bears a lot more features than mentioned as follows.

Multimedia Networks: Protocols, Design, and Applications, First Edition. Hans W. Barz and Gregory A. Bassett.
© 2016 John Wiley & Sons, Ltd. Published 2016 by John Wiley & Sons, Ltd.

Figure 2.1   Rotary dial telephone

Conferences: More than two persons are in a call. One can have up to thousands of people in a teleconference. Even for a mid-sized number of participants, users need to be muted to avoid background noises. There is a control mechanism to unmute participants. For example, a user dials a predefined number to signal his desire for input and a central control allows this.

Entering a conference call can be done by just dialing a number plus a code (Meet-me Conferences), by dialing to registered members or by calling an agent who adds the caller to a conference after an authorization check.

Distributed meetings: Today, conference calls are often accompanied by shared presentations. Such distributed meetings sharing slides and a conference call may exist in isolation, can be integrated at least for the reservation, or can be completely integrated. In a complete integration, one would either click a link or dial a phone number to enter the meeting on both media. For a video clip, implicitly showing features of a distributed meetings, see Ref. [4].

Emergency calls: When calling prespecified emergency numbers, laws in many different countries require that the location of the caller is communicated – commonly known in the United States as the (enhanced) E911 obligation [5].

Storing information: The system can store voice messages in a Voice-Mail system, Fax messages, and conference calls. For the management of Voice Mail, a rich and easy-to-use customer interface has to exist.

The duration of the storage is often governed by legal restrictions to avoid that courts can later use stored messages. The restrictions are necessary since people talk much less controlled than they do in writing.

Transfers: Of course, calls need to get transferred to another party with or without former consultation – blind or consultative transfer. Another option is "Call Parking" where the transfer goes to a virtual extension and later gets reused. If parked calls come back automatically, this feature is called "Orbiting".

Gateways: There are mobile gateways with built-in SIM cards to call mobile partners from a fixed phone, avoiding high provider costs. E-mail gateways translate Fax message as well as Voice Mails.

Directories: For easy searching, it is beneficial if all phone users are included in a company directory. The directory should be linked or integrated with the computer directory.

Presence: The information in directories and local contact lists is expanded by reachability information and how to best contact a person. Note that different viewers can get different reachability information; the boss might see more than a regular user. Other communication methods are included, such as short messages and E-mail. Figure 2.2 shows an example.

Music on hold: Music or information is played during call transfers or when waiting in hold queues.

Hunt groups: A phone call to a single extension gets distributed to a group of several extensions. In serial hunting, a call is handed to the next in the serial order when the former one is busy or not answering (Ring No-Answer Hunt Group). By using Hunt Groups, rudimentary service functions can be implemented.

Application integration: The most typical example is the integration of Customer Relation Management (CRM) applications. CRM systems store client information and consequently can prioritize calls, present customer information for the calling person as well as storing conversation details. Of course, we all know how to interact with automated attendants or Interactive Voice Response (IVR) systems that ask us to press numbers and use voice recognition for navigation.

Call center: Call centers are the customer interface for most companies that we have to pass when contacting a company. Call centers distribute calls to agents. Each agent uses applications that are integrated with the PBX. A Call center always includes an Automatic Call Distribution (ACD), which distributes each call based on different criteria. Criteria can be based on time of the day, availability of call agents, or an application integration. Call centers must support call agents at various locations seamlessly and also include telemarketing solutions for outgoing calls. Other communication methods such as chat or short message are integrated. Ongoing control and monitoring functions for call agents with at least one level of supervisor must exist.

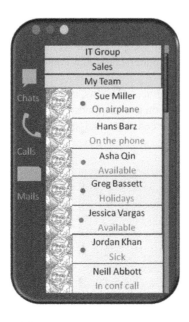

Figure 2.2 Presence example on a smartphone

2.2 Streaming

Media streaming to the households came into existence with affordable bandwidth, sound compression methods, and enough power in personal computers. In 1993, the band "Severe Tire Damage" played the first live audio stream from the Xerox site in Palo Alto. A year later, the "Rolling Stones" did the same but with more attention from the world press. Probably, the first Internet Radio started in 1993/1994 by the Georgia University student radio station WREK and the University of North Carolina WXYC. Video streaming commenced in 1997 with "shareworld.com", which allowed the user to upload files and share them with others – the so-called Video Hosting. This service was probably too early to succeed, but Vimeo and YouTube were successful when starting again in 2004 and 2005, respectively. A prerequisite for the use of video streaming is a good media player on computers; media players were introduced from 1999 onward by Microsoft, Apple, Real Networks, and finally Adobe.

The first Video-on-Demand (VOD or VoD) service already started in 1994 in Cambridge (United Kingdom) but ended in 1996. This service used the Acorn set-top box STB1 on an ATM network – Asynchronous Transfer Mode standard (ITU). Figure 2.3 shows the probably first Set-Top Box (STB) used for VoD. The initial speed of this service was 2 Mbit/s to homes and was later increased to 25 Mbit/s. Other companies in the United Kingdom followed on that road – for example, Kingston Communication, Home Choice, Sky, and BBC. In 1997, Netflix was founded in the United States but only offered a DVD mailing service. Netflix is now a very well known VoD service, but began only in 2007, the same time Hulu started. At the very same time, some European Telcos – for example, Swisscom, German Telekom – also began to offer VoD services as part of an overall IPTV offering. Apart from the United Kingdom, VoD services are a young business.

When we handle streaming in this book, we will solely look at video streaming since it is much more challenging for the network than audio streaming.

The user expectation for a VoD service, as well as Video Hosting, is uninterrupted play in HD quality (High Definition). Forward and backward jumps using thumbnails within the video should work fast, but the user will accept several seconds of delay. The same requirements hold for start-up times. The user understands that content is only available hours or weeks after creation.

Streaming should work on all devices, TV sets, and beamers. The devices should interoperate so that one can look at a video on a smartphone, but with a single command, send it to the TV set – the so-called multiple-screen or second-screen support. In addition, those platforms

Figure 2.3 Arcon historic set-top box STB1 – first VoD system (Source: Reproduced by permission of Centre for Computing History, Cambridge, UK)

should maintain favorite lists, recommendation of other videos based on already seen ones, and all videos should resume at the position where it was interrupted.

TV channels or Telcos in diverse countries stream live events in a restricted way over the Internet. Internet live sports events are rarely available outside of their paid customer base. Consequently, there will hardly be more than 10,000 users looking concurrently at the same media stream.

Mostly, streaming services deliver the content Over-the-Top (OTT); this means that the delivery runs over the Internet and in principle is accessible from everywhere. However, providers often have to restrict the access from other countries due to licensing conditions.

2.3 IPTV

The first regular TV programs started in 1935/1936 in the United Kingdom and Germany, visible in black-and-white on huge TV sets – see Figure 2.4. The transmission in the beginning was terrestrial, from the 1940s, also by cable (CATV: Community Access Television), and in the 1960s, satellite transmission became available. TV transmission is expected to work without any breaks and quality problems, very rare short outages – 1 or 2 seconds – will be accepted without a mass storm of complaints. Quality issues may exist with antennas and the receipt of satellite signals.

Figure 2.4 Telefunken TV set from 1936 (Source: Licensed under CC BY 2.0 via Wikimedia Commons – https://commons.wikimedia.org/wiki/File:Telefunken_1936.jpg#mediaviewer/File:Telefunken _1936.jpg. Courtesy: Erik Newth)

Over the years, color was added, and quality increased up to HD. Years before the Internet got live, online text systems were integrated with TV sets in Europe – in the beginning, by a telephone line; the most successful example was Minitel in France. In the last decade, the typical TV picture included browser-like features such as footer and header lines.

Internet Protocol (IP) TV delivers to the user all what his current TV system offers plus additional services. The user has hundreds of channels available and can switch between them instantaneously. Most channels are in HD and of course offer, for example, news plus sports events. In principle, one can have over 100 million users in a media stream – for example, Super Bowl 2014 had 111.5 million viewers. Just to note, TV does not deliver information in real time as telephones – the information typically arrives 7–10 seconds after the event. Nobody is complaining about this since no interaction occurs, but it helps to control the transferred information and eases the transmission.

The term "IPTV" was first used by the company Precept, which was acquired by Cisco in 1998; so the trademark for IPTV is with Cisco. On the history side, it seems that there is some overlap with streaming because Kingston Communication in the United Kingdom did also offer TV over IP [6] from 2000 to 2006. However, the interpretation changed somewhat because, today, IPTV is not sent over the general Internet but over national Telcos provider networks to their domestic subscribers. So, IPTV is not an OTT service.

At the end of 2014, IPTV had more than 115 million subscribers worldwide in over 30 countries. According to Ref. [7], the 10 biggest countries (with subscribers in million) are China (33), France (15), the United States (13), South Korea (11), Japan (5), Russia (5), Germany (3), Thailand (3), the United Kingdom (2), and the Netherlands (2). The main providers are the incumbent Telcos such as China Telecom, Orange, German Telekom, and AT&T (brand: U-verse).

IPTV can include VoD, picture-in-picture for channel switching preview without leaving the current channel, electronic program guide, parental control, time-shift television, and recording. Some providers offer a recording control application on standard smartphones that works over the Internet. A Companion Screen Application (CSA) on another device operates in parallel with the content shown on the TV. The CSA can be used as a remote control but will offer additional information and links for the main program; an integration in home management systems is foreseeable. In essence, it is a way to connect the TV content with the Internet and the home network.

Another way to link the TV to the Internet is the Hybrid Broadcast Broadband TV (HbbTV), which does the integration on the STB, and the user can change between linked information formats. The ETSI standard on HbbTV [8] includes an example that we used as the basis for Figure 2.5. It shows from the upper left to the lower right:

- TV with a visual prompt of available information ("Star Button").
- TV with information overlaid and buttons to decide for options.
- Information with video, audio, or picture inset and buttons to decide for options.
- Just information without audio/video and buttons to decide for options.

2.4   High-End Videoconferences

Videoconference (VC) describes a conference that is based on video and audio transmission in both directions between the conference members. Each conference member hears and

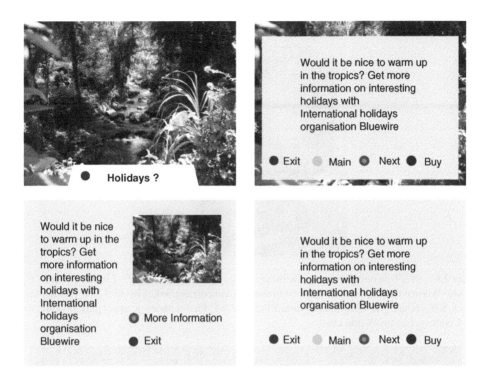

Figure 2.5   HbbTV: linking between TV and Internet information content

sees each other. The first commercial "See-each-other Telephone System" – as it was called initially – started in 1936 in Germany. The transmission used dedicated coax cables between the endpoints. The service was available in post offices of a few cities and a 3-minute call valued $20 in today's dollars. In the early 1970s, AT&T tried to broadly introduce the Picturephone service, which, however, ended in 1980 after a 500 million dollar investment disaster (see Figure 2.6). Until the beginning of the 1990s, all video conferences were only possible between two end systems [9].

The software CU-SeeMe was available from 1994 onward to run videoconferences over the Internet [10]. CU-SeeMe was common for many years but now has been replaced by software from Skype, Google, and others. All the videoconference software running over the Internet can definitely not be considered high-end, which we describe next.

From the mid-1990s onward, a few international companies regularly used Videoconferences in a decent quality over Integrated Services Digital Network (ISDN). They could afford the high costs and had available sufficiently trained personnel to support it. High-End Videoconferences became available at substantial costs from 2004 onward, and the word "Telepresence" was introduced. Please note that a science fiction story named "Waldo" invented the term Telepresence in 1942.

A Telepresence system as shown in Figure 2.7 has the following characteristics:

- The pictures are in HD.
- Each person is displayed in life-size – at least the people in the front row in case there are multiple rows.

Figure 2.6   AT&T picturephone – by Courtesy: LabguysWorld.com (Source: Licensed under CC BY 3.0 via Wikimedia Commons –https://commons.wikimedia.org/wiki/File:AT%26T_Picturephone_-_upper_RH_oblique_view.jpg#mediaviewer/File:AT%26T_Picturephone_-_upper_RH_oblique_view .jpg. Courtesy: LabguysWorld.com)

Figure 2.7   Telepresence system IX5000 (Source: Courtesy of Cisco Systems, Inc.)

- The room is set up with excellent lighting, which is delivered together with the system. In the best case, there is the illusion that the table continues to the other VC room.
- A regular interactive conversation style is possible without any delay or interferences.

Overall, one can read the emotions on other peoples' faces and a participant has to control himself, not to try to shake hands with the persons on the other end.

Since high-end VC systems are expensive, the users in companies expect that the systems are available and reliable similarly to phones. Many users are from senior management, and each small delay in using the system results in an escalation. The users expect that all connections are set up when they enter the room, or they are at best willing to press a single button to initiate a meeting. The expectation of the user is that all problems can be solved by an immediate remote support.

Additional important requirements for VCs are:

Slides: The participants of a meeting want to share slides on their laptop. Meeting members will connect their laptop VGA (Video Graphics Array) port to available connections, and one of the screens will be displayed in all video rooms. The Telepresence systems have displays either below the middle screen or within the tables. The slides can also go to a display screen if it is unused due to fewer participants – see the right screen in Figure 2.7. On rare occasions, a shared whiteboard instead of slides is used. Participants at the different sites can draw on a shared whiteboard and create together a drawing. Remote education systems also use whiteboards.

Audio: Per screen, there is one microphone and one loudspeaker. The sound of each microphone goes to the corresponding speaker. This way, participants hear other persons talking from the direction where they see them. Good echo cancelation need to be in place since the microphones as well receive the audio from the remote speakers. Also, the participants will place their mobile phones on the tables, but this should not cause interference.

Scheduling: Meeting rooms in companies are reserved using scheduling systems that are part of an E-mail system. When scheduling a meeting, a VC room is treated as any other meeting room. Of course, a VC room is required at all sites where the VC has a participant. The VC control system will use the information in the scheduling system on jointly booked VC rooms for an easy setup of a VC meeting.

Additional users: The system can also integrate users on a phone and give them access to the common slides – at best on the fly.

Other features: Additional features are accounting, recording, and encryption – see also Ref. [11].

2.5 Webcast

A Webcast is a multimedia presentation running on a browser that is distributed over a network. The included multimedia elements are always audio, video, slides, and possibly a feedback channel. It is not foreseen to do interactive discussion with viewers, but viewers' questions that are sent may be answered. Usually, Webcasts are recorded and can be viewed later on demand – of course, without asking questions. The term Webcast does not describe the form well: a better name could be Web-Presentation. Due to the uncertain name, the history is unclear, and sources mention audio-streaming or video-streaming events as Webcasts as well.

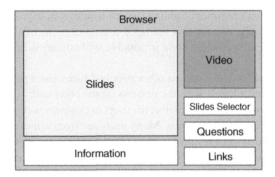

Figure 2.8   Typical browser setup for a Webcast

Webcasts are used for lectures, events, and meetings to up to multi-thousand viewers. From the typical Webcast setup on the browser shown in Figure 2.8, we see that the slides are the important part and have to be delivered in good visibility. The video is typically small because the audio of the presenter is more important than viewing him. In real presentations in conference centers, the speaker is sometimes hard to see but his slides are always visible. The information field contains at least the title, the name of the speaker(s), and other information.

A viewer can type questions that arrive at a central system. In big Webcasts, a communication person views the feedback and hands selected questions to the presenter at an agreed time. For the on-demand version, the viewer with the slide selector can choose at which slide the Webcast continues synchronized with audio and video.

A Webcast is not time-critical since the only interactions are text message questions. If a Webcast has delays – below 10 seconds – nobody will complain. The slides have to change when the presenter moves to the next one, but again this can lag a few seconds behind for the viewer.

2.6   Requirement Summary

Table 2.1 summarizes the network characteristics of the aforementioned different multimedia applications. We described the requirements in user language. We will later translate those user terms to measurable values, and in Section 12.5, we reuse this table and show the implementation characteristics that were elaborated in all the chapters in between.

In the table, those requirements in bold have significant implications for the network. In the following chapters, we will explain why those bold elements are challenging. Of course, the more bold items per service, the more complicated the overall service implementation will be.

Table 2.1 Requirements of typical multimedia applications – network view

	Telephony	Streaming	IPTV	Telepresence	Webcast
Availability	**Very high**	Normal	High	**Very high**	High
Stability	**Very high**	High	High	**Very high**	High
Simultaneity	**Almost**	No	Delayed	**Almost**	Delayed
Start-up delay media stream	Nationally: <1 second	>5 seconds	**<1 second**	Few seconds	<1 second
Delay within session	Not recognizable	For forward/ backward >5 seconds	Not recognizable	Not recognizable	1 second tolerable
Error	Not recognizable	Not recognizable	Not recognizable	Not recognizable	Not recognizable
Interacting devices	Yes – distrib. meetings	Yes	Yes	No	No
# users within same media stream	2 (seldom – 10,000)	VoD/Video-Host.: 1–5,000, Live: – 100,000	**100,000– 100,000,000**	2–50	20–10,000
Transported media	Audio	Audio (stereo, mult. languages), video – up to HD, opt. subtitle	Audio (stereo), **HD video**	Audio (mult. channels), slides, **HD video**	Audio, video, slides

3

Audio, Image, Video Coding, and Transmission

Everything starts with delivery of the analog media signal – video and voice. Figure 3.1 provides more detail about the scope of this chapter.

The analog audio and video signal $\sim$ is in all cases digitalized by an analog-to-digital converter $\boxed{A \setminus D}$. An uncompressed HD-video stream carries at minimum 1.485 Gb/second and clearly shows the need for compression.

We start with audio compression in Section 3.1 and introduce diverse mechanisms used for compression. Section 3.3 first discusses image encoding, since all video encoding is based on image encoding. We introduce different video compression methods in Sections 3.4 and 3.5. The network introduces errors and multimedia applications have to cope with this – see Section 3.7. Please note that transmission, in this chapter, relates to the application layer, while the following chapter will explain the underlying network function.

In the final section, we handle efficient changes in video formats – that is, transcoders.

This chapter describes the decompression and presentation in only a rudimentary way since the standards for compression do not include/comprise the decompression part. The decompression system must correctly re-create the compressed content. However, in the following chapter on synchronization, we will treat decompression challenges.

This chapter discusses the concepts to a depth that the reader can understand the network implications. For a deeper understanding, Li et al.'s book [12] gives much more detail but does not treat transcoding.

3.1 Audio

Analog-to-digital converters digitize analog signals by the following techniques:

Sampling: This works in the time dimension, and signal samples are taken based upon a given sampling frequency.

Quantization: In the amplitude dimension, the signal is converted to a digital value. It is a linear quantization if the amplitude dimension uses a linear scale. Otherwise, it is a nonlinear quantization.

Multimedia Networks: Protocols, Design, and Applications, First Edition. Hans W. Barz and Gregory A. Bassett.
© 2016 John Wiley & Sons, Ltd. Published 2016 by John Wiley & Sons, Ltd.

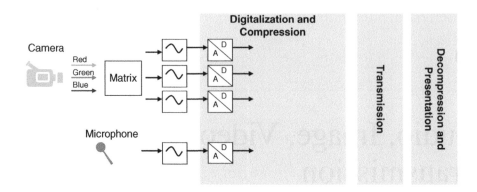

Figure 3.1   Scope

Figure 3.2 shows an example of a sampling along with a linear quantization at each dot. Here, this results in the following values: 0, 8, 21, 80, 139, 195, 229, 260, 257, · · ·. The sampling frequency must be at least twice the maximum frequency of the analyzed signal if we want to reconstruct the original signal later – this is the Nyquist–Shannon theorem. In the given example, we see that we have sampled roughly eight times more often than the original signal frequency.

If the quantization uses values out of a fixed-code alphabet, it is a Pulse-Code Modulation (PCM). Our example shown in Figure 3.2 is a PCM with an alphabet of 0–260.

The human hearing is restricted depending on the frequency and amplitude as shown in Figure 3.3 for pure tones. Humans can hear everything above the curve depending on the tone and loudness. Also, humans can differentiate 100 Hz at low frequencies but only 4 kHz at the highest tones. Phones use frequencies of 200–4000, and for a CD, the frequency range

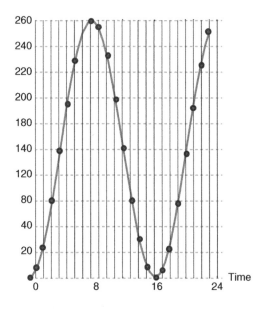

Figure 3.2   Example for sampling and quantization

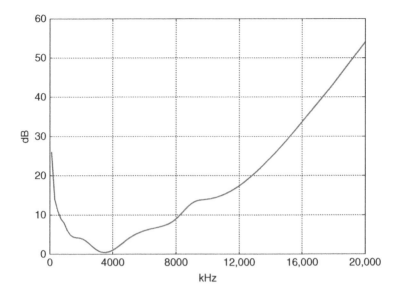

Figure 3.3   Human hearing range for pure tones

Table 3.1   Characteristics of sampling and data rate for audio

	Sampling rate (kHz)	Bits	Data rate (kB/second)	Frequency
Phone	8	8	8	200–4000
CD (stereo)	44.1	16	176.4	5–20,000

is 5–20,000. This frequency range defines the required sampling rate and in the end also the data rate – as shown in Table 3.1. With a sampling rate of 8 kHz for a phone and 2^8 quantized values, we obtain 8 kB information per second. Similarly, for a CD, we obtain information of 88 kB per second for each stereo channel.

3.1.1   Companding

There exists another restriction on our hearing: as stronger a signal already is as much proportionally more must be added to hear the difference (Weber's law) – that is,

$$\Delta \text{Response} \propto \Delta \text{Stimulus}/\text{Stimulus} \tag{3.1}$$

when we can tell the difference of an increase of 1 at a loudness level 10, then it needs 2 at a level of 20 to be recognized as a difference. The difference is the ΔResponse, the Stimulus is 10, and the ΔStimulus is 2. In other words, the more silent it is, the more differences we can hear. But this also means that we should use more bits during quantization in that area where the signal is rather quiet. In the following, we explain how this is integrated with telephony.

By rewriting Equation 3.1, we obtain a differential equation with r Response, s Stimulus, and k for the proportionality:

$$dr = k \cdot ds/s \tag{3.2}$$

By integration and replacement of the integration constant C for the situation $r = 0$, we arrive at

$$r = k \ln \frac{s}{s_0} \tag{3.3}$$

with s_0 being the lowest stimulus, which causes a response.

The developed logarithmic equation 3.3 was the reason for an adapted quantization in telephony. The standard ITU G.711 [13] defined this adjustment in the 1970s. The method carries the name "Companding." G.711 describes two alternative companding formulas: μ-law and A-law – used in the United States and in Europe, respectively:

$$\mu - \text{law} \; : \; r = \frac{\text{sgn}(s)}{\ln\,(1 + \mu)} \ln \left(1 + \mu \left| \frac{s}{s_{peak}} \right| \right) \tag{3.4}$$

$$A - \text{law} \; : \; r = \begin{cases} \dfrac{A}{1 + \ln A} & \text{if } \left| \dfrac{s}{s_{peak}} \right| \leq \dfrac{1}{A} \\ \dfrac{\text{sgn}(s)}{1 + \ln A} \left(1 + \ln A \left| \dfrac{s}{s_{peak}} \right| \right) & \text{if } \dfrac{1}{A} \leq \left| \dfrac{s}{s_{peak}} \right| \leq 1 \end{cases} \tag{3.5}$$

$$\text{where sgn}(s) = \begin{cases} 1 & \text{if } s > 0 \\ -1 & \text{otherwise} \end{cases}$$

The parameters are normally set to $\mu = 100$ (or 255) and $A = 87.6$; Figure 3.4 shows the curves with those values for Equations 3.4 and 3.5. s_{peak} is the peak stimulus value, and the fraction s/s_{peak} normalizes the values in the range $[-1{:}+1]$.

Before quantization, the signal s is transformed by the μ- or A-law to the response value r. So, a small signal value s leads to a value of s/s_{peak} close to 0, and those values are differentiated highly in the response. Larger signal values s lead to value of $s/s_{peak} \geq 0.5$ or ≤ -0.5, which are compressed in the possible response values r. After transmission during analog output generation, the reverse function needs to be applied.

The configuration of voice switches has to reflect that Europe and the United States use different companding functions to avoid distortions.

We mentioned before that the standard G.711 [13] describes companding. In addition, this standard also defines the transfer of voice without any compression apart from companding. It uses 8-bit samples and requires 8 kB/second transmission rate; here, no further IP overhead is included. Other elements are 10 millisecond sample interval and every 20 milliseconds a packet is sent by default.

The shorthand PCMA describes PCM (G.711) with the A-law and PCMU describes PCM (G.711) with the μ-law.

At the end of this section, it needs to be mentioned that some newer measurements exist [14], which contradict Weber's law for voice and so also the reasoning for the μ- or A-law. However, the μ- or A-law will stay.

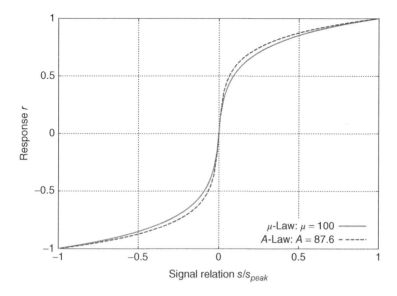

Figure 3.4   Companding function in telephony

3.1.2   *Differential Quantization*

Differential quantization methods use only the differences between consecutive samples – the so-called Differential Pulse-Code Modulation (DPCM). The differences are normally much smaller than the sample values, and this reduces the transmitted data. However, errors can propagate to later values and quality can drop.

When we look at the example for sampling and quantization – Figure 3.2, we see that the differences are quite small around the turning points of the function but can become quite larger during steep parts of the function. For that reason, only Adaptive Differential Pulse-Code Modulation (ADPCM) is used. ADPCM uses a prediction for the change of the next sample, and only this deviation is transmitted.

The schema for ADPCM is shown in Figure 3.5 and is described as follows:

- The encoders work sequentially on the sampled input signals $s_1, s_2, s_3, \cdots, s_n$ – let us assume that currently we are working on signal s_n.
- From the last iteration, we obtain a predicted signal s_n^{pr}.
- s_n is compared against s_n^{pr}, and the prediction error e_n is computed – typically just the difference.
- The adaptive quantization function Q is applied on e_n, and we obtain e_n^q. This value is transmitted but is also required in the feedback loop.
- The adaptive inverse quantization function Q^{-1} is applied on e_n^q. The receiving side acts on the transmitted value accordingly. So, the prediction is based on a commonly reconstructed value s_n^q.
- The adaptive predictor function A is applied on s_n^q together with the last prediction s_{n-1}^{pr} and creates a new predicted signal value of s_{n+1}^{pr}. Since the schema shows a whole iteration, the predicted value s_{n+1}^{pr} is already obtained for the next iteration.

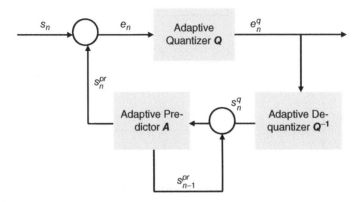

Figure 3.5   Schema ADPCM Encoder

In a short form, the whole schema is:

Create: $s_n^{pr} = A(s_n^q, s_{n-1}^{pr})$, $e_n = s_n - s_n^{pr}$, $e_n^q = Q(e_n)$,
Transmit: e_n^q,
Reconstruct: $s_n^q = Q^{-1}(e_n^q, s_{n-1}^{pr})$.

A broadly used ADPCM algorithm was defined by the Interactive Multimedia Association (IMA) in 1992. This organization ceased to operate in 1998, but the specified algorithm is still in use and is accessible [15]. Different names are used for this algorithm – for example, the IETF uses the name DVI4 [16], and Microsoft uses it for compressions of files with the extensions .avi and .asf. A variant is the Linear Pulse-Code Modulation (LPCM) used by Microsoft for .wav files. The ADPCM algorithm reduces the signal size from 16 bit to 4 bit. The C-alike pseudocode in Algorithm 3.1 uses the notation shown in Figure 3.5 as much as possible. Note that the original algorithm works with bit-shift operations for division by 2, but we changed this to a more readable version here.

Example for Algorithm 3.1 As an example for the code, we use the quantized values from
Figure 3.2 as our signals – that is, $s_i = 0, 8, 21, 80, 139, 195, 229, 260, \cdots$:

- We obtain in lines 5–9 $e = 0$ and then fill the last bit of the prediction error with $e\_q[3] = 0$. The last bit characterizes the sign of the difference.
- In lines 5–19, the size of e versus the *stepsize* is checked, and the higher the value, the more bits are set. Here, we obtain $e\_q = 0000$, and $diff = 0$ indicates the required change of the *stepsize*.
- Line 22 can only deliver values in the range $[-32767:32767]$, which corresponds to 2^{16} as the possible input values. Hence, the next predicted value s_pr stays unchanged. In general, s_pr can change maximally by $1.75 \cdot stepsize$.
- Line 24 can only deliver values in the range $[0:88]$, which corresponds to the size of the *ima_step_table* – in this case, *index* is not changed. In line 26, the *index* selects from the *ima_step_table* the new *stepsize* – also the *stepsize* is unchanged. The *stepsize* is a measure of the currently observed steepness of the function or said differently for the associated multiplicator for each bit in e_q.

- In the second iteration for the value 8, we have the following values in the end: $e\_q = 0000$, $diff = 0$, $s\_pr = 7$, $index = 0$, $stepsize = 7$.
- In the third iteration for the value 21, we have the following values in the end: $e\_q = 0111$, $diff = 11$, $s\_pr = 18$, $index = 8$, $stepsize = 16$.
- In the fourth iteration for the value 80, we have the following values in the end: $e\_q = 0111$, $diff = 30$, $s\_pr = 48$, $index = 16$, $stepsize = 34$.
- In the fifth iteration for the value 139, we have the following values in the end: $e\_q = 0111$, $diff = 63$, $s\_pr = 111$, $index = 24$, $stepsize = 66$.
- In the sixth iteration for the value 195, we have the following values in the end: $e\_q = 0101$, $diff = 91$, $s\_pr = 202$, $index = 30$, $stepsize = 118$.

Algorithm 3.1: IMA ADPCM algorithm (also: DVI4, .wav, .avi, .asf)

Input:
ima_step_table[0,...,88] := [7,8,9,10,11,12,13,14,16,17,19,21,23,25,28,31,34,37,41,45,50,55,60, 66,73,80,88,97,107,118,130,143,157,173,190,209,230,253,279,307,337,371,408,449,494,544, 598,658,724,...,32767];
adjust_index[0,...,7]:=[−1,−1,−1,−1,2,4,6,8];
s: Next signal; /*16 bit*/
Output: e_q[0,...,3] Binary array: quantized prediction error;

```
1  index := 0 ; /* Index of current interval */
2  stepsize := 7; /* Quantizer stepsize */
3  s_pr := 0; /*Predicted signal */
4  while not eof input do
5      e:= s-s_pr; diff:=0; e_q:=0;
6      if e < 0 then
7          e_q[3]:=1; e:=−e; help:=−1
8      else
9          e_q[3]:=0; help:=1
10     end
11     if e >= stepsize then
12         e_q[2]:=1; e:=e-stepsize; diff:=stepsize
13     end
14     if e >= stepsize/2 then
15         e_q[1]:=1; e:=e-(stepsize/2); diff:=diff+(stepsize/2)
16     end
17     if e >= stepsize/4 then
18         e_q[0]:=1; diff:=diff+(stepsize/4)
19     end
20     /* Transmit e_q */
21     diff:=diff+(stepsize/8);
22     s_pr:= min(max(s_pr + help*diff,−32767),32767) ;
23     /* guarantees predicted signal is in range*/
24     index:= min(max(index+adjust_index[e_q[1:3]],0),88);
25     /* guarantees adjusted index is in range*/
26     stepsize:= ima_step_table[index];
27 end
```

The dequantifier receives the sequence of values $e^q_{1,2,3,4,\cdots}$ and also knows the initial values of the variables $s\_pr, index, stepsize$. Based on each e^q_i, the dequantifier maintains the analog values of $s\_pr, index, stepsize$. The values s_pr are output values of the transmission. It is obvious that a lost e^q is never recovered in this transmission method and all the state variables would stay at a wrong value for the following transmitted values.

Note that diff*help is the s^q_n from the schema definition of the ADPCM Encoder (Figure 3.5) and the adaptive predictor A is just $s^{pr}_{n-1} + s^q_n$ to form s^{pr}_n.

The ITU issued in 1984 the ADPCM G.726 [17] standard. It is much more complicated then IMA since, for example, the predictor function is based on a combination of six quantified differences from six previous signals. In addition to the IMA standard, it has the option to compress to 16, 24, 32, and 40 kbit/second. The compression to 32 kbit/second is the most used one and is the standard for DECT phones (Digital Enhanced Cordless Telecommunications – European standard for in-house phones). G.726 samples every 5 milliseconds, and every 20 milliseconds a packet is sent by default.

3.1.3 Vocoders

So far, all presented methods for voice compression could be used for any type of audio. Vocoders – that is, Voice Coders – are optimized for compression of human voice – see, for a broad overview, Ref. [18]. These codecs can, however, not be used to encode music or tones generated by operating the buttons on the phone.

In the 1930s, subband filtering was introduced for speech transmission over telegraph lines in the Bell Labs. Here, the signal band is divided into frequency subbands based on band filters. The signals from each band are sampled, encoded, and transmitted. The receiver decodes, demultiplexes, and by building the sum, creates the output. The bands are quantized with different numbers of bits – the more the number of bits, the better this spectrum can be heard. For example, the previous ITU standard G.722 quantizes a 50 Hz–3.5 kHz subband with 48 kbit/second and a 3.5–7 kHz subband with 16 kbit/second.

Subbanding aids the vocoders, but the newest method is to understand the letters or a combination of those and transfer, in essence, the letters and re-create the speech by an audio generator. The speech sounds are formed in the mouth, the nose, and the throat. The important frequencies that form a letter are called formants. We can associate in each language particular base frequencies to a letter – see Table 3.2. F_1 is the formant with the lowest frequency of a letter, F_2 is the formant with the next frequency of a letter, and so on. Typically, the first two formants are enough to disambiguate a vowel. Formants have a length of 10–40 milliseconds.

Table 3.2   Vowel formants in English

Vowel	Formant F_1 (Hz)	Formant F_2 (Hz)
I	240	2400
E	390	2300
O	360	640

The Linear Predictive Coding (LPC) analyzes the waveform and not the frequency. The algorithm describes the waveform by a set of LP coefficients (gain, pitch, voiced/unvoiced, etc.) and sends an analysis of the sound but not the sound itself. The name linear prediction results from the fact that each sample of the speech sequence is predicted from a linear combination of p past samples with $s(n)$ output. a_i is a coefficient in the linear prediction model, $f(n)$ is the output of the vocal-tract model, and G gain:

$$s(n) = \sum_{i=1}^{p} a_i s(n - i) + Gf(n) \qquad (3.6)$$

The Internet Low Bit Rate Codec (ILBC) [19] uses LPC with blocks of 30 milliseconds. Each block is independent, and no predictions from blocks before are used. Since each block describes a number of formants, losing data packets does not result in any error propagation. ILBC 30 (ILBC 20) compresses to 13.33 kbit/second (15.2 kbit/second) and samples every 30 (20) milliseconds and also sends every 30 (20) milliseconds a packet by default.

LCP was further improved by the Code(-book) Exited Linear Prediction (CELP) methods. This method adds Long-Term Prediction (LTP) based on codebooks. The methods of LPC are described here as adaptive codebooks. The achieved quality is usually acceptable also for audio conferences, but the additional time for LTP requires a lot of processing power. The Algebraic Code Exited Linear Prediction (ACELP) is a patented expansion that is better performing.

The G.723.1 [20] standard uses ACELP but can switch at every 30 milliseconds boundary also to Multipulse Maximum Likelihood Quantization (MP-MLQ). In the case of packet loss at the beginning of a voiced segment this algorithm is severely impacted [21]. G.723.1 compresses to 6.3 kbit/second (5.3 kbit/second) and samples every 30 milliseconds and packets are sent each 30 milliseconds by default.

The G.729 [22] standard also uses ACELP and exists in multiple versions; the important versions are lower complexity (G.729a), silence compression (G.729b), and low complexity with silence compression (G.729ab). Silence compression is the technique to cut out silence for the transmission. G.729 compresses to 8 kbit/second and samples every 10 milliseconds and packets are sent every 20 milliseconds by default.

Table 3.3 summarizes the standards that are typically usable for Voice Compression. We will again treat this in Section 6.8.3.

3.1.3.1 Audio MPEG

The Moving Pictures Experts Group (MPEG) started at the end of 1980s to create numerous audio compression standards. The standards are based on the principle of an asymmetric coding system – see Figure 3.6. This principle works well if the encoding process does not occur in an interactive communication and encoding is executed in a batch modus. The decoder is thought to be dumb since it has less processing power and needs to decode on the fly. Note that these coders are not vocoders – that is, the coders work for all audio-inclusive music.

Since the encoder in MPEG is smart, it can use more elaborated characteristics of the ear – the so-called psychoacoustic or perceptual model – see Figure 3.7. This model decides what

Table 3.3 Typical voice compression standards

Codec	Description	Rate (kbit/second)	Frame (millisecond)
G.711	Uncompressed Voice Codec apart from A/μ-Law	64	20
G.723.1	MPC-MLQ or ACELP	5.3,6.3	30
G.726	ADPCM	40, 32, 24, 16	var.
G.729	ACELP (G.729a lower complexity, G.729b silence compression scheme, G.729ab reduced complexity G.729b)	8	10
ILBC	Internet Low Bit Rate Codec uses LPC, stable at packet loss	15.2,13.3	20,30

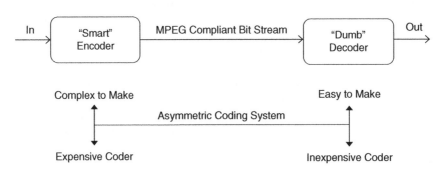

Figure 3.6   Overview of MPEG audio compression

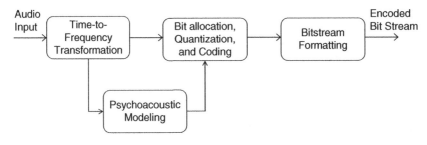

Figure 3.7   Basic MPEG audio coding

gets quantized and transferred. This model describes how tones influence the hearing of other frequencies close by:

Frequency masking: Tones at a particular frequency can mask other tones with a slightly different frequency. As louder a tone is, the more impact it has in its surrounding. Lower tones can well mask higher frequencies, but the reverse is not true.

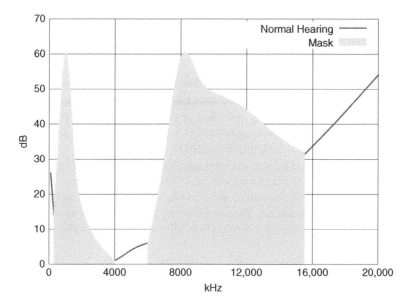

Figure 3.8 Frequency masking

In Figure 3.8, we have added two 60 dB loud tones at 1 and 8 kHz to the overall hearing curve (Figure 3.3). The shaded area below those two frequencies is masked and cannot be heard at the same time.

Temporal masking: After the end of a loud tone, the masked tone needs some time to be heard. The time span until the masking disappears depends on the length of time the original tone existed.

The psychoacoustic model is an excellent example of the asymmetric coding system shown in Figure 3.6 since it has a high effort when encoding but even decreases the work for the decoder.

Note that MPEG specifies the use of the psychoacoustic model, but it does not specify what the model exactly looks like and which measures need to be implemented. This results in the fact that the quality of the generated MPEG audio stream can be quite different.

All MPEG standards split the input into different frequency bands – typically 32 – as we have seen this already with the vocoders. Note that these bands are partially overlapping because no filters exist that can perfectly differentiate frequency bands.

The most important MPEG audio standards are:

MPEG-1 The standard was issued in 1992/1993 and covers three layers [23]. Supported bit rates are in the range of 32–448 kbit/second. The possible sampling rates are 32, 44.1, and 48. The standard supports mono and stereo. The different layers are:

Layer 1: It is abbreviated to MP1 with a file extension .mp1. It uses 32 subbands for coding. It is outdated.

Layer 2: It is abbreviated to MP2 with a file extension .mp2. Joint-stereo mode removes redundancies between channels. It is known to be fault-tolerant and is used by broadcasters. In addition to Layer 1, it uses the psychoacoustic model.

Layer 3: It is abbreviated to MP3 with a file extension .mp3. Of course, it is the most famous one of all MPEG audio compression standards. In July 1994, the Fraunhofer Society released the first MP3 player, and in 1998, the ISO publicized a freely accessible reference software.

In addition to MP2, this method uses a modified Discrete Cosine Transform (DCT), which we treat in Section 3.2.4. Further, MP3 uses Huffman Encoding – see Section 3.2.3.

MPEG-2 The standard was originally published in 1994 and also covers three layers [24]. In essence, the methods are the same as of MPEG-1 but with higher bit and sampling rates. All methods are backward compatible with MPEG-1.

MPEG-2 Part 3/7: The standard was released initially in 1977 [25] and is named Advanced Audio Coding (AAC). File extensions are .mp4, .m4a, .aac, and others. It comes in three profiles: low complexity, main profile, and scalable sampling rate. Supported bit rates are in the range of 8–160 kbit/second. The sampling rates are 8–96, and up to 48 channels are possible. It includes improvements in relation to MP3 in all areas and also covers prediction methods.

MPEG-4 Part 3: The standard was published originally in 1999 [26] and is named Audio Lossless Coding (ALS). Supported bit rates are in the range up to 384 kbit/second and arbitrary sampling rates. Multichannel is possible. It is the only lossless coding standard in MPEG.

MPEG-7 The standard was issued in 2001/2002 [27] and is a content description for multimedia.

MPEG-21 The standard was published in three parts since 2004 (first part: [28]). It describes a multimedia framework.

For a complete overview, see Refs [12, 29, 30].

3.2 Basics of Video Coding

Figure 3.9 uses Figure 3.1 as a basis but provides much more detail. While the light arrives at the camera in red (R), green (G), or blue (B), conversion takes place in different color elements

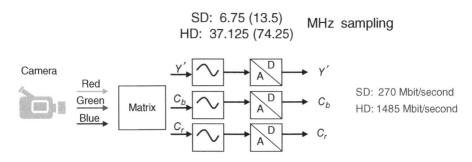

Figure 3.9   Overview of video sampling

Y, C_b, C_r. After that, it passes through the analog–digital converter. Also, a gamma conversion is run to adapt the input to our human eye that has a greater sensitivity to relative differences between darker tones than between lighter ones. The gamma conversion is denoted by $'$.

Y' is the luminance or luma and C_b, C_r are color differences. The colors are also named chrominance or chroma. The R.601 color space YCbCr [31] defines this as:

$$Y' = 0.299R' + 0.587G' + 0.144B'$$

$$C_b = 0.564(B' - Y') + 0.5$$

$$C_r = 0.713(R' - Y') + 0.5$$

Note that there exist many more color models, but MPEG and JPEG commonly use the R.601 space. Consequently, we restrict ourselves to "Rec. 601." Note that only PAL (Phase Alternating Line) uses $Y'UV$ (mostly one YUV) with $Y' = 0.3R' + 0.6G' + 0.1B'$, Hue $U = 0.493(B' - Y')$, and Tint $V = 0.877(R' - Y')$. NTSC (National Television System Committee) uses the color space YIQ with in-phase chrominance I and quadrature chrominance Q instead of U/V.

Figure 3.10 gives an impression of the color range defined by C_b, C_r with a luminance value of $Y' = 0.5$: $C_r = 1, C_b = 1$ delivers purple; $C_r = 0, C_b = 0$ dark green; $C_r = 1, C_b = 0$ orange; and finally, $C_r = 0, C_b = 1$ light blue.

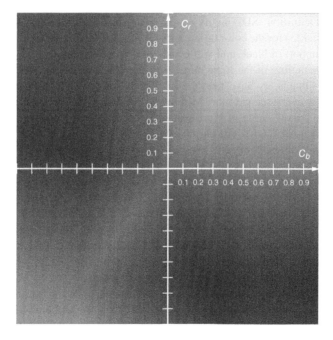

Figure 3.10 $C_b C_r$ value range for $Y' = 0.5$ (Source: Licensed under CC BY 2.0 via Wikimedia Commons – http://upload.wikimedia.org/wikipedia/commons/3/34/YCbCr-CbCr_Scaled_Y50.png. Courtesy: Simon A. Eugster)

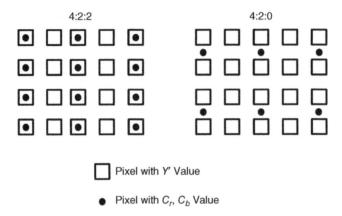

Figure 3.11   Subsampling schemas

Common to all standards is, however, the separation of luminance and color difference. The separation of luminance is essential since the eye can exhibit visual acuity in black and white much better than in any color. Consequently, the luminance needs a higher sampling compared to the color components. As illustrated already in Figure 3.9, the sampling value for the Standard Definition (SD) quality is 13.5 MHz (HD: 74.25) for the luminance Y' and 6.75 MHz (HD: 37.125) for the color differences C_b, C_r. Another reason to separate luminance was the parallel support of black-and-white and color-receiver sets.

The increased sampling of the luma means subsampling for chroma. R.601 subsample by the 4:2:2 schema – that is, we have, per iteration, 4 pixel values of Y, but the 2 pixel values for C_b, C_r are sent only every second time. MPEG and JPEG use the schema 4:2:0, which means that, per iteration, 4 pixel values of Y are sent and a transmission occurs at every second time for the 2 pixel values for C_b, C_r. The chroma subsampling is shown in Figure 3.11.

The screen output can be painted in two basic ways:

Interlacing: The output writes first the uneven rows and then the even rows on the screen. This technique was invented at 1930 by Telefunken – see the old TV shown in Figure 2.4 from the same company – to overcome technology limitations for TV. Cathode ray tubes (CRTs) could only be written with the AC line frequency (Europe: 50 Hz or United States: 60 Hz), but this rate was too high for the delivered input. In the motion screen displays, the same screen was shown multiple times, but this was not feasible on TV because of insufficient memory availability. As a consequence, we have still today interlacing on TV.

Due to the separate writes – even and uneven – of the screen, interlacing results in display of two pictures on the screen at the same time. Interlacing can result in flicker when showing fast movements or in shimmering when showing finely stripped patterns. Consequently, TV professionals avoid wearing finely striped clothes.

Sometimes, an interlaced video requires conversion to a noninterlaced video – that is, deinterlacing. Deinterlacing delivers lower resolution and can create artifacts. The best way is to use the original before interlacing.

Progressive: The output writes all rows in one run. Progressive is the standard on all computer screens and also on most screens based on LCD (Liquid Crystal Display) technology.

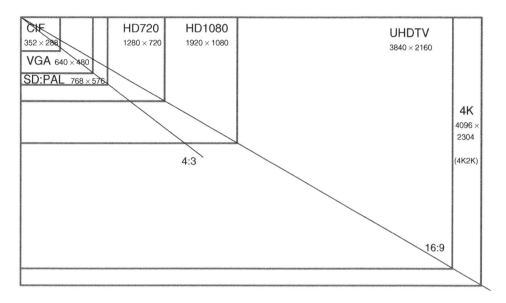

Figure 3.12   Display size and resolution

Table 3.4   Characteristics of display types

Overall resolution	Sampling	Aspect ratio	Picture rate	Data Rate (MB/second) – Raw (10 bit resolution)
Standard-definition TV (SDTV/SD)				
720 × 480	4:2:2	4:3	60/50 I	26
High-definition TV (HDTV/HD)				
1280 × 720	4:2:0	16:9	60 P	211
1920 × 1080	4:2:0	16:9	60 I	237
Ultra-high-definition TV (UHDTV/UltraHD)				
3840 × 2160	4:2:0	16:9	≤120 P	3000

Screen displays have increased resolution over the past few decades, and Figure 3.12 provides an overview of size.

Table 3.4 summarizes the main characteristics of the display types. While these numbers are impressive, it is worthwhile to note that only UHDTV has a higher resolution compared to good old 135 (United States) and 35 mm (Europe) camera films. Only with a new UHDTV projector, the old slide projector becomes obsolete. The required raw data rates are, however, good reasons to look at compression mechanisms in the following sections.

In audio compression, the mechanism to achieve compression has been to remove redundancy or remove parts the human eye does not recognize as missing. When removing redundancy within videos, this could be spatial or temporal. Spatial redundancy is reducing within single images of a video while temporal redundancy describes reduction across multiple pictures that follow each other in time.

The techniques to execute video compression can be classified as:

Simple: Simple mechanisms incorporate truncation, Color Look-Up Tables, (CLUT) and Run Length enCoding (RLC). Section 3.2.1 treats these techniques.

Interpolative: In those mechanisms, an interpolation between values takes place. We have already discussed the typical method with subsampling and will not elaborate more on this.

Predictive: These mechanisms predict values based on other earlier or later values. We treat Motion Compensation or Motion Estimation as the typical method in Section 3.2.2.

Statistical: The statistical methods decrease the required information transfer based on statistical characteristics of the treated data – this is handled in Section 3.2.3.

Transform: Transform methods use transform functions to concentrate relevant information in certain areas and provide the opportunity to drop or partly drop nonrelevant information. Section 3.2.4 treats the most prominent mechanism Discrete Cosine Function (DCT).

3.2.1   Simple Compression

Truncation is a very basic method. Here, the least significant bits from a pixel value are truncated.

In the 1980s, the Graphical Interchange Format (GIF) [32] used CLUT for pictures; this was supported by CompuServe. Here, a color pixel is mapped to the closest color in an array. The indexes in the color array were transferred as replacement for the pixel. There existed standard color tables for 8 bit (named Pseudocolor), for 16 bit (Hicolor) and 24 bit (Truecolor). Also, a color array could be generated only for one image – for example, one that is visible in Figure 3.13.

RLC replaces a sequence of pixels with the same value by the pair ($< count >, < value >$). For example, the sequence $\boxed{11,22,34,34,34,77,11,22}$ will be replaced by $\boxed{11,22,(3,34),77,11,22}$. For computer-generated graphics, this works well, but less with natural images. Also, other compression algorithms use RLC in appropriate situations; it will first appear in JPEG – Section 3.3.

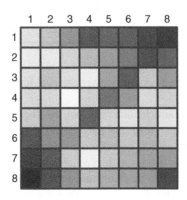

Figure 3.13   Eight bit CLUT

3.2.2   Motion Estimation

This method is solely usable with videos. When looking at a video, it is obvious that apart from scene cuts, successive pictures from a video change slightly. In the following, we will not use the word picture within a video but it is a frame. The changes from frame to frame often are small movements of individual elements. So, if we can describe the changes from frame to frame, we can avoid sending the new frame at all or send only the 1% that has changed.

Motion estimation or motion compensation is the method to find out where certain parts of the frame have moved. Typically, if one has identified the movement of one piece to exactly another position, the most parts of the rest of the frame will have moved in an identical way.

Frames are split into quadratic (macro)blocks. Different standards define different block sizes, but for the principle of the motion compensation method, the size is only a parameter. Here, we assume a block size of N. Between frames $n - 1$ and n, we need to look at two different blocks: C from frame $n - 1$ and D from frame n. C and D have a certain distance from each other, which is measured by the pair (i, j). Figure 3.14 illustrates this situation.

The Motion Vector (MV) is the pair (i, j) where Minimal Absolute Difference (MAD) from Equation 3.7 has its minimum. MAD just computes the differences between the pixel luminance values. Of course, only the luminance values are used because higher accuracy is required:

$$\text{MAD} = \frac{1}{N^2} \sum_{k=0}^{N-1} \sum_{l=0}^{N-1} |C(x + k, y + l) - D(x + k + i, y + l + j)| \qquad (3.7)$$

As a full search of the whole frame is not feasible, one concentrates on a search window that is seen in Figure 3.15. Within this window, a 2D logarithmic search can be used. The algorithm starts with nine well-distributed points in the search window – here marked as a *. For each * position, MAD is calculated. Around * with the minimum MAD, another nine new points are generated with half of the distance as in the last iteration, and the search continues.

There are many more options to make the quest more efficient – for example, one can down-sample the resolution and search first a more coarse resolution and then higher resolutions. Many of the later provided standards also define which type of motion vector is acceptable – for example, which directions of movements are allowed.

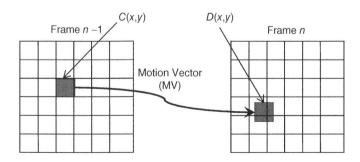

Figure 3.14   Displaced blocks for treatment by motion compensation

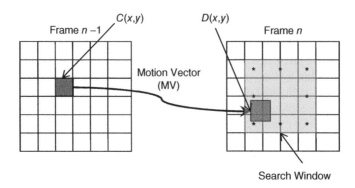

Figure 3.15   Searching the motion vector MV

3.2.3   *Statistical Compression*

The methods in this space use the fact that sample values are not equally likely. So, more common values are encoded with fewer bits and less common with more bits. Shannon [33] has shown how many bits are required to encode information. The *entropy η* of an information source with alphabet $S = s_1, s_2, \cdots, s_N$ is given by

$$\eta = \sum_{i=1}^{N} p_i \log_2 \frac{1}{p_i} \tag{3.8}$$

with p_i probability that sample s_i is part of N samples. $\log_2 \frac{1}{p_i}$ indicates the minimum number of bits required to encode s_i. η represents the average amount of information contained in $s_1, \cdots, s_N$.

Each statistical method is measured how close it gets in relation to η. It is clear that all methods require a codebook that translates a sample value to its encoding. Such a codebook can be static or adaptive. In the adaptive case, the codebook changes when new values arrive and need encoding. Obviously, adaptive codebooks will be more efficient since they can represent the actual probabilities better. However, sender and receiver need to adapt their codebooks in sync. Adaptive codebooks run the risk of getting out of sync when nonrecoverable losses occur during a transmission.

In detail, we will see how the Huffman coding [34] generates the codes. It belongs to the class of Variable Length Coding (VLC) algorithms. Algorithm 3.2 produces codes that are close to the optimum since ($\eta \le$ *average code length* $< \eta + 1$) – that is, the code either has the optimal size or is at most one greater than the optimum.

The algorithm creates a binary tree where each leaf node is a symbol value out of the set of samples. During the algorithm, the tree is built so that in the end, one can read the codebook values by passing from the root to a leaf. The creation process uses the probabilities of the symbols to move nodes with a higher probability closer to the root. When a node is closer to the root, the generated code for the symbol in this node becomes shorter.

Example of Algorithm 3.2 As an example, we use the following tuples: $(E, 148)$ $(T, 93)$ $(A, 88)(O, 72), (R, 68)$. The five letters E, T, A, O, and R are the most probable letters in English texts, and the second value in each tuple denotes the frequency of occurrence for 1000 letters.

Algorithm 3.2: Huffman coding
Input: N number of input tuples Tuples s_i, p_i for all N /* Symbol and their probability of appearance in alphabet */ **Output**: codebook

```
 1  char * codebook [N+1]; /* Stores optimal code for input symbols*/
 2  struct node char symbol; int count; struct node* left; struct node* right;;
 3  /* tree element node with symbol value, count, and two pointers left,right*/
 4  node new_root;
 5  node coding [N]; /*array of tree elements node */
 6  Procedure Sort (node, k);
 7  /*sort decreasingly the node array between 1,...,k based on the value count of each element*/
 8  Function Gen_Code ( node,code)
 9  if node!=null then
10  │   Gen_Code (left, code ‖ 0); Gen_Code (right, code ‖ 1);
11  │   codebook[node.symbol] := code
12  end
13  /* Initialization */
14  for i := 1 by 1 to N do
15  │   coding[i] := NEW node; coding[i].symbol := s_i; coding[i].count := p_i;
16  │   coding[i].right := coding[i].left := NIL
17  endfor
18  /* Create Huffman tree, Sorting can be optimized by swapping sub-trees*/
19  for  i := N-1 by -1 to 1 do
20  │   Sort (coding,i+1); new_root := NEW node;
21  │   new_root.count := coding[i].count + coding[i+1].count;
22  │   new_root.left := coding[i];
23  │   new_root.right := coding[i+1];
24  │   coding[i]:=new_root
25  endfor
26  Gen_Code (new_root,"");
```

- **Initialization:** In lines 14–17, we fill the tuples into nodes accessed via the array coding. Each node has pointers to the left and the right that are at this stage empty or NIL. This results in the structure [A] shown in Figure 3.16.
- $i = 4$: We now continue with the loop from lines 19 to 25 and the first value for $i = 4$. The array is already sorted decreasingly, and no change is required. Now, we build, above the last two elements, a new node with the accumulated probability of 140 and pointing to – non-dotted – two elements containing O, R. This is visible in [B] as shown in Figure 3.16.
- $i = 3$: The array is not sorted when using the non-dotted arcs. A sort results in the array with the dotted arcs. Again, a new node is created above the last two elements containing T, A with the value of 181. This is visible in [C] as shown in Figure 3.16.
- The steps for $i = 2$ and $i = 1$ are left out, but the reader can easily execute them and will arrive at [D] as shown in Figure 3.16. Here, the array *coding* with all the connecting arcs is not shown any longer because the algorithm does not require it. We have arranged the nodes anew to avoid crossover arcs. It is now obvious how the tree is arranged and that symbols with a higher probability tend to be closer to the root.

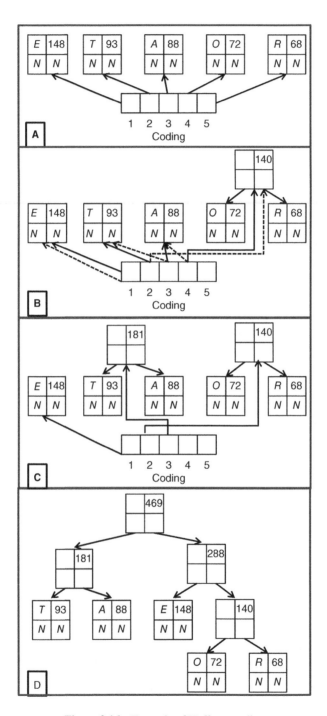

Figure 3.16   Example of Huffman coding

The final step is to create the codes out of the tree $\boxed{\text{D}}$ as shown in Figure 3.16. The recursive procedure *gencode* in lines 8–12 generates the codes. Initial parameters for *gencode* are (*new_root*,""). The process goes down through the tree and the code gets a "1" appended when moving to the left and a "0" when moving to the right. In the end, the following code values have been assigned to the codebook char array: $T = 11, A = 10, E = 01, O = 001, R = 000$. The observant reader will see that the procedure also assigns symbols at intermediate nodes when moving up the tree, but this is always an assignment of an empty character.

η for those five values yields 1.57, and the average code length for those five values is 2.4. So, we are in the expected range, but the example range is, of course, rather small. For the complete English alphabet, $\eta = 4.7$.

The adaptive Huffman algorithm builds on the presented static variant. In any adaptive variant, the statistical values are not given, but with each new value, adaptation takes place. Updates in the statistics lead to updates in the count values of the nodes of the Huffman tree. If this requires a change in the tree structure, the code will change.

Another well-known system is Lempel–Ziv–Welch (LZW), which was patented twice in the beginning of 1980s; it is a Dictionary-Based Coding (DBC). DBC have fixed-length codewords, but a codeword represents a variable-length string of symbols. LZW is used as an adaptive method since the efficiency depends on intense use of longer strings in the dictionary. The algorithm assigns on the fly codewords to seen letters and strings. The longer a string for a codeword is, the better the algorithm works. When strings cannot be extended any more, the dictionary is full, and then the method is static.

Many newer standards use Arithmetic Coding. Here, we will only sketch the main underlying idea but will not detail any solution. Assume that we have five symbols A, B, C, D, E with an occurrence probability of $0.2, 0.16, 0.34, 0.22, 0.08$, respectively. A string of these characters constantly builds smaller arithmetic ranges between $[0 : 1]$ – see Figure 3.17. From the

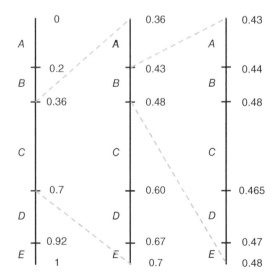

Figure 3.17   Example to arithmetic coding

figure, we could see that the character string C, B, B is uniquely represented by the interval [0.44:0.448]. It is clear that this method could be continuously repeated for larger strings, resulting in even smaller real number intervals. Now, a good encoding of real numbers is required, but we will not go further in this area – see Refs [35, 36] for good overviews on arithmetic coding.

3.2.4   Transform Functions

The idea behind all transformation functions is to "decorrelate" values and compact signals in a reversible way. Consecutive values in a video are correlated – that is, they are close to each other, and one can somehow predict them. The transform functions create layers, and each layer contains more details of the input. All these functions do not compress, but they "order" the information differently. However, the functions open new ways to compress further.

We do not handle the selection of transform functions here. The presented DCT is quite prominent, but there also exist others such as Slant transform [37] or Karhunen–Loève transform [38]. The DCT is a special case of the Discrete Fourier Transform. Note that the treated DCT discussed in this section is one of the existing DCTs, but it is sufficient if we only treat one. For further details on DCT, see Ref. [39].

First, we provide the definition of the two-dimensional DCT function 3.9 but start with the explanation of the one-dimensional DCT function 3.10. Later, we will come back to the two-dimensional DCT. The 2-DCT of $s(i,j)$ is defined with M, N size of the matrix, $S(u, v)$ the transformed matrix, and $C(u), C(v)$ constants as

$$C(\alpha) = \begin{cases} \frac{\sqrt{2}}{2} & \text{if } \alpha = 0 \\ 1 & \text{otherwise} \end{cases}$$

$$S(u, v) = \frac{2C(u)C(v)}{\sqrt{MN}} \sum_{i=0}^{M-1} \sum_{j=0}^{N-1} \cos \frac{(2i+1)u\pi}{2M} \cos \frac{(2j+1)v\pi}{2N} s(i,j) \qquad (3.9)$$

The one-dimensional DCT can be easily derived from Equation 3.9, and additionally, we restrict ourselves to $N = 8$ (block size) as

$$S(u) = \frac{C(u)}{2} \sum_{i=0}^{7} \boxed{\cos \frac{(2i+1)u\pi}{16}} s(i) \quad \text{with } u = 0, \cdots, 7 \qquad (3.10)$$

The boxed part of Equation 3.10 are the base functions of the one-dimensional DCT for all $u = 0, \cdots, 7$. All these base functions are shown in Figure 3.18 for all u, i values. The zeroth base function is a constant and is also named DC (Direct Current). All the other seven base functions are named AC (Alternative Current). Some characteristics of these 1-DCT base functions are:

- If for all i $s(i) = K$ with constant K, then only $S(0)$ is unequal to 0
  $S(0)$ will be roughly $2.83 * K$ since all cos values will be 1. Now, let us verify for $S(1)$:

$$S(1) = \frac{1}{2} \left(\cos \frac{\pi}{16} K + \cos \frac{3\pi}{16} K + \cos \frac{5\pi}{16} K + \cos \frac{7\pi}{16} K \right.$$
$$\left. + \cos \frac{9\pi}{16} K + \cos \frac{11\pi}{16} K + \cos \frac{13\pi}{16} K + \cos \frac{15\pi}{16} K \right) \qquad (3.11)$$

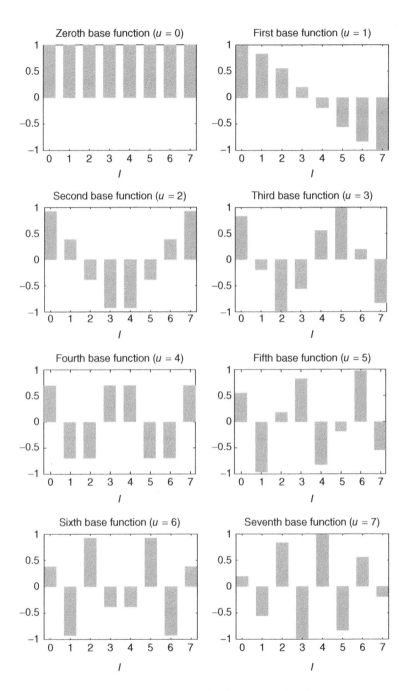

Figure 3.18　One-dimensional DCT base functions

It holds now that $\cos \frac{a\pi}{16} = -\cos \frac{(16-a)\pi}{16}$ since a cos function repeats after 2π; so after one π difference, its value turns negative. This leads to the result $S(1) = 0$ since the summation elements cancel each other.

One can easily see that the same holds for all $S(u)$ with $u > 1$. This shows that the zeroth base function takes "care" of constant input values.

- If $s(i)$ is equal to the uth base function, then only $S(u)$ is unequal 0.

Let us consider $u = 1$, which then results in $s(i) = \cos \frac{(2i+1)\pi}{16}$. We claim that $S(1) \neq 0$. We have calculated $S(1)$ already in Equation 3.11 and need only to replace each K by $s(i)$, which results in:

$$S(1) = \frac{1}{2} \left(\cos^2 \frac{\pi}{16} + \cos^2 \frac{3\pi}{16} + \cos^2 \frac{5\pi}{16} + \cos^2 \frac{7\pi}{16} \right.$$
$$\left. + \cos^2 \frac{9\pi}{16} + \cos^2 \frac{11\pi}{16} + \cos^2 \frac{13\pi}{16} + \cos^2 \frac{15\pi}{16} \right) \qquad (3.12)$$

We know that $\sin \frac{\pi}{16} = \cos \frac{9\pi}{16}$ because sin and cos are shifted by $\frac{\pi}{2}$. Also known is $\sin^2 x + \cos^2 x = 1$ and so $\cos^2 \frac{\pi}{16} + \cos^2 \frac{9\pi}{16} = 1$ holds. The analog holds for the pairs $(3\pi/16, 11\pi/16)$, $(5\pi/16, 13\pi/16)$, and $(7\pi/16, 15\pi/16)$. So, we get in Equation 3.12

$$S(1) = \frac{1}{2}(1 + 1 + 1 + 1) = 2$$

We will not calculate the other values $S(u)$ with $u > 1$ but all will result in 0 because, similarly as in Equation 3.11, the terms cancel each other.

We have seen that $S(0)$ equals the magnitude of the signal $s(i)$ apart from a constant factor. The other seven DCT coefficients reflect the signal $s(i)$ at different frequencies. Upon checking the base functions, it is obvious that the first base function does half a cos cycle, the second base function does a full cos cycle, and each other one half of a cos cycle more. If the input function $s(i)$ has the same form of one base function j, the corresponding $S(j)$ will yield a high value.

The inverse function 1D IDCT (Inverse DCT) is – noted as s^q since the transform is not lossless:

$$s^q(i) = \sum_{i=0}^{7} \frac{C(u)}{2} \cos \frac{(2i+1)u\pi}{16} S(u) \qquad (3.13)$$

Now, let us return to the two-dimensional DCT as in Equation 3.9 with $N, M = 8$. Here we also have two-dimensional base functions as shown in Figure 3.19. The signal is now indexed in space. Each column step and line step are an increase in frequency by $\frac{1}{2}$ cycle in a way similar to that seen for the 1-DCT as shown in Figure 3.18.

The DCT coefficients are created by inner products of the base functions with 8×8 blocks of the image data – as shown in Figure 3.20. As the elements move closer to the bottom right, less additional improvement is seen. The element in the left upper corner is still named DC, and the other elements are named AC. Looking at the two-dimensional DCT base function in Figure 3.19, one visually recognizes the transformation of Lea in Figure 3.20.

However, be aware that the transformation shown in Figure 3.20 used the whole picture of Lea. This is not the case in future use. In the future, DCT will be applied to small 8×8 or possibly 16×16 pixel blocks out of a picture.

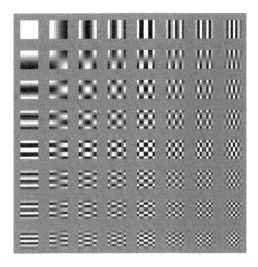

Figure 3.19   Two-dimensional 8×8 DCT base functions

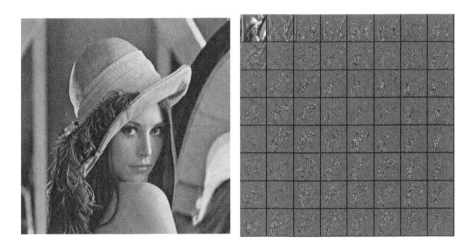

Figure 3.20   Lea transformed by 8×8 DCT coefficients

3.3   JPEG

The Joint Photographic Expert Group (JPEG) created this standard in 1992 [40], and the standard took over the group name. This group is a collaboration between the ITU and ISO/IEC.

In the previous sections, we introduced all elements that will allow us to understand how JPEG works. Figure 3.21 gives an overview of JPEG encoder.

The details of the JPEG encoder are:

- The input picture is split into 8×8 pixel blocks out of different color spaces *YIQ* or *YUV*. Each of these 8×8 blocks is individually handled by a 2-DCT.

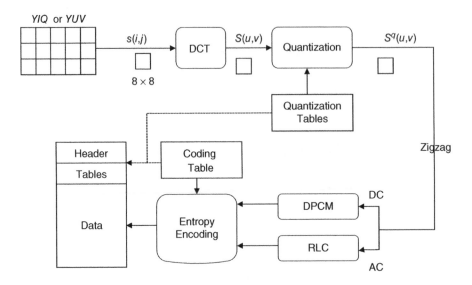

Figure 3.21 JPEG encoder overview

Table 3.5 (a) Luminance Q_T and   (b) Chroma Q_T

16	11	10	16	24	40	51	61
12	12	14	19	26	58	60	55
14	13	16	24	40	57	69	56
14	17	22	29	51	87	80	62
18	22	37	56	68	109	103	77
24	35	55	64	81	104	113	92
49	64	78	87	103	121	120	101
72	92	95	98	112	100	103	99

17	18	24	47	99	99	99	99
18	21	26	66	99	99	99	99
24	26	56	99	99	99	99	99
47	66	99	99	99	99	99	99
99	99	99	99	99	99	99	99
99	99	99	99	99	99	99	99
99	99	99	99	99	99	99	99
99	99	99	99	99	99	99	99

- The output is then quantized based upon quantization tables – Table 3.5 – with

$$S^q(u, v) = round\left(\frac{S(u, v)}{Q_T(u, v)}\right)$$

The JPEG standard defines the two quantization tables as shown. However, other tables can be used. Note that quantization tables are also named Stepsize Table. The higher the value is in the quantization table for a given u, v, the smaller $S^q(u, v)$ will become. Many values in the lower right part will typically become 0. Of course, the chroma values are even more impacted by their quantization table.
- So far, the values were handled in small blocks. By Zigzag, each block gets enumerated in a sequence from top left to bottom right as follows: $S_{00}^q(DC), S_{01}^q, S_{10}^q, S_{20}^q, S_{11}^q, S_{03}^q, S_{04}^q,$ $\dots, S_{77}^q$. Zigzag is used in variants in all video compression algorithms.

- Since many of the ACs will have the same value or even be 0, RLC seems the right solution. The DCs for multiple blocks are combined to run DPCM on them. Typically, DC values are large but consecutive ones should have values close to each other since following blocks are not entirely different. Consequently, DPCM seems to be a reasonable choice.
- In the next step, the algorithm encodes the DC and AC values by an adapted Huffman coding. Huffman is changed in the sense that for the output tuple $<$ *count, value* $>$ of RLC, only in the *count* field a coding makes sense. The coding tables provide the information of the entropy encoding.
- The encoded data is written in the JPEG File Interchange Format (JFIF). JFIF might contain changed coding tables or quantization tables if changes occurred.

JPEG has different compression modes:

Sequential: The presented implementation is this mode. A "Motion JPEG" that applies this method for each separate picture in a stream of images also exists. However, "Motion JPEG" is highly inefficient and is not used.

Progressive: This mode is used mainly in Web browsers. Here, the picture quality increases over time. It starts with a low resolution and gets better until the final situation. There exist two options to accomplish this:

- The algorithm first sends only DC plus a few AC components and afterward sends chunks of more AC components. This algorithm carries the name Spectral Selection.
- The other algorithm sends the Most Significant Bits (MSBs) of all DCT coefficients first and afterward sends chunks of more bits. This method is the Successive Approximation.

Hierarchical: The picture is coded in multiple resolutions. So, it could replace progressive mode requirements. This variant is seldom used.

Lossless: The lossless JPEG replaces DCT by DPCM. DCT has particular problems at the border between blocks. These border artifacts are also an issue in video compression; we will treat these methods later during video compression. Lossless JPEG avoids this problem too, but the file size typically doubles.

JPEG is not the right choice for digital plots or drawings since there exist a lot of hard transitions that get smeared. Also, for black-and-white pictures as well as for newspapers, JPEG is not the first choice. Here, formats such as GIF (Graphic Interchange Format) or PNG (Portable Network Graphics) achieve better results – see here in Ref. [41]. Additional standards to address the aforementioned disadvantages are JPEG2000, JPEG-LS for medical images, and JBIG (Joint Bi-level Image Processing Group) – see for all those standards [12, 42].

3.4 MPEG/H.26x Video Compression

Since 1990, the Moving Pictures Expert Group (MPEG) of ISO and ITU-T have published diverse video compression standards. Since 2003, the two standard bodies work together and issued common standards. Note that some of the video compression standards contain audio parts – see Section 3.1.3.1.

We will first provide the reader with an overview of the different standards. All compression methods write their information in special packets for either transmission or storage; the first subsection treats the MPEG transmission packets. In the following subsections, we address different relevant standards in more detail.

The most important video compression ISO MPEG standards are:

MPEG-1 These standards were issued from 1992 onward after working for over 4 years (numbered 11172– ...). Here, only Part 2 [23] is important. ITU H.261 heavily influenced this standard, but it never reached the coverage of H.261 since it covered storing and playing videos on computers at a low bit rate – that is, not supported interlaced pictures on TV.

MPEG-2 From 1995 onward, these standards were issued (numbered 13818– ...). Part 1 [43] is the extension of MPEG-1, Part 1 and specifies the system layer that includes the Packetized Elementary Stream (PES) and Transport Stream (TS) layer. It treats additional synchronization, buffering, and timing. We treat this in Section 3.4.1.

Part 2 [44] is the extension of MPEG-1, Part 2. The standard is the same as H.262. DVD and Blu-ray use this standard but also other H.26x standards. Some countries use it for Broadcast TV and Cable TV.

MPEG-4 The committees issued another wave of standards from 1999 onward (numbered 14496– ...). Part 2 [45] covered video, still textures, and synthetic images. The guideline also includes H.263. The standard has some interesting concepts, which we treat in Section 3.4.3. However, the concepts never worked and also the industry stepped back. Part 10 is identical to the very successful ITU H.264 [46].

MPEG-H ISO issued this set of standards from 2013 onward (numbered 23008– ...). Here, only Part 2 is important and is identical to ITU H.265 [47]. Under development are guidelines for reference software and a conformance specification.

The most important video compression ITU H.26x standards are:

H.261 This was the breakthrough video standard in 1990 [48] – treated in Section 3.4.2. It was developed to work on multiple narrow-band ISDN (64 kbit/second) lines.

The intended use was for interactive video applications, but this requires some additional related standards. For multiplexing audio and video (H.221 [49]), for call setup and negotiation (H.230 [50]), and for conference control procedures (H.242 [51], H.245 [52]) is necessary. We treated the additionally required audio standards in Section 3.1.

H.262 This standard was already mentioned under MPEG-2, Part 2.

H.263 In 1996, this standard [53] was released for low bit rate video compression for data networks. Also, MPEG4, Part 2 included this standard.

H.263 can replace the ISDN-based H.261 for video applications and can use the same conference control, multiplexing (here: H.223 [54]), and setup and negotiation control. We treated the recommended audio standard G.723 in Section 3.1.

H.264 As of today, this is the most used video compression standard [46], which we address in Section 3.4.4. This was created along with ISO (MPEG-4, Part 10) and was released in 2003. Since 2003, the ITU released eight new versions, not counting corrigendum. Also, an implementer's guide [55] is available.

H.265 The standard "High Efficiency Video Coding" (HEVC) [47] is the outcome of 9 years of work in the international bodies. In 2004, the goal was to create a new compression system that delivers the same quality at doubled compression rate – and they made it! The author is not aware of any other standard that took as long or must have been as hard.

However, the goal was right since a new standard is only acceptable for a global implementation if it brings substantial benefits. When a new standard brings only 20% improvements, it will not be implemented because it is not worth the effort.

The best experts from around the world fill the teams in the ITU/ISO. Members send in proposals that are then analyzed and also tested. The new standard was not the result of a single contribution. The result was a combination of many, many proposals that merged together over those years. The team released the standard in April 2013, and an update is available from October 2014.

With high chances, no new standard will come within the next generation, unless someone has an outstanding new idea for compression that no one has thought of in the past 20 years. We treat H.265 in Section 3.4.6.

3.4.1   MPEG Data Streams

Originally, each compression method in the ISO or ITU world stored the result of a compression in packets of a PES. Afterward, these packets are inserted into either an MPEG Transport Stream or an MPEG Program Stream. The Program Stream packets are for storing the data – for example, on a DVD – while the transport stream is for sending the data over a network. With H.264, the strict use of MPEG data stream ended, but it is still an option. This is discussed in Section 3.4.4.2.

In addition to the encoded video, the MPEG Data Stream also transports the following information:

Time: A video is decoded along a time axis. The play-out of video and audio needs synchronization – lip-synced. Consequently, at multiple places, time information needs to be stored.

Encryption: Certain video might get scrambled and end users need to have the right decryption key.

Entitlements: Certain videos might only be visible to individual entitled users that may have paid an additional fee.

Program information A video can be part of a whole program information – for example, channel, frequencies, transponders. There may also exist a verbal description of the content.

Length: The lengths of data fields need to be mentioned.

The necessity to exchange this information is obvious, but one would not have expected that all this go in the MPEG data stream.

Some of the mentioned information can be highly structured, and today, an XML-like description could be chosen. However, the definition of TS was provided at the beginning of the 1990s, and at that time, only a rather compressed description was acceptable. At that time there existed a rather compressed but also well descriptive possibility in the ISO communication standards named Abstract Syntax Notation 1 (ASN.1) – see also Section 7.1. Instead, this specification was created in complex structures with pointers and tables. In essence, we need an analyzer to understand it thoroughly. The following will give us some understanding of the structure of those packets.

The encoder writes its output in PES packets as shown in Figure 3.22.

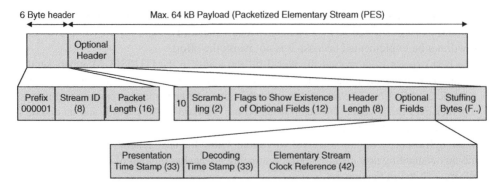

Figure 3.22 MPEG packetized elementary stream (PES)

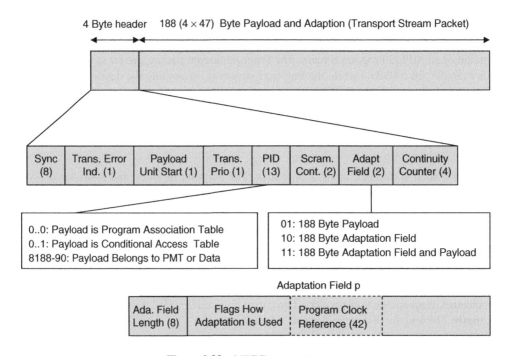

Figure 3.23 MPEG transport stream

The *Stream ID* describes the type of stream – audio, video, and so on. We see multiple clock information in the packet. We explain the *Presentation Time Stamp* (PTS) and *Decoding Time Stamp* (DTS) in Section 5.4. The *Elementary Stream Clock Reference* is the base clock for the stream and generates the system clock for the decoder. Timing information is sent every 40 milliseconds.

PES packets are placed in Transport Stream packets – typically, they cover multiple TS packets. TS packets, as shown in Figure 3.23, have the size of 192 Bytes. The length fits for four ATM packets or slots (Asynchronous Transfer Mode). At the beginning of the 1990s, ATM was expected to be the network technology of the future despite the inherent complexity. The TS Packet was not part of the original MPEG-1, Part 1 standard, but was introduced in MPEG-2. TS stays unchanged in MPEG-3 and MPEG-4. The MPEG-TS is also named MPEG2-TS.

Let us look at some details of the packet in Figure 3.23. The Transmission Error Indicator (TEI) is set if information arrives with error and could not be corrected. The Packet IDentifier (PID) specifies the content and program content for the payload. Multiple options are possible, mostly using tables:

PID = 0: The payload is a Program Association Table (PAT), which contains a list of *PID*s of Program Map Tables (PMT). Or to say it in programming terms, it is an array of pointer to PMTs.

PID = 1: The payload contains Conditional Access Tables (CAT). CAT is used for the management of cipher keys and privately defined data.

PID = 10: The payload contains a Network Information Table (NIT). NIT assigns frequencies to channels and transponders. In addition, it may include program guide information. The standard has not specified this in detail.

$8188 \leq PID \leq 8190$: The payload is a PMT or data. A PMT holds a list of PIDs that are either Video or Audio Elementary Streams. In other words, a PMT is an array of pointers to video or audio elements.

The *Adaptation Field* marks the existence of an adaptation. In case of an adaptation, the payload is shortened by the *Adaptation Field Length*. As shown in Figure 3.23, the reference clock is part of the adaption field.

If an MPEG-TS is transported for only one stream, it is a Single Program Transport Stream (SPTS); otherwise, it is a Multiple Program Transport Stream (MPTS). Further information on MPEG data streams is available in Ref. [2].

For some time, the standard bodies work on a replacement of MPEG-TS. MPEG Media Transport (MMT) is the name for the replacement [56]. The corresponding standard is still in draft form.

3.4.2   H.261

Even if H.261 [48] is rarely used today, it deserves a section because of its historic role. Before the existence of H.261, interactive videoconferencing was not possible. Also, all other H.26x video compression standards still retain the same basic H.261 model but of course advanced it in many, many directions. So, it is worthwhile to understand H.261 before moving on to the modern ones.

The base characteristics of H.261 are:

Encoding delay: The design was driven by a maximum encoding delay of 150 milliseconds. In Section 4.5, we discuss in detail why 150 milliseconds was and still is a critical time limit for all interactive multimedia applications.

ISDN: The transmission should be scalable for $p \times 64$ kbit/second with $1 \leq p \leq 30$ – that is, 128 kbit/second until 2 Mbit/second. $p \geq 2$ is required minimally for videoconferencing.

Macroblocks: The specification states a $16 \times 16/8 \times 8$ luminance macroblock size, and each video frame is split into those blocks.

Subsampling: The standard specifies a 4:2:0 sampling.

Output: Two frame sizes CIF (352×288 luma and 176×144 chroma) – see also Figure 3.12 – and an even smaller one Quarter CIF (QCIF) were possible. The specification mentions 30, 15, 10, or 7.5 frames per second. This translates into maximum uncompressed bit rates of 36 Mbit/second.

Figure 3.24 shows the overall scheme. First, let us look at the two modes the system can use.

Intraframe: This means we are working within a frame. A frame is one complete picture out of the video stream. However, when we work within a frame, we always restrict ourselves to the macroblocks of that frame at one time.

The intraframe method is shown in lighter gray at the bottom of the figure. Note that there exists an area in the middle where intraframe and interframe are overlapping.

Interframe: This means we are working on consecutive frames – that is, two consecutive complete pictures out of the video stream. When we work on two consecutive frames, we always restrict ourselves to a selected macroblock of the current frame. However, when working on the previous frame, we look at multiple macroblocks at a position close to the current one.

The intraframe part of H.261 always works on 8×8 chroma macroblocks. Due to the different sampling, this corresponds to a 16×16 luminance macroblock. The intraframe method works as JPEG – see Section 3.3 – apart from the stepsize or the quantizer matrix can be

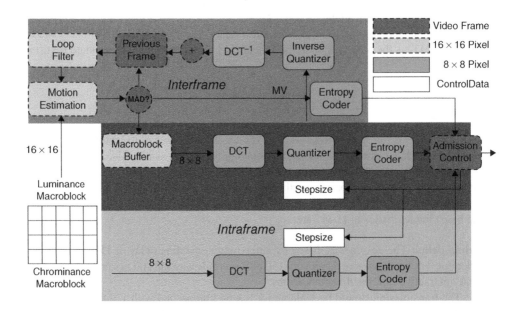

Figure 3.24   H.261 schema

influenced by the admission control at the end of the process. The stepsize may change the AC values in a range of $2, \cdots, 62$.

The luma macroblock has a size of 16×16 and is bigger than the chroma macroblock. Here, the motion estimation is performed as explained in Section 3.2.2. The loop filter is a means to smoothen the previous frame and to minimize prediction errors; the filter was, however, removed in later H.26x versions [57]. If the motion estimation found a matching block – that is, MAD was good – an MV is created. The motion vector contains the shifts from the previous frame to the current frame. The MV is sent out after applying entropy encoding. If the motion estimation found no matching block, the current block is cut into four pieces, and for each such block, the intraframe procedure is applied.

For each quantized luma macroblock, the inverse quantizer generates the inverse quantified luma macroblock. This is the macroblock the decoder would generate at the receiver side. These regenerated blocks are used as comparisons within the motion estimation. The comparison with the regenerated blocks ensures the quality of the estimation process.

The admission control assembles the packets and manages the stepsize. How the control process works is explained in a later section – see Section 5.4. The bit stream syntax of each standard governs the assembly of packets. Here, we use a syntax for a superset of H.261 avoiding some historical formats. So, PES packets are filled with the following information from inside to the outside:

Block: The block contains DC and the ACs in the further encoded form

Macroblock: A macroblock has up to six blocks – four Y, one C_b, one C_Y. A Coded Block Pattern (CBP) tells if a block is instead represented by an MV.

Slice: The slice contains a variable part of macroblocks so that a decoder can resynchronize after a slice.

Frames: Each frame consists of slices. There exist I-Frames, P-Frames, and B-Frames as shown in Figure 3.25.

An I-Frame contains no prediction at all. I-Frames are sent at certain intervals to guarantee that transmission may resynchronize again. If pictures with predicted information are lost, any follow-up image cannot be reconstructed, but such a problem ends with an I-Frame.

P-Frame is predicted from previous I-frame or another P-Frame. A bidirectional (B)-frame contains predictions from the past and the next frame. H.262 introduced B-frames, and they exist in all subsequent H.26x standards.

Group of pictures: A GOP is a number of pictures with at least one I-Picture.

Sequence: A Sequence contain GOPs with equal aspect ratios and sizing.

Traffic pattern of compressed videos

The structure as shown in Figure 3.25 means that H.261 and all other H.26x compression mechanisms exhibit a Variable Bit Rate (VBR). Traffic peaks occur when transmitting I-frames, and P- or B-frames only generate a small portion of traffic. Also, the peak traffic contains the most relevant information. However, after a peak, there is for some milliseconds no other peak. Note that there are ways to smoothen the peak somewhat – see later under GDR in Section 3.4.4.

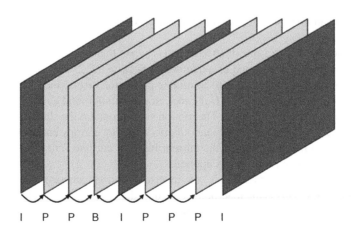

I   P   P   B   I   P   P   P   I

Figure 3.25   Possible frames

3.4.3   MPEG-4

The standard MPEG 4, Part 2 [45] used entirely another approach to video compression. It is still a great idea, but so far no effective implementation was possible.

The idea is close to how persons describe a video to each other. We start with characterizing the objects, and we create a chronicle of events where these objects are acting.

The standard uses Video Objects (VO), which are visible in many Video Object Planes (VOP). A VOP is a snapshot of video objects at a particular point in time. In Figure 3.26, we see two objects that we name *person* and *car*. The *person* moves while the *car* stands. The algorithms for analysis and description of motion can be the same as for H.26x. However, the challenge is to find the objects. Note that we also need a three-dimensional description of the objects because they do not only move horizontally or vertically.

This idea was so far not implementable for the analysis and compression of videos. However, more and more modern motion pictures are created today electronically based on 3D objects. So far, they were all delivered in a H.26x format to our television set.

For further reading on this topic, a worthwhile recent article is Ref. [58].

3.4.4   H.264

We will explain the standard H.264 [46] by providing the main additions to the previously introduced H.261 compression – see also [59, 60]. Sometimes, H.264 is described simply by

Figure 3.26   Two video objects in three VOPs

the term Advanced Video Coding (AVC) or also H.264 AVC. In Section 3.4.4.2, we address the byte stream in H.264. We will look at the achieved efficiency improvements of different H.26x standards at the end of the Section of H.265 – Section 3.4.6.

The base unit for any prediction is no longer a frame but a slice. A slice is a part of a frame or a complete frame. The standard has many new predictions:

Bidirectional: The prediction supports maximum five past slices and five future slices. Consequently, an implementation needs more buffering of slices and a higher delay is possible.

Intra: A block can now predict other blocks within the same slice. Three luma sizes and one chroma size are possible. The prediction supports nine different modes. Each mode describes a particular direction between the predicted block and the reference block – for example, vertical, horizontal, and vertical right.

Before H.264, an I-slice and I-frame had no predictions included. Now, an I-slice is still not dependent on other frames but may contain predictions within the slice.

Inter: Nine different Prediction Block (PB) sizes exist now (one 16×16, two 8×16, two 16×8, four 8×8). Since the motion estimation algorithms have become so efficient over the years [61, 62], it was feasible to subdivide or interpolate luma pixels. Two refinement levels are possible: Half Pixel (HPEL) or Quarter Pixel (QPEL).

MV: Motion Vectors are highly correlated for consecutive blocks or even within a whole slice. So, Motion Vectors are also predicted.

Weighted: Multiple blocks are the basis for the prediction. This is optimal for fading.

Of course, H.264 has to support interlacing, but this is a challenge for the different prediction methods. We have two options on how to handle interlacing. In one option, we keep the frame with all the interlaced lines together – see Figure 3.27(a). The other options are to keep Top and Bottom together. If we keep Top and Bottom together, the prediction lines look like that in Figure 3.27(b). Here, the B-frame is predicted from the I- and P-frame and the P-frame only from the I-frame.

As shown in Figure 3.27(b), some of the predictions must now point from top to bottom or vice versa. In both cases, particular methods for predictions need to be added – for example, Picture-Adaptive Frame/Field (PAFF) or Macroblock PAFF (MBAFF) [63]. Also, H.264 applies a particular entropy encoding method – Context Adaptive Binary Arithmetic Coding (CABAC) – to cope with Interlacing [64]. CABAC has different probability models available and can switch between them. CABAC is an arithmetic coding method as explained in Section 3.2.3.

Additional features from H.264 are:

SP/SI-slices: The "Switching P Picture" (SP) and "Switching I Picture" (SI) slices set synchronization points for the transmitted information. Those synchronization points allow the decoder to switch between different video streams and efficient random access. SP/SI then allows switching between the two videos during the playing. The random access is required for the "fast-forward" or "rewind" operation of players. This operation dates back to the analog systems. However, digital decoders cannot decode faster to deliver the same experience. So, decoders jump over a multitude of pictures to the next one. For this reason, this operation also carries the name "trick mode" or "trick play."

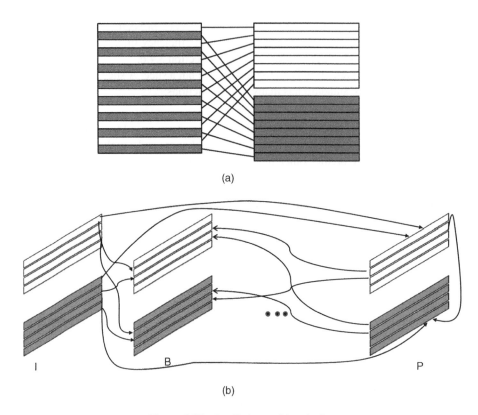

(a)

(b)

Figure 3.27 Predicting and interlacing

Variable order: The system has options of where to place the slices or any data in the stream –
named Data Partitioning. Arbitrary Slice Order (ASO) is the term for moving a slice at
another position. However, the internal order in a slice is also flexible but controlled by a
MacroBlock Allocation map (MBAmap). These ordering options are discussed in Section
3.4.4.2 for the generation of groups of equal importance when sending the data over the
network.

Gradual decoding refresh: Instead of sending a complete I-frame for refreshing the picture,
multiple GDRs can be sent, which always include only a part of the overall picture. This
mechanism splits the otherwise big I-Frame information into smaller pieces.

Textures: The standard understands picture elements such as water, grass, sand, and clouds as
textures. These textures generate a lot of analysis work, but the detail is not necessary for
the viewer. After the detection of textures, they are en-/decoded in a special way and save
up to 20% of the bit rate [65].

Improved entropy encoding: The system uses a Context-Adaptive Variable-Length Coding
(CAVLC). CAVLC has multiple contexts that are built into two runs over the input data
while also incorporating neighbor blocks' contexts [66].

Diverse: The standard does not only support Zigzag for scanning the quantized matrix but also
horizontal or vertical scanning. Intra-respective interprediction uses different quantization
methods. The system can include redundant slices to improve robustness.

Figure 3.28 Block boundaries treated by deblocking filter (Source: Licensed under CC BY 2.0 via Wikimedia Commons "Deblock1" – http://commons.wikimedia.org/wiki/File:Deblock1.JPG# mediaviewer/File:Deblock1.JPG. Courtesy: Rausler Raddell)

The standard also addresses the block boundary visibilities. In the left picture in Figure 3.28, the macroblocks are quite visible since they create explicit boundaries to the neighbor block. Due to the low resolution, the macroblock boundaries are easily visible. H.264 has a deblocking filter that smoothes the boundary transitions as can be seen on the right side of Figure 3.28.

3.4.4.1   H.264 Profiles

Overall, H.264 has 17 profiles, but the most important ones are Baseline, Main, and Extended as shown in Figure 3.29.

Each profile has a typical application area:

Videoconferencing: Baseline is foreseen for videoconferencing. Interactive stability is a primary requirement for videoconferencing as well as fast encoding and decoding. The reader may cross-check these demands with the requirements summarized in Table 2.1. Stability is increased by redundant slices and grouping the essential elements. Fast decoding is eased by avoiding B-slices. Since no picture fading exists in videoconferencing, there is no need for a weighted prediction. SP/SI-slices are required in streaming for fast-forward or rewind operations.

TV broadcast and video-storage: Main covers TV broadcast and video storage. TV sets play interlaced videos and CABAC helps here. Because of missing interactivity, we can add B-slices and remove redundancy elements. SP- and SI-slices do not make sense since there is no need to change the resolution during play-out.

Streaming media: The extended profile serves streaming media best. End devices in streaming media do not require interlacing. Since interactivity is not necessary, stability elements are not included in the profile.

3.4.4.2   Network Abstraction Layer

In the Network Abstraction Layer (NAL), the byte stream is formed, which is handed over to the transmission layer. Figure 3.30 shows the interfaces of the NAL. NAL can hand over its output to an MPEG data Stream (PES/TS) as described in Section 3.4.1 or to the Real-Time

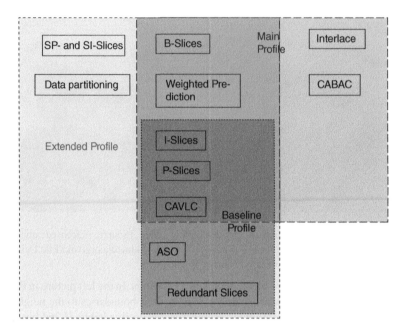

Figure 3.29   H.264 profiles: main, extended, and baseline

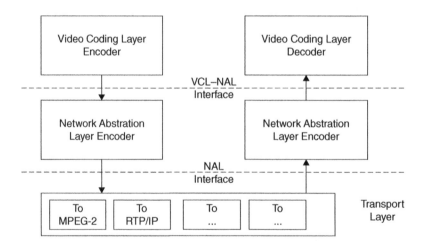

Figure 3.30   Overview of layering NAL

Protocol (RTP) as outlined in Section 4.1. Section 4.1 explains the embedding into RTP. Since we concentrate on IP-Transport in this book, we could assume that we will never see the MPEG data stream anymore. However, we discussed in Section 3.4.1 the usage of the MPEG data stream for additional purposes: encryption, entitlement, program information, and timing. RTP is not able to serve these other uses, and this would need additional protocols. If the particular protocols do not exist, applications may fall back to MPEG-2.

The information sent over the NAL Interface is a NAL Unit (NALU) – see Figure 3.31.

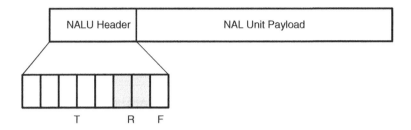

Figure 3.31 NAL header

The following elements are part of the header:

T: The five-bit NALU **T**ype characterizes the included information. Thirty-two different NALU types exist. Some examples are: eight for Picture parameter set, nine for Access Unit Delimiter, and five for Slice of an Instantaneous Decoder Refresh (IDR) picture – that is, I-frames, which compose a picture.

R: The two-bit **R**eference Id specifies the importance of the NALU. The higher the value of the Reference Id is, the more significant the NALU is for the picture quality. The standard does form rules for the values – that is, IDR > 0 or Access Unit Delimiter = 0.

F: This field contains a 1 if the NALU has an error, else the value is 0.

H.264 bundles information due to the Reference Id. The bundling was the reason for the variable order or data partitioning before. Note that the decoding normally has still to follow the original encoding order and has to reassemble the incoming information.

Overall, parameter sets are normally valid for multiple frames or the whole video. This data is also named non-"Video Coding Layer" (VCL). The correctness of the parameter sets has the highest priority for the overall transmission. H.264 even mentions out-of-the-band transfer for the transmission – that is, out of the regular transfer by more safe methods. Non-VCL NALUs transport the parameter sets.

A set of VCL NALUs is an Access Unit (AU). The decoding of an AU generates one picture. The structure of an AU is visible in Figure 3.32. The Primary Coded Picture (PCP) contains all compression information for decoding under a given parameter set. The Supplemental

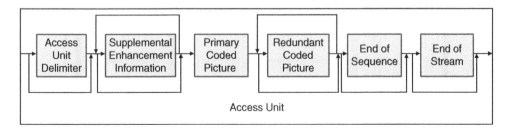

Figure 3.32 Video coding layer NAL Access Unit

Enhancement Information (SEI) provides noncritical addition to the overall parameter. The redundant coded picture contains slices that are transmitted multiple times for redundancy.

Video compression built for proper network handling
A router can check the two bits in the header of an NALU packet to decide on drop probability in overload situations. The new video compression standards were built to align better with the network. If an NALU is embedded in MPEG-2, this is hardly possible.

3.4.5 Scalable Video Codec

The H.264 standard contains the Scalable Video Codec (SVC) since the introduction of version 8 in 2007 [67, 68]. A video is scalable when it is offered in different scales that are predicted from a base variant. So, scalable video compressions are not a Simulcast, which means that all scales are sent independently of each other. A downscaling can occur by quality reduction, by temporal subsampling – that is, reduced frame rate, or by spatial subsampling – that is, smaller frame size. Scalable video coding standards are not new, so H.262 and a later version of H.263 included all three aforementioned downscaling methods. However, the market had, at that time, no interest in these features. In those times, there were no handheld or tablet devices that required support for other screen sizes, slower wireless links, or reduced power consumption.

A bit stream is *temporal scalable* if it can be separated into layers $T_0, T_1, \cdots, T_k$ so that by removing AU's of layers $> k$, the bit stream is still a valid bit stream for the decoder. The base layer is T_0. Figure 3.33 shows such a temporal scalable stream of access units. Note that the figure also shows prediction dependencies, but this is not essential for the temporal scalability. The adaptation of the temporal scalability is straightforward in H.264 since the multiple reference picture concept allows the construction of a frame by various inputs.

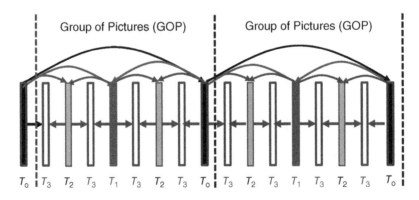

Figure 3.33 Temporal scalability

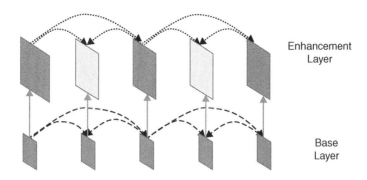

Figure 3.34   Spatial scalability

Figure 3.34 shows the spatial scalability that also used an interlayer prediction. This interlayer prediction helps to avoid transmission of both sizes in parallel – the so-called Simulcast. The interlayer prediction includes a particular motion prediction whereby a scaling of the motion vector occurs. Also, the intraprediction of the base layer is adjusted by the interlayer prediction. When increasing the size of a picture macroblock, boundaries become more visible. The interlayer prediction has to maintain information for the upsampling of a macroblock for use within a deblocking filter.

The Quality Scalability is a special case of the spatial scalability with identical picture sizes for base and enhancement layers. However, only a few parts of the interlayer prediction are used. When using only the base layer prediction, the enhancement layer prediction is not available. The loss of the enhancement layer prediction leads to a drift problem since the dotted arcs in Figure 3.34 are not available, and the prediction is not entirely correct until a next I-frame occurs. The drift is not an issue if predictions are only allowed in the base layer. This solution is named Fine-Grained quality Scalable (FGS), but it decreases the efficiency of the enhancement layer coding.

SVC can be used in the many situations, and examples are:

Different receivers: When a video is encoded for receivers with different capabilities – bandwidth, available codecs, screen resolution – all receivers can use the same SVC stream.
Variable bandwidth: If transmission bandwidth varies, the number of used SVC streams can be adapted to the current bandwidth.

We will further treat the application of SVC transmission in the following chapters.

For SVC transmission, the NALU gets expanded by a 3 Byte header extension. The header extension specifies, for example, the use of interlayer prediction, layer characteristics, temporal resolution, or the necessity of this NALU for decoding higher layers. The signaling of the scalability structure occurs within the SEIs of the AU.

3.4.6   H.265

As stated before, the standard "High Efficiency Video Coding" (HEVC) [47] is the result of nearly 10 years of work. In this section, we discuss the most significant differences in H.265 in comparison to the previous standards – for more details, see [69–73]. While we will highlight

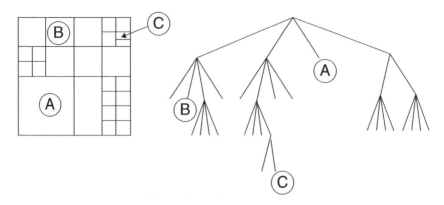

Figure 3.35   Partitioning of the 64 × 64 macroblock by the Coding Tree Block (CTB)

the main differences, it needs to be noted that H.265 has refined many, many details of the preceding standards. However, this cannot be easily summarized. At the end of this section, a performance overview of different H.26x standards is presented.

The size of the macroblocks was in all video compression standards an important refinement element. Here, the overall macroblock of 64 × 64 can be split hierarchically down to 4 × 4 coding blocks (CB) if required. A Coding Tree Block (CTB) defines how the 64 × 64 macroblock is split. After the split, some CBs are bigger, and other ones are smaller – see Figure 3.35. All CBs are leaves in the tree. Each CB can be a Prediction Block (PB) or Transform Block (TB).

A Transform Unit (TU) combines the luma and chroma values of the TBs. A Prediction Unit (PU) incorporates the luma and chroma PBs and associated elements such as motion information. One luma CB and ordinarily two Chroma CBs form a Coding Unit (CU). A Coding Tree Unit (CTU) consists of a luma CTB and the corresponding chroma CTB and some additional elements.

Tests [69] show that the coding efficiency increases with larger CTUs. The test evaluated CTU sizes of 64 × 64, 32 × 32, and 16 × 16.

Based on the introduced terminology, the following characteristics are notable:

Intraprediction: This prediction mode now supports 33 directions compared to eight in H.264 for PUs within a slice. However, there exist restrictions on the size of the PUs in one of the two supported modes (DC intra, Planar).

Interprediction: The PU can also be based on asymmetric split CTBs. The CTBs in different pictures at the same positions are split into dissimilar trees. The widely open configurable trees create problems when referencing them. A merge mode allows the reference to a number of CUs. A skip mode refers a PU to another PU, which contains motion vector for usage.

Motion vector: The precision of the vector was significantly increased. The advanced mode also includes several probable candidates. Also, the inheritance from temporal or spatial can be expressed.

In-loop filtering: The potentially small TB sizes introduce the so-called ring artifacts. The artifacts can be dissolved by Sample Adaptive Offset (SAO) or Adaptive Loop Filtering (ALF). SAO adjusts the adjacent pixel on the surrounding information. ALF merges adjacent pixel information.

Parallel processing: The specification allows two options for parallel processing. Tiles are rectangular parts of a picture that are independently decodable. A typical configuration consists of tiles of a similar size. On each of those tiles, a process can work independently.

In the other option, a slice is described as a matrix of its CTUs. Per row of this matrix, a parallel process runs in thread 1, thread 2, and so on. Each process starts two CTUs after the process that works on the row above – see Figure 3.36. The delay in starting results from context dependencies of the entropy encode.

Profiles: The first version of the standard in 2013 contained three profiles: Main, Main 10, and Main Still Picture. The second version of the norm in October 2014 added 21 range extension profiles, 2 scalable and 1 multiview profiles. The last overall ITU H.265 document [47] contains those profiles. The profile Main will typically be used with end-user devices (bit depth 8, 4:2:0 chroma sampling). The profile Main10 supports up to 10 bits per sample, which UHDTV requires. Note that the standard also differentiates the available bit rates by main and high tiers. Both tiers are specified for different levels (screen and resolution sizes). The main tier is the usual one.

Interlaced: The norm reduces the support for interlaced and does not contain the tools MBAFF and PAFF – as mentioned in H.264.

H.265 will not be in mass-market roll-out in 2015. Some products are on the market since 2014, and major providers have stated in which release of their products it will be included. However, forecast expects the beginning of the broader use in 2016.

Figure 3.37 shows a comparison of H.265, H.264, H.263, and H.261. One axis of the graph is for PSNR. PSNR stands for Peak Signal-to-Noise Ratio, which measures the ratio between the maximum possible signal and the noise. PSNR is an approximation for the human perception of quality [74]. For a bit depth of 8, values range between 30 and 50 dB – logarithmic unit $\log_{10}(x)$. The other axis gives the rate in kilobit per second. Here, it is only important to compare the result with same PSNR values.

Most parts of Figure 3.37 were built on the data described by Ohm et al. [75] (Kimono1 1920×1080, 24 Hz). However, the curve for H.261 is projected using results presented in Ref. [76]. We needed to project the H.261 curve since normally all basic data has to be the same to compare PSNR values.

Figure 3.37 shows that the coding efficiency over the past 25 years has improved by a factor of 10. However, the improvements have slowed down.

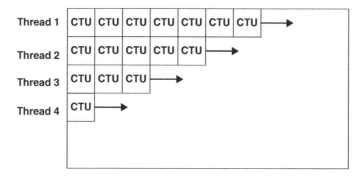

Figure 3.36   Parallel operation on CTUs of a slice

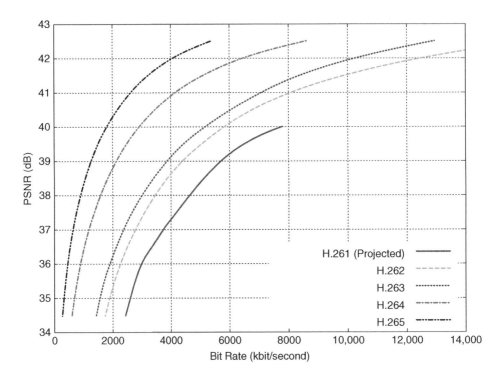

Figure 3.37   Comparison of H.265–H.261 compression efficiency since 1990

Video compression efficiency improvements will get small

In the past decades, compression standards delivered every few years an efficiency increase by a factor of 2. Technology is now in such a state that a next standard with an efficiency increase of 2 will probably be available at the earliest in 20 years. Incremental improvements of a few percent may still happen with full backward compatibility but no jump to a new level.

3.5   Other Video Compression Standards

Other main video compression standards are:

VC-1: Video Codec 1 (VC1) is a product from Microsoft, which is well documented in Refs [77–79]. SMPTE (Society of Motion Picture and Television Engineers) released this video compression system as a standard in April 2006. Note that this standard is not open-source software. VC-1 is today found in Blu-ray Discs, Windows Media, Xbox, and Silverlight.

VC-2: Dirac developed this standard in the United Kingdom [80] under the support of the BBC. SPMTE standardized it in 2008. It is an open-source standard. Since 2011, no new release was issued.

VC-2 is used occasionally within the BBC.

VP8: VP8 was standardized in 2011 by Google [81] using the technologies of a bought company On2.

VP9: VP9 aims for a 50% improvement over VP8. The Internet Draft [82] documents a first version of VP9 and is available as license-free software. Other names for the product are VP-Next or Next Gen Open Video (NGOV). Google runs this project. Also, Google mentioned in 2014 the beginning of the development of VP10.

At the time of writing this book, VP9 was not finally released and no actual draft is available. The developers still try to get close to H.265 performance.. The Chrome Browser, Mozilla Firefox, and Opera include the latest development versions of VP9.

China AVS: China has, over the years, standardized its own video and audio compression algorithms. All the Chinese standards carry the name Audio Video coding Standard (AVS) [83, 84]. From 2003 onward, different profiles for the first version of AVS – called AVS1 – have been published, covering broadcasting, cinema, surveillance, and HDTV.

Currently, the new version AVS2 is close to publication [84].

AVS1 is used broadly in China, for example, in broadcasting.

Table 3.6 gives an overview of the features of the other main standards – see also Ref. [85]: Many comparisons exist to show the superiority of other methods above H.264 and the other way around as well. For example, Google claims that their standard VP9 is close to H.265, but all recent tests show that they are closer to H.264 [86, 87]. However, the problem with all tests

Table 3.6 Comparison of other standards in relation to H.264

Feature	H.264	VC-1	VP-9	AVS
Picture type	I, P, B, SI, SP	I, P, B, BI, Skip	I, P	I, P, B
Reference frames	Max 16	Max 2	Max 8	Max 2
Transform	Similar to DCT (2 sizes)	Similar to DCT (4 sizes)	DCT (4 sizes)	Similar to DCT
Intraprediction	4×4, 16×16, 9 modes	8×8, 2 modes	4×4–64×64, 10 modes	4×4, 8×8
Interprediction	Y	Y	Y	Y
Motion compensation	7 sizes	2 sizes	>150 sizes similar H.265	6 sizes
MV resolution	$1, \frac{1}{2}, \frac{1}{4}$ pel	$1, \frac{1}{2}, \frac{1}{4}$ pel	$1, \frac{1}{2}, \frac{1}{4}, \frac{1}{8}$ pel	$\frac{1}{4}$ pel
Reference frames	Max 16	Max 2	3 out of 8 stored ones	Max 2
Entropy coding	CAVLC, CABAC	Adaptive VLC	Arithmetic BoolCoder	VLC, CABAC
Deblocking	Yes	Yes	Yes	Yes
Error resilience built-in	Yes	No	Yes	No

is that the superiority depends on the used video. Overall reliable data would require an agreed set of videos for comparison. An agreed testing suite exists within the H.26x developments, but this was not performed in the same way as other proposed video codecs. Consequently, a reliable comparison as shown in Figure 3.37 is not available.

So far, no broad move to one of those other standards has been seen. Again, the reason is that another standard must be at least 50% better than an existing one to make a migration attractive. However, such an improvement has never been seen in relation to the last published international standard and is not very probable to be seen in the future.

The market can accept new codecs faster if they are not used in hardware. For encoding of live events, HD or UltraHD hardware implementations are required. Even after a standard is ready, the chip production will take time. Nonstandardized codecs such as VP9 are only available in software and can be used to decode videos in a browser. Only slow software encoders produce such a video. For live event transmission in HD/UltraHD, only well-accepted video standards are applied.

Google and others state that the reason for their own developments of encoding standards is that they are license-free. However, the licenses for HEVC seem to be quite moderate: US $0.20 per unit after first 100,000 units each year [88]. Below 100,000 units/year, there are no royalty fees. So, there might be additional reasons for development and distribution of alternative standards.

3.6   Three-Dimensional Video

The idea for 3D (3DTV) started with early photography 150 years back. So, in each stage of image and video creation, stereoscopy was immediately included. However, stereoscopy always raised a curiosity and not suitable for daily video or image use. The use of stereoscopy in cinemas happened in waves. The first wave was at the beginning of the 1950s when people reduced their time in the cinemas since there was a TV set at home. On a black-and-white TV set, 3D cannot be generated, but it was possible for color films in the cinema theaters. Recently, another wave happened.

From H.262 onward, all video coding standards included a Multiview Coding (MVC), which includes stereoscopy as a special case. For H.264 and H.265, a Free Viewpoint Video (FVV) was standardized. With FVV, the user can choose his viewpoint when watching a video that includes multiple views [89, 90].

When transmitting pictures from different cameras, the easiest way is to send one stream per camera – again called Simulcast because parallel send-outs take place. However, MVC compresses the stream by using redundancy between the streams. We treated so many forms of predictions already that a compression by an Inter-View Prediction is apparent. This kind of prediction can compress spatial and temporal redundancy but does not create any issues with lower compression elements.

There arise problems solely when there are graphical overlays on the picture – for example, subtitles or stock exchange prices at the bottom of a video. There are specific standards that treat this – see Ref. [91].

MVC is part of the Blu-ray standard. However, broadcasting MVC is still considered to be too complicated, and the corresponding standard bodies (DVB for Europe, SMPTE for the United States, etc.) recommend sending two streams – that is, Simulcast. SMPTE stands for

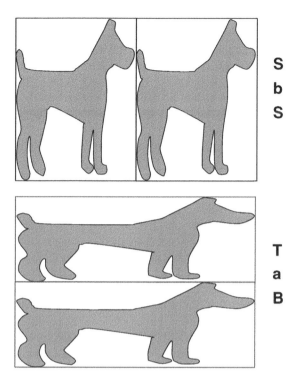

Figure 3.38   SbS and TaB

Society of Motion Picture & Television Engineers. When two streams are sent, the picture must still fit in the standard format. There are few options to make the pictures frame-compatible:

Top-and-bottom: Top-and-Bottom (TaB) is to place the squashed pictures above each other. See also Figure 3.38.

Side-by-side: Here, the two channels – for example, the original right and left pictures – are placed in an squashed form on the right and left parts of the sent picture (SbS). The receiver separates the two images and prepares them for display by superimposing on the screen. Today's HD TV services use this option, and DVB issued the standard in 2011 [92]. SbS was preferred over TaB because it better fits for interlaced devices.

Line sequential: The stereo pairs are woven into each other.

The production of 3D videos is more challenging – especially for live events. Cameras need to be equipped with two lenses, and the picture should be sharp for all parts. The latter criterion is especially very hard to achieve in a life event. This is treated in additional literature [93, 94]. If all works correctly, there are still around 5% of the users who have problems in fusing the two different pictures on the screen. Those users will stop using 3D after a short time. Time will show if users accept the discomfort when using lenses or use the special screens with a restricted viewing angle.

For a broader perspective on three-dimensional video, see Ref. [95].

3.7 Error Resilience

Only the decoder can initiate the error handling, and this is the first section that treats the receiving side exclusively. The H.264 decoder finds an error in the following situations:

- Intraprediction mode: Decoder cannot find pixels in intraprediction coding mode.
- Slice run-on: Decoder is not in the state to move to next slice after reading the last byte in the current slice.
- Slice fragment: Decoder is instructed to move to next slice, but unprocessed bytes are still available.
- Macroblock overrun: The number of macroblocks exceeds the expected number.

All other potential errors are nondetectable errors for the decoder. Tests showed that those non-detectable errors are also not observable by a viewer [96] and are below 0.1% of all mistakes. The decoder does not detect the error at the real position of the error but with some delay as shown in Figure 3.39 based on numbers in Ref. [96]. When the decoder detects the error condition, either it can skip all input until the next resynchronization point or it may try some of the measures listed below. The next resynchronization can always happen at the next GOP.

Here, we list those measures that can be taken to avoid or restrict errors:

Reversible Variable-Length Codes (RVLC): When the decoder finds an error, it moves forward until the next synchronization point. The decoder cannot work backward to see what is still correct because it cannot be sure if the read information is correct. Here, codes are reverse readable too – that is, one can find a maximum valid string until a last position. Codes that are readable forward and backward are in essence similar to a mirror image [97].

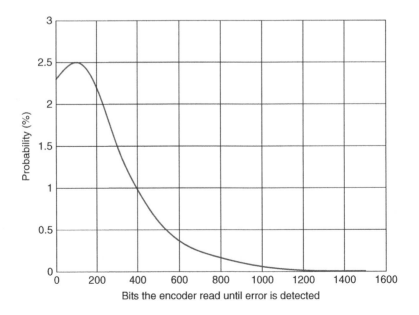

Figure 3.39 Probability depending on bits passing through decoder until error is detected

Table 3.7 Comparison of RVLC to
Huffman coding (English alphabet)

	Probability	Huffman	RVLC
E	0.148	001	000
T	0.093	110	111
A	0.088	0000	0110
O	0.072	0100	1001
R	0.068	0101	00100
N	0.064	0110	11011
H	0.058	1000	01010
	. . .		

Table 3.7 summarizes an example code for the most probable English letters in comparison
to the Huffman coding – see Section 3.2.

Data partitioning: As discussed in Section 3.4.4.2, the encoder can combine data of equal
importance into common NALUs. In this way, less drop probability for the relevant parts of
the transmission is possible.

Forward Error Correction (FEC): The sender adds redundant information to the transmitted
messages. This redundancy allows the sender to recover a limited number of errors. A very
simple FEC can send each bit three times. In the case of a bit error, the other two bits decide
on the correct value. For much more efficient codecs, see Ref. [98]

Duplication of frames: H.264 duplicated frames if configured this way. The Header Extension
Code (HEC) informs if this option is in use for an NALU.

The decoder has some techniques available to recover macroblocks if one is missing –
named as Decoder Error Concealment:

Spatial: The decoder uses the surrounding existing macroblocks to calculate the missing one
and most simplistically by averaging the pixel values of the other blocks.

Temporal: Here, the decoder uses the corresponding macroblock of a previous or following
frame to replace the missing one – this has the name Previous Frame Concealment (PFC).
The decoder finds the corresponding macroblock by looking at the surrounding motion vectors and calculates the mean and the Sum of Squared Differences (SSD) to decide on an
appropriate value.

Hybrid: The decoder could apply a hybrid concealment technique if surrounding macroblocks
were also intrapredicted. The motion vectors of the intrapredicted surrounding macroblocks
are used to predict the missing macroblock. If the surrounding blocks were interpredicted,
the temporal method is used instead. This technique carries the name Adaptive Error Concealment (AEC).

For error avoidance, the application could use a network protocol that guarantees the delivery. In many cases, the multimedia application still use protocols that do not guarantee the

delivery but are faster – see Section 4.1. This choice will be discussed later for individual applications. One video compression standard – H.263, Annex N [53, 99] – has proposed an unusual combination named NEWPRED of interworking with network protocols. The underlying transport protocols send Acknowledgments (ACK) and Nonacknowledgments (NAK) after receipt of packets back to the sender. These messages do not initiate any transport layer follow-up activities. However, the encoder reacts to those messages as follows:

- The encoder can only use acknowledged frames in any further predictions.
- When a NAK is received, the encoder sends an I-frame as soon as possible.
- When a NAK is received, the encoder sends I-frames for all so far nonacknowledged frames.

This technique is a kind of interactive error control combining the transport and application layers. The effectiveness of this procedure depends on the possible play-out delay of the decoder – see Section 5.4 – and on the fast availability of differently encoded frames by the encoder. Implementations of this technique are rare but were discussed for mobile and wireless networks [100]. For simulation results, see Ref. [101].

For overall error handling, see Ref. [102], and for error control for audio, see Ref. [103].

3.8   Transcoder

Transcoding is the conversion of a video stream from one format or bit rate into another format or bit rate. We have already seen in Section 3.4.5 that the encoder can create videos in scalable format that allows the viewing device to adapt the scale – that is, quality and bit rate, temporal subsampling, spatial subsampling. There are, however, situations when this is not available or hardly possible. For example, in an interactive videoconference, any delay in creating the video stream by a sender changes the arrival times for all receivers. If only a few receivers need, for instance, a downscaled resolution, a transcoding unit can create this in the transmission path on the fly.

A system that supports transcoding is the so-called Multipoint Control Unit (MCU) in video-conferencing. An MCU's primary task is to take the input from one sender and distribute it to a number of receivers. Since all receivers may not have all the same characteristics, the MCU has to transcode the scale for a receiver. In an interactive multipoint videoconference, the sender and the associated receivers change over time since only the pictures from the currently speaking partner are sent. Consequently, the transcoding of the MCU also has to change dynamically depending on the incoming and outgoing scales. It is also part of the optional functionality of an MCU to assemble many receivers in a single multiparted picture. In that way, all participants are visible at once but in a much smaller size. Also, this functionality obviously requires transcoding.

A change of the scale is also necessary if only one compressed video exists and needs to be made available in other scales or additions such as watermarks and labels need to be added. However, in such cases, one can fully decode the video and encode it again with the additional information. Full decoding and encoding in another scale are not feasible in interactive videoconferences due to delay restrictions.

When SVC is used, Transcoding often becomes superfluous. Still, one stream needs to be distributed to many receivers. Such a forwarding unit carries the name Single Forwarding Unit (SFU).

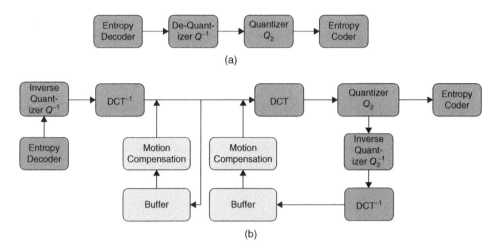

Figure 3.40   Open-loop (a) and CPDT (b) for quality adaptation

Transcoding is also a possibility for adaptation of a video stream to multiple previously unknown end-devices. This adjustment is required for the Companion Screen Applications (CSA) as described in the requirements – Section 2.3. CSAs will also be handled in Section 11.7.

Main transcoding techniques are [104–106]:

Quality adaptation: The open-loop technique drops the high-frequency DCT values. Dropping is performed by first dequantization Q^{-1} and executing, afterward a stricter quantization Q_2 as shown in Figure 3.40(a). The motion information is transported unchanged. Note that Figure 3.40 fits exactly with Figure 3.24 from the H.261 compression.

This method leads to the drift problem as already mentioned in Section 3.4.5. Initially, an I-frame is reduced by higher DCs, and afterward, all sent differences are not exact, and slowly the picture accuracy drifts away. The open-loop mechanism has no feedback loop on the error, and so, only with a new I-frame, the image is adjusted again.

The improved method for quality adaptation is the so-called closed-loop technology as shown in Figure 3.40(b). Here, the frame is also treated by DCT^{-1} plus the motion compensation, and so a complete unencoded picture is available. After the stricter quantization Q_2, the inverse Q_2^{-1} and DCT^{-1} a correct motion estimation for the newly quantized Q_2 exist. This method carries the name Cascaded Pixel-Domain Transcoding (CPDT). The effort is now of course much higher, and variants exist, which do not completely encode but calculate the new motion vector by a combination of the two quantized sets.

Spatial adaptation: The typical approach is to use the CPDT architecture and then take multiple macroblocks and combine them into a single one. The single macroblock is created by averaging the data from the multiple ones. When combining the macroblocks, the associated motion vectors also need to be combined. The combination can get tricky if motion vectors have a 1:n association with macroblocks. A combination of macroblocks is also difficult to accomplish when we have a video stream for an interlaced display. Finally, a set of adjacent macroblocks can be inter- or intrapredicted from different locations. So, it is not as easy as first described.

Temporal adaptation: This adaptation is the easiest one since we can just drop frames. However, the encoder has to create new motion vectors that are an aggregation of the motion vectors of all removed frames. This technique is called the Telescopic Vector Composition (TVC), but other methods exist as well.

Since transcoders are sitting in the transmission path, they are also able to act as a network proxy. For instance, they can execute an interactive error control – see Section 3.7.

MCUs contain the most powerful transcoders and their providers – for example, Cisco, Polycom, Ericsson – have extensive experiences with the technology.

The recent book [107] concentrates on H.264 transcoding.

4

Underlying Network Functions

This chapter covers key network protocols and capabilities used to transfer multimedia content over a packet network. Understanding what these capabilities and protocols provide is critical to understanding multimedia network services. It describes key protocols at both the transport and the application level, such as Real-Time Protocol (RTP) in Section 4.1 and Session Description Protocol (SDP) in Section 4.2, while Section 4.3 covers Real-Time Streaming Protocol (RTSP). The required infrastructure services are also discussed, such as Multicast in Section 4.4, Quality of Service (QoS/RSVP) in Section 4.5, and Network Time Protocol (NTP) in Section 4.6. Finally, we treat the Caching of content to be delivered in Section 4.7.

In Figure 4.1, we see a summary of some of the components used to deliver multimedia content. All of these items are IETF/IEEE/ISO standards. The **bold** items are covered this chapter. Session Initiation Protocol (SIP) and Conferencing are covered in Chapter 6. H.323 and H.248 are covered in Chapter 7. Service Location Protocol (SLP) is covered in Chapter 10. For Audio/Video and Shared Applications, more detailed protocols are covered in Chapters 7, 9, and 11.

4.1 Real-Time Protocol (RTP)

From the beginning of packet networks, there has been a constant stream of new ideas, services, and capabilities envisioned and implemented. Many of these new capabilities require standardized ways to exchange data and other information. For example, the need to exchange large files efficiently led to the creation of the File Transfer Protocol (FTP). Similarly, the advent of the World Wide Web began with the continuing development of the HyperText Transfer Protocol (HTTP). In many ways, the transmission of real-time data, such as voice and video, differs from that of other data types such as File Transfer. In the transfer of bulk data, the primary goals are high throughput/efficiency and data integrity. In the case of voice and video, the end goal is the delivery of time-sensitive data streams.

Several points are interesting to note when comparing UDP/TCP and RTP for transport.

- UDP provides only simple, unreliable transport. The packet either arrives or does not arrive. The upper levels have no visibility into the transport layer.

Multimedia Networks: Protocols, Design, and Applications, First Edition. Hans W. Barz and Gregory A. Bassett.
© 2016 John Wiley & Sons, Ltd. Published 2016 by John Wiley & Sons, Ltd.

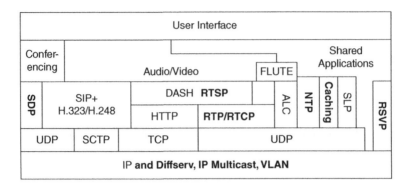

Figure 4.1   The multimedia network stack

- TCP offers reliable end-to-end service but reacts to congestion and packet loss by slowing the transfer of data. This is often not possible with real-time data that has a constant flow rate. Also, there is no visibility into the transport layer. For example, the upper layer protocols are not informed about varying network conditions. TCP does not support multicast.
- The design of RTP was based on the requirement to provide end-to-end, real-time delivery of traffic over an unreliable packet network.
- RTP provides visibility into its transport layer. This permits applications to understand how RTP is delivering the information and intelligently react to any network problems.
- RTP also provides a number of key capabilities to support real-time data transport. For example, RTP provides transport, timestamps, sequence numbers, payload source and identification, reception quality feedback, media synchronization, and membership management.
- RTP relies on the underlying network layers to provide delivery of its packets. Although RTP does not specify a transport, IP/UDP is almost always used to deliver RTP packets.

RTP was first defined in 1996 by RFC 1889 [108]. That RFC was superseded in 2003 by RFC 3550 [109], which remains in effect up to this day. Although RTP was first developed to support audio- and videoconferencing, it was intentionally designed to be flexible and extensible. It was clear during the development of RTP that it might be used in a wide variety of environments from videoconferencing to voice transmission to commercial broadcasting. This required the design to be a framework that would support a wide variety of applications.

It can be used in both Unicast (point-to-point) and Multicast (one-to-many or many-to-many) environments. It can scale from one to many thousands of endpoints. It can transport streams encoded with virtually any current or future encoding scheme. RFC 3550 also defines and supports the concept of mixers and translators. Mixers generally receive streams from one or more sources and mix them together to create a new, combined stream. It is responsible for synchronizing the streams. Therefore, the mixer generates its own source identifier and timecode for the stream it outputs. Translators forward RTP packets with the synchronization intact. They can change the encoding, replicate packets for multicast to unicast transmission, and provide filtering for firewalls.

In summary, although other transport mechanisms, such as UDP and TCP, can provide some of the services listed, RTP provides the widest and most complete support for these capabilities.

It is not perfect for every application, but its flexibility makes it the protocol of choice for many multimedia implementations.

This chapter provides an overview of RTP, but much more extensive information exists. For detailed information on RTP, we refer the reader to the book by Perkins [110]. Perkins was heavily involved in the development of RTP.

4.1.1   Elements of RTP

RFC 3550 defines two separate protocols. RTP is used for the transport of media streams. RTCP, which is defined in the same RFC, provides a control channel for monitoring, limited control, and identification capabilities. RTP is best thought of as a framework for the transfer of real-time data. The specification is intentionally incomplete in two ways. First, the algorithms that apply to the media, such as synchronization, play-out, and regeneration must be specified by the application. The second part is the ability of the application to influence the transport such as resolution of the timestamps and the marking of interesting packets. Therefore, additional information about the actual payload of the RTP packets is required. The specification required to complete an RTP implementation is the *profile*. The profile defines the application-specific information such as codecs used to encode the data and the payload format in the RTP packets. RFC 3551 [16] was published concurrently with RFC 3550 and defines a default profile for audio and video data transfer using RTP. Many other profiles, such as the H.264 Profile [111, 112] are now defined.

IP/UDP   provides packet delivery services.

RTP packets   transport real-time data, including timestamp, type of data, and source of streams.

RTCP packets   provide quality feedback, identification of receivers, and stream synchronization.

Profiles   define how RTP header fields should be interpreted and define the mapping from Payload Type (PT) to data encoding specifications.

4.1.2   Details of RTP

Figure 4.2 shows the layout of the RTP packet.

Version (V):   Version of RTP Protocol. Value is 2 as of 2003.

Padding (P):   If the padding bit is set, the last octet contains a count of how many octets should be ignored; used for encryption algorithms or underlying transport protocols.

Extension (X):   If the extension bit is set, the fixed header must be followed by a header extension. This is very rarely used. Normally, extensions are included in the payload.

CSRC count (CC):   The CSRC count contains the number of CSRC identifiers that follow the fixed header.

Marker (M):   The interpretation of the marker is defined by a profile. It is intended to allow significant events such as frame boundaries to be marked in the packet stream. For example, MPEG-4 ES [111] uses the Marker to indicate the *end of Access Unit* (end of decoded picture). For more information, see Section 3.4.1.

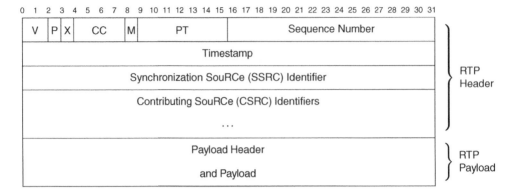

Figure 4.2   RTP packet

Payload type (PT):   This field identifies the format of the RTP payload and determines its interpretation by the application. For information on values of the PT field, see Section 4.1.3.

Sequence number: The initial value of the sequence number should be random and increments by one for each RTP data packet sent. If the received sequence number is more than 100 out of order, a restart of the stream is assumed.

Timestamp: This is the encoding time of the first octet of the media data in this packet. The timestamp does not need to be the actual time (wall clock). The resolution of the clock should allow for good jitter measurement and media synchronization. Different RTP sessions may have different timestamp clocks and resolutions.

Synchronization source identifier (SSRC):   The SSRC uniquely identifies the source of the stream being carried in the payload. The SSRC should be randomly chosen.

Contributing source identifiers (CSRC):   This is a list of contributing sources for the payload. There can be up to 15 CSRC entries. For example, if a number of audio streams are combined to a single stream by a mixer, the SSRC in the RTP packet will be the mixer and each of the original audio sources will have their SSRCs entered as CSRCs in this list.

Optional header extension (not shown):   This will exist if defined by a profile and the Extension bit is set.

Payload: See Section 4.1.3.

4.1.3   RTP Payload

The payload of the RTP packet is one of the most critical components of RTP. The specification of the layout and interpretation of the payload are defined by profiles. The profiles also define the possible values for the PT described. The original PT values were defined in the first profile standard [16], but are now maintained by IANA.[1] Table 4.1 lists some of the most common values for RT. Additional payload type codes may be defined dynamically through non-RTP means. An RTP source may change the payload type during a session, but this triggers a new SSRC [113]. For multiple, concurrent streams, such as voice and video, parallel RTP sessions are used [114].

---

[1] http://www.iana.org/assignments/rtp-parameters.

Table 4.1 Common
payload type (PT) values

Payload type (PT)	Value
5	DVI4
26	JPEG
31	H.261
32	MPEG I/II

Network abstraction layer

An important concept in the transfer of video over a packet network is the *Network Abstraction Layer* (*NAL*). This is a method and format for sending encoded video in a way that the content can be properly handled by a network, storage devices, and more. Part of the format describes segmenting the video content into transportable blocks called *Network Abstraction Layer Units* (*NALU*). NALUs were designed for network transmission and not for use on storage devices. The MPEG Program Elementary Stream is used for stored video. See Section 3.4.1.

The content of the RTP payload is specified in the profile associated with the session. For example, the H.264 Profile [111, 112] and the SVC Profile [114, 115] define the following items. See also Sections 3.4.4 and 3.4.5.

Network Abstraction Layer Unit (NALU) octet: This octet contains three fields: F, NRI, and Type. It is used to identify what type of payload structure follows. "F" is the Forbidden bit and should be 0 if no bit or syntax errors exist in the payload. Decoders will ignore packets with $F = 1$. The NAL Reference ID (NRI) bits indicate if this NALU is used to generate reference pictures. If NRI = 0, then the NALU can be discarded without impacting the generation of reference pictures. The larger the number, the greater the impact of the NALU on the reference picture. The Type field provides additional information such as whether the NALU is a single packet, an aggregation packet, or fragmentation unit. For single NALUs, it also indicates the type of packet (e.g., End-of-Sequence).

Figure 4.3 shows a simplified flow of video traffic from the coding layer, through the NAL packetization, and finally into the RTP payload. Some important points to notice are that the NALU size will vary depending on the video stream. Also, note that the RTP payload can contain one or more NALUs. Although not shown in this figure, multiple RTP packets can be used to transport a single NAL unit. This is shown in Figure 4.5.

RTP can transport NALU also using MPEG-2 data stream as described in Section 3.4.1 using RFC 2250 [116]. Older video compression information (H.263/2/1) was always transported in an MPEG-2 data stream. Section 3.4.4.2 lists the reasons why MPEG-2 is still, sometimes, used instead of direct embedding into RTP.

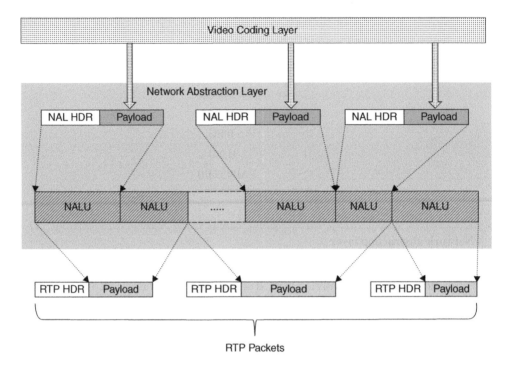

Figure 4.3   Mapping video streams to RTP using NALUs

Packetization modes:

- **Single NALU (Type 1–23):**   In this case, the entire NALU is contained in the payload.
- **Noninterleaved mode:**   Multiple NALUs are sent in strict decoding order.
- **Interleaved mode:**   Multiple NALUs are sent, but not in strict order. In the case of a single stream, there is an additional field that describes the decoding order, *Decoding Order Number (DON)*. For ordering multiple RTP sessions with an SVC Profile, *Cross-Session DON (CS-DON)* is used.[2]

Aggregation modes for interleaved/noninterleaved sessions: There are use cases where the NALUs are relatively small, sometimes as small as a few bytes. In these cases, it is helpful to aggregate them with other NALUs in a single RTP packet. Two types of NALU aggregation exist:

Single-time aggregation packets (STAP):   This type is generally used in low-delay environment. In these cases, interleaving of information is not desired, and therefore, all NALUs share a single timestamp. The format is simple with a single byte with the same format as that of the NALU Octet, followed by one or more STAP units.

Multiple-time aggregation packets (MTAP):   In high-delay environments, such as streaming, interleaving is more common, resulting in the requirements that NALUs have different timestamps. The timestamp of each MTAP unit is coded relative to the RTP timestamp to save bits and contains a DON that indicates the decoding sequence of the NALUs in the MTAP.

---

[2] The NALUs can span multiple RTP streams. For example, each layer of an SVC stream is sent with a separate RTP session and ordered by the CS-DON value.

Fragmentation modes: This field is used to segment a single NALU over multiple RTP packets. There are two modes: FU-A and FU-B. The fragmentation header contains the fields: *start_indication, end_indication, reserved, type.*

Combinations of H.264 profile modes: Table 4.2 lists which modes of traffic can be transported by different Packet Type (PTs).

An example of an RTP packet carrying NALUs is shown in Figure 4.4. This is an MTAP16 packet aggregating two NALUs. Note that the DON appears here with the DON Base (DONB) value in the RTP packet and the DON Difference (DOND) in each NAL unit.

Figure 4.5 illustrates many of the concepts discussed so far. It shows a multilayer SVC bit stream with both aggregation and fragmentation using three different RTP streams.

Table 4.2 H.264 profile modes

Type	Packet	Single NALU mode	Noninterleaved mode	Interleaved mode
1–23	NALU	Yes	Yes	No
24	STAP-A	No	Yes	No
25	STAP-B	No	No	Yes
26	MTAP16	No	No	Yes
27	MTAP24	No	No	Yes
28	FU-A	No	Yes	Yes
29	FU-B	No	No	Yes

Figure 4.4 Sample RTP packet with MTAP16

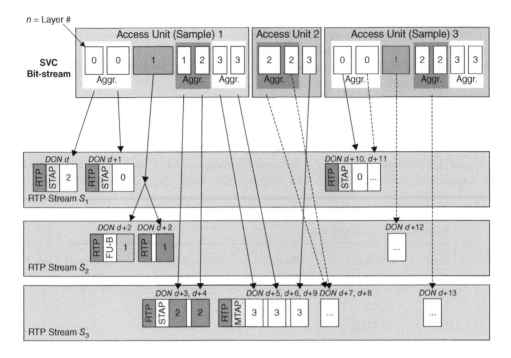

Figure 4.5   Multilayer synchronization with SVC

As mentioned earlier, the DON in each packet is used to determine how the packets should be ordered after they are received. The following is a summary of the packetization [117]:

- This SVC bit stream consists of four layers; the base layer (0) and three enhancement layers (1–3).
- The sender groups together subsets of different layers into Access Units (AU). The figure shows three of these AUs.
- The first two NALUs (layer 0) are placed in separate, unfragmented packets in RTP stream 1.
- The NALU for layer 1 is fragmented into two separate RTP packets in stream 2.
- Note that the DON used to reassemble the data stream is incremented for each of these operations.
- Next two *sequential* NALUs for layer 2 are placed in a single-time RTP packet in stream 3.
- Although two sequential NALUs for layer 3 are placed in the same packet, the addition of a third packet from a different AU (with a different time) requires the use of a multi-time RTP packet.

Many combinations of packet types are possible. The use of multiplexed video streams has been shown to reduce the packets required to send the data by 60% as compared to nonmultiplexing methods [118]. The multiplexed streams also showed a reduction of 5% in damaged frames [119]. Note that the NALU concept for H264 is also valid, with adaptations, for the new H.265 standard and is documented in a draft H.265 RTP payload format document [120].

There are also extensions such as Secure Real-Time Protocol (SRTP) [121] and Header Compression. See Section 4.1.4.5.

Simplifications to RTP have also been developed, such as Reduced Complexity Decoding Operation (RCDO) [122], Reduced RTCP size [123], and the control of port numbers to ease NAT transversal [124].

Clocks and timestamps

Perhaps one of the most important topics when discussing real-time multimedia is *time*. Although timestamps have already been discussed, the subsequent sections deal with time in much more detail. Some packets, such as the RTCP Sender Report (SR), can easily have five or more time-related fields. For an example, see Section 4.1.4.1. Some key points to remember:

- Timestamps can be based on a variety of clocks or, in some cases, no clock at all. These different origins will be noted for each type of timestamp.
- There are different lengths for timestamp, offset, and clock resolution fields. Watch for them.
- The clock resolution – the length of each clock tick – is an extremely important value. It will depend on the system, the media, the sample rate of the media, the encoding bit rate, and several other possible factors. There are situations when the clock rate will change.

4.1.4   Details of RTCP

The RTP Control Protocol (RTCP) is part of an overall RTP implementation and can provide feedback on the quality of transmission as well as maintaining a persistent identifier to track senders in the event of restarts or excessive packet loss. RTCP provides this service by the periodic transmission of control packets to all session participants. Feedback on the quality of transmission can be used by adaptive encoding schemes to select a stream with an appropriate data rate. It can also provide critical information to isolate and diagnose distribution faults, especially in multicast environments. Conflicts or program restarts can cause the SSRC of a data stream to change. The use of a Canonical Name (CNAME) by RTCP permits the receiver to track the stream source if the SSRC changes. The CNAME can also be used to associate multiple streams, such as voice and video, to a single source.

The standard requires RTCP packets be sent to all session participants. To enable RTCP to scale to a large number of participants, it is necessary to control the rate at which these packets are sent. RFC 3550 provides timing rules for the calculation of the RTCP transmission interval. The goal is that RTCP should use no more than 5% of the total bandwidth used by the RTP session. Each participant should maintain a table of all other active participants in order to calculate the transmission interval of the RTCP packets. Translators that modify RTP packets must also modify the associated RTCP packets. Mixers generate their own reception reports (SR, RR) but do not forward RTCP packets from others. Any expansions to RTCP by profiles are governed by Ott and Perkins [125].

There are five RTCP packet types.

Sender Report (SR): The SR packet is sent by participants that both send and receive data and provides both transmission and reception statistics. If this participant has not sent any data packets since the previous report was issued, an Receiver Report (RR) is sent instead.

Receiver Report (RR): The RR packet sends reception statistics from participants that only receive data. There should be one reception report for each of the sources from which this participant has received packets.

Source Description (SDES) items: Source Description items, including CNAME.

BYE: Session Membership Management.

APP: Application-defined Functions.

The beginning of each RTCP packet is a fixed header similar to the RTP packet. What follows are structured elements of variable length, but always ending on a 32-bit boundary. This is defined to make RTCP packets "stackable." Multiple RTCP packets should always be concatenated to form a compound RTCP packet sent as a single packet by the lower layer protocol, such as UDP. In order for the protocol to function properly and to simplify identifying damaged packets, RFC 3550 has imposed constraints on which packets must be sent and which packet types must be first in the compound RTCP packet. The details of these constraints are outside the scope of this book and can be found in the RFC. There are two important implications of these constraints. First, virtually every RTCP packet is a compound packet containing, at a minimum, a reception report (SR or RR) and an SDES CNAME packet. The reception report must be the first packet in the compound packet.

4.1.4.1   RTCP Sender Report Packet (SR)

The SR packet shown in Figure 4.6 is comprised of three sections, with an additional extension if defined by a profile. The first section is the RTCP header, which all packet types have in common.

Reception report count (RC): This contains the number of reception report blocks contained in this packet. Zero is a valid value if no data packets have been received.

PT: A Packet Type of 200 specifies this as an RTCP SR packet.

Length: The length of this RTCP packet is 32-bit words minus one, including the header and any padding.

SSRC: The SSRC of the originator of this SR packet.

The second section, the sender information, is 20 octets long and is present in every sender report packet. It summarizes the data transmissions from this sender.

NTP timestamp: Contains the wall clock time when this report was sent.

RTP timestamp: This corresponds to the same point in time as the NTP timestamp, but is based on the offset and units used by the timestamp in RTP data packets. If the NTP timestamps from different senders are based on the same clock, these timestamps can be used to synchronize multiple data streams.

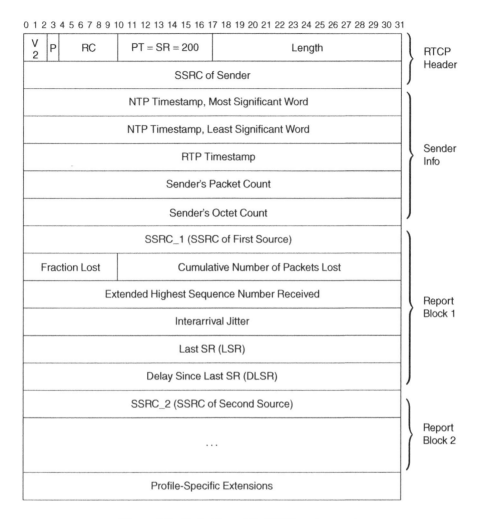

0 1 2 3 4 5 6 7 8 9 10 11 12 13 14 15 16 17 18 19 20 21 22 23 24 25 26 27 28 29 30 31

V 2	P	RC	PT = SR = 200	Length

SSRC of Sender

RTCP Header

NTP Timestamp, Most Significant Word

NTP Timestamp, Least Significant Word

RTP Timestamp

Sender's Packet Count

Sender's Octet Count

Sender Info

SSRC_1 (SSRC of First Source)

Fraction Lost	Cumulative Number of Packets Lost

Extended Highest Sequence Number Received

Interarrival Jitter

Last SR (LSR)

Delay Since Last SR (DLSR)

Report Block 1

SSRC_2 (SSRC of Second Source)

. . .

Report Block 2

Profile-Specific Extensions

Figure 4.6 Packet format for RTCP sender report

Sender's packet count: The total number of RTP data packets sent since starting transmission.

Sender's octet count: The total number of payload octets sent since starting transmission (excluding header and padding).

The third section contains one reception report block for each source heard since the last report. Zero is a valid value. Each block contains statistics on the reception of RTP data packets from that source.

SSRC_n (source identifier): The SSRC of the source to which this report pertains.

Fraction lost: This field reports the fraction of packets lost since the last report. It is a fixed-point integer with the decimal point on the left edge of the field. It is calculated by multiplying the loss fraction by 256. Therefore, if 25% of the packets were lost since the last

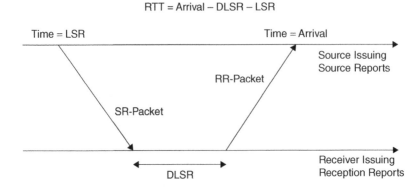

Figure 4.7   Calculation of RTT using RTCP reports

report, the value of this field would be 64. If the number of packets is greater than expected due to duplicate or late packets, the value is set to zero. Note that if the loss is 100%, that is, no packets were received, a reception report will not be generated at all.

Cumulative number of packets lost: This field is a signed integer with the total number of RTP data packets from this source that have been lost since the beginning of this session. The total loss may be negative due to duplicate packets.

Extended highest sequence number received: The low 16 bits contain the highest sequence number received in an RTP data packet from the source. Since this number is 16 bits and may start with any number, this number will sometimes wrap (cycle). The high 16 bits are the number of sequence number cycles.

Interarrival jitter: This field is an estimate of the variation in arrival times of RTP data packets, expressed in timestamp units. The exact formula is defined in RFC 3550 and is used for all implementations to permit profile-independent monitoring of jitter. See Section 5.2.

Last Sender Report (LSR): This value is the middle 32 bits out of 64 in the NTP timestamp from the most recent RTCP sender report (SR) from the source.

Delay Since Last Sender Report (DLSR): The delay, expressed in units of 1/65536 seconds, between receiving the last SR packet from the source and sending this reception report block. If no SR packet has been received yet from this source, the DLSR field is set to zero. DLSR, together with LSR, can be used by the sender to calculate approximate round-trip time to this receiver.

The Round-Trip Time (RTT) between the sender and the receiver can be calculated as shown as follows and in Figure 4.7.

$$RTT = Arrival - DLSR - LSR$$

4.1.4.2   RTCP Receiver Report Packet (RR)

The format of the receiver report (RR) packet is the same as that of the SR packet except that the packet type field contains the value 201 and the five words of sender information are omitted (these are the NTP and RTP timestamps and sender's packet and octet counts). The remaining fields have the same meaning as for the SR packet. Both SR and RR reports can be extended with additional information when required by a profile. This extension adds a fourth section to the report packet after the sender/receiver report sections. For both SR and SS packets, if the SSRC identifier changes, all the statistics are reset. In addition, an SSRC change frequently results in a changed RTP clock rate.

4.1.4.3 RTCP Source Description Packet (SDES)

As the name implies, the purpose of SDES packets is to provide identification and additional information about each participant. Normally, the packet will describe one or more items for a single source. Mixers and translators that are aggregating multiple data sources will also aggregate the SDES information. The SDES packet shown in Figure 4.8 has an RTCP header followed by one packet for each source. The packet includes the SSRC of the source followed by a list of identification or information items. Standard SDES items are CNAME, NAME, EMAIL, PHONE, LOC, TOOL, NOTE, and PRIV.

Packet Type (PT): The value of 202 identifies this packet as an RTCP SDES packet.

SDS Item – CNAME: This value normally has the format *user@host*. The host can be a Fully Qualified Domain Name (FQDN), a dotted decimal IPv4 number, or an IPv6 address. This value remains the same even if the data stream is restarted (with a corresponding new SSRC). This value can also represent more than one data stream. This permits a receiver to identify, for example, the audio and video streams used in a simple multimedia broadcast.

SDS items – NAME, EMAIL, PHONE, LOC, TOOL, NOTE, PRIV: These optional values provide additional information about the session to the receiver. Name, E-mail, Phone,

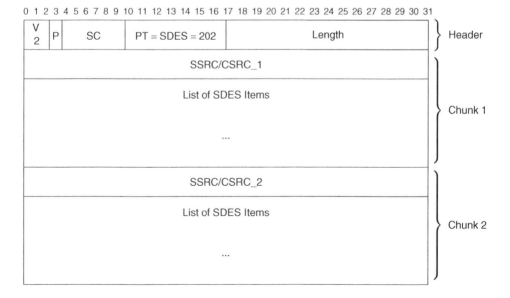

Figure 4.8 Packet format for RTCP Source Description (SDES) packet

Figure 4.9 Packet format for RTCP goodbye packet (BYE)

Location, and Note are self-explanatory. The TOOL field can be used to identify the application used to generate the stream. This may help the receiver debug problems with the stream. Priv is a private, profile-specific extension to the SDS packet.

4.1.4.4 RTCP Goodbye Packet (BYE)

RTP supports relaxed membership control. The RTCP BYE packet shown in Figure 4.9 informs the recipient that the participant is terminating the session and all future RTP/RTCP packets related to the listed SSRCs should be ignored. Since there is no guarantee of delivery, the BYE packet is not the only method to determine that a participant has terminated the session. An optional "Reason for Leaving" field can provide a displayable text message to the user stating the cause of the termination.

4.1.4.5 RTP Header Compression

To enable RTP usage with low bandwidth links, header compression is used. This was especially important when low-bandwidth dial-up network connections were more common but can still provide significant benefit with today's networking technologies. In most cases, the normal IP/UDP/RTP header of 40 Bytes can be reduced to 2 Bytes. See Figure 4.10. Two basic standards have been developed to provide this capability; Compressed RTP (CRTP) and Robust Header Compression (ROHC) [126, 127]. CRTP was originally designed for use on low-speed serial links that generally had low error rates. It remains the protocol of choice for these links because it is efficient and easy to implement. ROHC was developed to work in environments with longer delays and higher error rates, such as cellular data networks. Although it is a more complex protocol, it is able to recover more quickly with less impact in the event of lost packets. Both compressions take place on a hop-by-hop basis.

4.1.4.6 Compressed RTP (CRTP)

CRTP is defined in Ref. [128] and compresses combined IP/UDP/RTP headers as shown in Figure 4.10. Since many IP/UDP/RTP header fields are either constant or normally change in a predictable way, these can be eliminated to create a compressed version. The mechanism used to identify the packets varies based on the type of link. In the case of PPP, the mechanism used

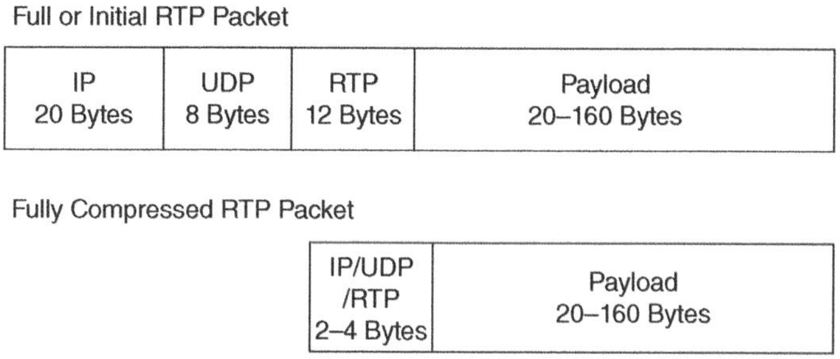

Full or Initial RTP Packet

IP 20 Bytes	UDP 8 Bytes	RTP 12 Bytes	Payload 20–160 Bytes

Fully Compressed RTP Packet

IP/UDP /RTP 2–4 Bytes	Payload 20–160 Bytes

Figure 4.10 Example of CRTP header compression

is defined by Koren et al. [129]. The following new packet types are defined in Ref. [128] and used to identify compressed packets to the link-layer protocol:

Full header: A packet with full headers including Context ID and Sequence Number.
Compressed UDP: Only IP/UDP headers are compressed [130].
Context state: This is a special packet with a list of Context IDs that have lost or may have lost synchronization.
Compress RTP: Only differences that cannot be predicted are sent. This can happen for the following fields; IPv4 Packet ID (bit:I), UDP checksum, RTP Marker bit (M), RTP Sequence Number (S), RTP Timestamp (T), RTP CSRC count and list. If these fields are not present in the packet, the information in the following decompressor context table is used.

Initially, CRTP packets with full headers are sent to establish context for subsequent packets. The initial packets include two additional fields; Context Identifier (CID, 8 or 16 bit) and Sequence Number (4 bits). These additional fields are stored in the length fields of the IP and UDP headers since this information can be obtained from the link-layer protocol. This initial packet is used by the decompressor to populate a context table entry for this session. Subsequent CRTP headers can then be derived from the compressed header and the context table. Some compressed headers will also contain information that will update the context table.

The CRTP packet contents can be seen in Figure 4.11. The first 2 or 3 Bytes (depending on when an 8 or a 16 bit CID is used) and the RTP payload are in every packet. The UDP checksum fields, Random Fields, RTP Header Extension, and Padding each will be in either every packet or none. The existence of each of these fields is determined at the beginning of each session and does not change. The status bits M, S, T, and I determine whether each of the associated fields will exist. There is a special case in the event that the CSRC list changes. Since there is no single status bit for this case, all status bits (MSTI) are set to 1. In this case, the CSRC list and count will be included and a second set of status bits (M$'$, S$'$, T$'$, and I$'$) are added to reflect the actual status of those fields.

When packets are lost during a session, a number of compressed packets must be discarded until the protocol resynchronizes. As CRTP was designed for low-loss circuits, this can result in significant disruption of the data stream. Enhanced CRTP (ECRTP) was developed with improvements in the area of error recovery and results in few lost packets with a minor increase in overhead [131]. One method uses ECRTP to accomplish this is by transmitting changes to compressed fields in the next N packets. Therefore, if less than N packets are lost, the decompressor is still able to maintain context [132].

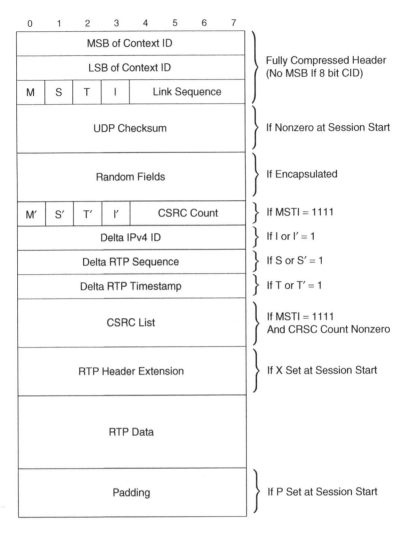

Figure 4.11   CRTP packet

4.2   Session Description Protocol (SDP)

4.2.1   SDP Overview

The SDP is actually a message format for describing the initialization parameters of a multimedia stream. The initial standard was proposed in 1998 [133]. This was superseded in 2006 [134]. It was originally developed together with the Session Announcement Protocol (SAP) for multicast broadcasts. However, it has found use as a general purpose description that is used by a variety of additional network environments, such as SIP (Chapter 6) and HTTP. It is not designed for the negotiation of the actual content or encoding of the media. Applications will use SDP to announce a session, invite participants, and negotiate parameters such as media type, format, and other properties.

A session description is a series of fields, each with a single <type>=<value> pair terminated with a CR/LF. The <type> is a single case-sensitive character. The <value> is structured text with a format defined by the type. The <value> is usually either one or more text strings separated by a single space or a free-form text string. The description has three sections. The first section is a session description that contains information such as originator, title, and contact information. The second section describes the time(s) the session is active, including a repeat count (if appropriate). The last section describes the media. Each message can contain multiple timing and media sections. Fields must be sent in the order defined in the specification to simplify the parser and more easily identify damaged packets.

In Figure 4.12, we can see a sample SDP message. Some key items of information are:

c=IN IP4 224.2.17.12/127	This is a multicast session (TTL=127) originating from the listed IP address.
t=2873397496 2873404696	Here are the start and stop times (in NTP format). There are no repeats listed (no r=).
m=audio ...	Audio stream using the RTP/AVP (RTP Audio Visual Profile) protocol.
m=video ...	Video stream using the RTP/AVP protocol.
a=rtpmap:99 h263-1998/90000	The *rtpmap* attribute is a key element in defining what media tools are required to participate in a session. In the example, it maps an RTP Payload Type number (99, same as m = number) to the encoding name (h263-1998) with a clock rate of 90,000 [135].

Table 4.3 describes the fields in the SDP [134] – * means the field is optional:

```
┌─────────────┐
│ SDP Sample  │
├─────────────┘
v=0
o=jdoe 2890844526 2890842807 IN IP4 10.47.16.5
s=SDP Seminar
i=A Seminar on the session description protocol
u=http://www.example.com/seminars/sdp.pdf
e=j.doe@example.com (Jane Doe)
c=IN IP4 224.2.17.12/127
t=2873397496 2873404696
a=recvonly
m=audio 49170 RTP/AVP 0
m=video 51372 RTP/AVP 99
a=rtpmap:99 h263-1998/90000
```

Figure 4.12   SDP sample message

Table 4.3   Description of SDP fields

Session description

---

v=	Protocol version = 0
o=	Originator and session identifier comprised of *"username session_id session_vers nettype addrtype unicast_addr"*
nettype =	IN for Internet, addrtype = IP4 or IP6, unicast_addr is either IP name or dotted decimal address
s=	Session name – must not be empty
i=*	Session information
u=*	URI of description
e=*	E-mail address
p=*	Phone number
c=*	Connection information – not required if included in all media comprised of *"nettype addrtype connection_addr"*. If session is multicast, TTL is added at the end
b=*	Zero or more bandwidth information lines – contains proposed bandwidth usage
	One or more time descriptions (see Time description below)
z=*	Time zone adjustments
k=*	Encryption key
a=*	Zero or more session attribute lines. Some attributes, such as *rtpmap* are defined in the SDP RFC, and others are defined in other RFCs to extend SDP capabilities. See Section 4.2.2
	Zero or more media descriptions (see Media description below)

Time description

---

t=	Time the session is active – Defines start and stop time in Network Time Protocol format. See Section 4.6
r=*	Zero or more repeat times – Defines repeat interval, duration, and offsets from start time

Media description

---

m=	Media name and transport address – This field defines the media type, port used, transport protocol, and media format
media type =	Audio, video, text, application, or message
transport protocol =	udp (unspecified), RTP/AVP (RTP for A/V conferences) [16], or SRTP/SAVP (Secure RTP) [121]
media format =	Interpretation depends on the transport protocol field
i=*	Media title
c=*	Connection information – optional if included at session level
b=*	Zero or more bandwidth information lines
k=*	Encryption key
a=*	Zero or more media attribute lines

4.2.2   Extending SDP

Since SDP was originally defined, it has been deployed in many more multimedia environments than anticipated. This has led to the development of an number of new standards and extensions to the original specification. Several of these are listed as follows.

Participants in a session can use SDP together with an Offer/Answer model as defined in Ref. [136] to reach a common view and agreement of the session parameters. This method is most often used in unicast sessions by protocols such as SIP. The offerer sends a proposal using the SDP message format and is answered in the same way by the other participant approving or replacing the session parameters. A number of examples can be found in Ref. [137].

While the Offer/Answer model [136] allows the participants to exchange and negotiate certain session parameters such as transport protocols and codecs, it does not include the ability to negotiate the capabilities of the participants. To provide this ability, SDP is extended by Andreasen [138] to provide a general Capability Negotiation framework. This standard was updated in 2013 by Gilman et al. [139]. This capabilities exchange sends offers of actual and potential configurations encoded in SDP attributes using the Offer/Answer model and receives answers potentially accepting one of the offered configurations. Examples of this expanded capabilities negotiation include the ability to offer multiple transport protocols (RTP, SRTP, SRTP with RTCP feedback, etc.) and the ability to negotiate the security parameters for SRTP.

Another standard that extends the capabilities of SDP is Ref. [140], which describes a method for Signaling Media Dependency. This capability is required to support media encoding, such as Scalable Video Coding (SVC) [114], which uses multiple, related media streams to provide different overall transmission quality levels. While SDP provides the method to describe the various media streams, it does not provide any mechanism to define the dependencies between them. This is achieved by grouping the media streams using the media identification attribute [141] and a media-level attribute "depend." With these additional attributes, the receiver is able to determine the relationship between the multiple streams in the session.

In addition to SDP extensions that further support negotiation and defining capabilities or media stream dependencies, underlying or related protocols can use SDP to modify or enhance their own capabilities. For example, in some use cases, the default bandwidth allocation for RTCP of 5% may not be optimal. Casner [142] provides a mechanism, using an SDP attribute, to modify the bandwidth allocation. Similarly, the ports used by RTCP can be defined using an SDP attribute defined in Ref. [143].

Another example is the compression of SIP signaling for use over low-bandwidth network links. An SDP extension [144] together with the Signaling Compression scheme [145] provides greatly reduced setup time in these types of network links. There are many more examples of SDP extensions in areas such as Security, QoS, and more.

Ott et al. [146] uses SDP to indicate the ability to provide media stream feedback using RTCP. With this extension to the Audio-Visual Profile, receivers can provide more immediate feedback to the sender on the reception of a stream. SDP parameters are used to define the operational details.

4.2.3   Javascript Session Establishment Protocol (JSEP)

The increased focus on the Web browser as the primary human interface has triggered the development of Web Real-Time Communication (WebRTC). This is a set of APIs that support direct browser-to-browser communications for voice, video, files sharing, and more. To enable browsers to establish real-time sessions, a new mechanism is being defined as a component of WebRTC. Javascript Session Establishment Protocol (JSEP) enables a Javascript to control the signaling plane of a multimedia session. It uses, at the moment, SDP to describe the session. At the time of this writing, JSEP is an Internet Draft [147]. This is treated in more detail in Chapter 8.

4.3   Streaming

In the preceding sections, we have discussed protocols and message formats responsible for the transport of media, quality feedback mechanisms, and the description/negotiation of media streams. In this section, we describe protocols used for controlling the broadcast of these media streams. For additional information on deploying streaming services, refer to Ref. [148].

There are generally three ways to stream media to clients:

Progressive download: The content is downloaded to the client using HTTP. Playback can begin before the download is complete. From the server viewpoint, this is a simple file download without special controls or feedback. The client chooses the buffering size and playback style. This is a simple method but usually results in significant delays before playback can begin.

RTP/UDP: The previous sections described streaming media over RTP/UDP with control of the stream provided by RTSP. The RTP/UDP method of streaming has the advantages of low latency, efficient use of available bandwidth, no local (client) storage of media, and the ability of a server to monitor the streams that the client consumes [149]. Live time-sensitive broadcasts use mostly RTP/UDP for streaming. However, the use of UDP for transport can frequently cause problems due to packets being blocked by firewalls.

HTTP streaming: To avoid these issues and to take advantage of the rapidly increasing use of HTTP infrastructure, new techniques and standards have been developed using HTTP as a transport to provide relatively low latency and firewall traversal. One very important additional capability that often is a part of HTTP Streaming is *adaptive streaming*. With this approach, the server maintains multiple versions of the content with different quality/bit rates. Since the client requests relatively small segments, different versions of the stream can be requested over time to adapt to varying network conditions. For in-depth information about HTTP Streaming, see Chapter 9.

4.3.1   Real-Time Streaming Protocol (RTSP)

RTSP [150] is an application-level protocol that enables a client to control the broadcast of multiple media streams sent from a server. The media streams can be either stored data or live feeds. RTSP also controls the delivery channels (UDP, Multicast UDP, and TCP) as well as the delivery mechanisms used by RTP [108]. This can be thought of as a *remote control* for streaming media with the ability to start, stop, pause, and many other operations.

Since RTSP is a connectionless protocol (regardless of the underlying transport protocol such as TCP), state is maintained on the server using a session identifier. The server states are: *Setup (reserve resources), Play/Record, Pause,* and *Teardown (release resources).* During a media session, RTSP may use either UDP or a number of TCP connections.

Although RTSP most frequently controls streams sent over RTP, other media transports can be used.

The syntax and operation of RTSP are intentionally similar to those of HTTP/1.1 [151]. This permits easier parsing using existing code and allows extensions to HTTP to be added to RTSP. Note that many of the specific field names and message structures are copied from the HTTP/1.1 specification. SDP [133] may be used with RTSP to describe media streams. Every message includes a session identifier (*session-id*, supplied by the server), which is maintained from SETUP to TEARDOWN, and a sequence number (*Cseq*), which is incremented for each new request.

The entire sequence is normally initiated using HTTP with the following URL:

[rtsp:|rtspu:]//host [:port] path

rtsp is specified for TCP transport and *rtspu* for UDP transport. The default port for both protocols is 554.

The exchange of RTSP messages consists of a number of *Requests* and *Responses*. A Request message begins with a *request line* and can be followed by a general header (not common), a request header, and an entity header. Similarly, the Response message can contain a general header, a response header, and an entity header.

4.3.1.1   RTSP Sample

To demonstrate a typical sequence of commands, please refer to Figure 4.13. On the left side of the figure, we find the user. This user is interested in (1) starting a video, (2) pausing the video, and (3) stopping the video. The commands are shown with the same symbols we find on a normal remote control: a triangle – play, a double line – pause, and a small circle – stop.

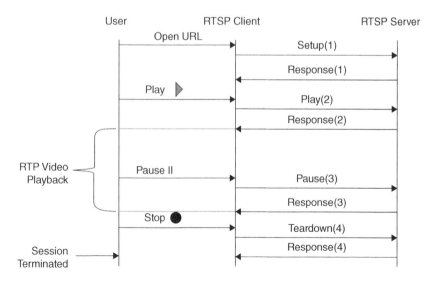

Figure 4.13   Sample RTSP session

Table 4.4   Sample RTSP session

Client sends	Server answers
User requests media	
SETUP rtsp://foo/twister RTSP/1.0 Cseq: 1 Transport: RTP/ATP; unicast; client_port=4588-4589	RTSP/1.0 200 OK Cseq: 1 Session: 567890 Transport: RTP/ATP; unicast; client_port=4588-4589; server-port=6256-6257
User clicks on PLAY starting at 10 seconds into recording	
PLAY rtsp://foo/twister RTSP/1.0 Cseq: 2 Session: 567890 Range: smpte=0:10:00-	RTSP/1.0 Cseq: 2 Session: 567890 Range: smpte=0:10:00-
Stream starts and playback begins	
User clicks on PAUSE	
PAUSE rtsp://foo/twister RTSP/1.0 Cseq: 3 Session: 567890	RTSP/1.0 Cseq: 3 Session: 567890 Range: smpte=0:15:00-0.30:00
Stream and playback pauses	
User clicks stop or exits program	
TEARDOWN rtsp://foo/twister RTSP/1.0 Cseq: 4 Session: 567890	RTSP/1.0 Cseq: 4
Session is shutdown	

Table 4.4 summarizes the same exchange as in the figure but displays the messages that are transferred between the client and the server. The exchange begins with the user entering a URL into his client media player – may also be a click on a Web link.

Although the details on the message format are provided in Section 4.3.1.3, review this simple exchange now and look for the following important points:

- URL of the desired media – rtsp://foo/twister.
- Sequence number *Cseq* that increments after each request/response pair.
- Unique session number provided by the server and listed in each subsequent operation.
- Request from client for a particular *transport* and the server's response on what it will send.
- Request from client to begin playing at 10 seconds after the start of the clip and the server's response on what it will send.

4.3.1.2   RTSP Request Message

The Request message begins with the request line containing three fields:

- Method – This is essentially the request or command to the other participant.
  - DESCRIBE – Request information about the specified media object (URI).
  - GET_PARAMETER – Retrieve the value of a specific parameter.
  - PAUSE – Halt all streams temporarily.
  - PLAY – Begin transmission of media streams.
  - RECORD – Begin recording of media streams.
  - SETUP – Establish an RTSP session with transport specifications.
  - TEARDOWN – Terminate an RTSP session.
- URI – The RTSP URL for the media object. For example,
    rtsp://server.example.com/media_stream.
- "RTSP/1.0" – This identifies the protocol and version. See Section 4.3.1.5 for information on future versions.

The Request message might also include a request header. While there are several parameters, the most common is *Accept*. This is used by the client to define acceptable responses from the server. For example, if the client is only capable of decoding certain content types, here is where those acceptable types would be listed.

- Example: "Accept: application/rtsl, application/sdp;level=2"

The entity header must always contain the sequence number "Cseq:" that allows the server to maintain state. This number is unique to each REQUEST/RESPONSE pair and is incremented for each new request. The remaining lines in the Request depend on the "Method" of the request.

4.3.1.3   RTSP Response Message

The Response message begins with a status line. This line consists of three fields: the protocol version, a numeric status code, and the textual phrase of the status code. Most of the status codes are adopted from HTTP/1.1.

- Example: "RTSP/1.0 200 OK"

Depending of the original request and the response type, additional information may be included in response header.

4.3.1.4   RTSP Message – Entity Information

Both Request and Response messages use the entity header and entity body to describe aspects of the media session. This can include transport information, content type, conference identity, playback speed, and the portions of the stream to play back.

Although there are a very large number of parameters, the following are the most commonly used fields in the entity header. As with the other elements of RTSP, many of these parameters are taken from the HTTP/1.1 specification.

- Transport: "RTP/AVP" ["TCP"|"UDP"] [;Parameters (see list)]
  - unicast | multicast
  - destination ["=" address]
  - port = port["–"port] (Used for multicast)
  - client_port = port["–"port] (Used for unicast)
  - server_port = port["–"port] (Used for unicast)
- RANGE – The range is used by the client to propose time ranges to be played. For example, "start – end" or "start-." The response from the server confirms this request with the possibility of changes. At the end of the time range, the server stops the stream. There are three different methods for specifying the time.
  - *smpte* is a timecode format that is designed for frame-level accuracy. It specifies the relative time since the beginning of the clip. For example, "smpte=10:07:05-" indicates a range staring at 10 minutes, 7 seconds, and 5 frames after the start of the clip and continuing to the end.
  - "Normal Play Time"[3] specifies time in hours, minutes, seconds, and fractions of a second. It also shows relative time since the start of a presentation or container (set of multiple streams). If the playback speed is doubled (scale = 2), then the NPT will advance at twice the normal rate. Similarly, npt will decrement when the stream is placed in reverse.
    - Test 2
    - "npt=123.45-125" – Send stream starting a 123.45 seconds after the start and continue until 125 seconds after the start.
    - "npt=12:05:35.3-" – Send stream starting at 12 hours, 5 minutes, and 35.3 seconds after the start and continue to the end of the stream.
    - "npt=now-" – (Used for live broadcasts) Start sending stream from the current time and continue to the end.
  - Absolute time uses the actual time (wall clock). It is specified as Universal Coordinated Time (UTC) [152, 153].
    - "time=20151105T152333.15Z-" – Send the stream starting at 5 November 2015 15:23 and 33.15 seconds and continue to the end of the Stream.
- RTP-info – This information is sent by the server in response to the "PLAY" request. It contains information about the RTP stream that will permit the client to map the incoming RTP packets (which contain the RTP timecode) to the RANGE request sent to the server. The fields are:
  - url – URL of the RTP stream.
  - seq – Sequence number of the first packet of the stream. This allows the client to identify packets outside of the requested seek range.
  - rtptime – This field is the RTP timecode that corresponds to the time value the client specified in the RANGE parameter. The client then uses this value to map the RTP time value to the RTSP RANGE time.

---

[3] "Digital storage media command and control" (DSM-CC) of MPEG-2, part 6–see Section DSM-CC.

- Example – "RTP-Info: url=rtsp://example.com/stream.avi; seq=12345; rtptime=1043292."
- Scale – This parameter specifies the playback speed of the media stream. 1 = Normal play speed, 2 = twice the normal viewing speed, and negative numbers indicate playback in the reverse direction. The specific implementation of scale will depend on the client/server implementations and the media type. Scale can be used with PLAY and RECORD requests.
- Conferences – Conferences are created outside of RTSP (e.g., with SIP). The "conference-id" is used within the SETUP request, consists of any octet value, and can be used by RTSP to contain parameters related to a conference stream.
- Content-length – See HTTP/1.1 [151] for details.
- Content-type – Values are defined by IANA.[4] If value is *application/sdp*, an SDP description follows.
- Session – This is a unique RTSP session identifier defined by the server and entered into the Response message to a SETUP Request from a client. This is optional if the server has an alternative method to identify separate media sessions, such as uniquely generated URLs.

The first commercial implementation of RTSP was the SureStream system from RealNetworks in 1998 [154]. This implementation also included RTP for media transport and adaptive streaming to adjust the media session to different or changing network conditions.

4.3.1.5 RTSP V2.0 – Draft

RTSP has become a heavily used protocol since it was first defined in 1998. During that time, experience and several technology changes and enhancements have uncovered limitations or unused portions of the initial RTSP specification. Some of these are:

- Network Address Translation traversal – Private (nonroutable on the Internet) IP addresses are now commonly used in companies and homes. The translation of these addresses causes issues with UDP-based RTP. The current version of RTSP does not address it.
- IPv6 is not supported in RTSP V1.0.
- Several features such as RTSP over UDP and Record were never utilized.
- The use of profiles in RTP has expanded into applications that were not anticipated and RTSP support needs to be extended.

Currently under development is an update to RTSP, which addresses the topics listed. RTSP 2.0 will replace RTSP 1.0 and is, at the time of this writing, a draft RFC [155]. If approved, it will obsolete the current standard. RTSP 2.0 has a large number of changes in the framework and syntax. Therefore, except for the version negotiation, it is not backward compatible with V1.0.

4.3.1.6 Media Fragments

An important capability of RTSP allows a receiver to specify to the server the segment of a media stream to transmit. In addition, the use of audio and video embedded in Web content has increased dramatically. However, none of the standards discussed so far define a standardized

---

[4] http://www.iana.org/assignments/media-types/media-types.xhtml.

way to describe portions of media streams or *media fragments* in URIs. The RFC for RTSP specifically states: *"fragment and query identifiers do not have a well-defined meaning at this time, with the interpretation left to the RTSP server."* Although a few registered media type definitions include fragment formats, most do not [156].

To fill this gap, World Wide Web Consortium (W3C) proposed a specification for media fragments within URIs [157]. This proposed standard includes encoding for both a time range and a display segment of a media stream. It also supports selection of tracks to be sent. For example:

http://www.mediaserver.com/sample.mp4#t=15,25#xywh=pixel:247,156,129,206

$t = 15, 25$:   Stream media starting 15 seconds from the beginning until 25 seconds from the beginning.

$xywh = $ **pixel:247,156,129,206:**   Display a portion of the video: <left corner *x*><left corner *y*><width><height>.

The interpretation of these media fragment URIs is possible in the client, on the server, or in an intermediate proxy.

4.4   Multicast

4.4.1   Multicast Overview

Up to this point, we have discussed mostly individual client – server or unicast connections over IP networks. As the number of participants increases, the network bandwidth and over-all overhead to maintain all of the individual unicast connections can quickly overwhelm the available resources. To provide a transport mechanism for one-to-many or many-to-many environments, *multicast* is used. The general term *multicast* refers to any method that provides group communication, whether it be over a packet network at the IP level, sometimes called network-assisted multicast, or further up the protocol stack at the application level. *Application Layer Multicast* (ALM) is discussed in Section 4.4.7. In this section, the terms multicast and IP multicast are used interchangeably. Although this section addresses both IPv4 and IPv6 multicast, the bulk of the information presented concentrates on IPv4.

IP Multicast was first standardized in 1986 [158] with augmentations in 2006 for Group Management [159] and in 2010 for administratively scoped addresses [160]. IP Multicast provides massive scalability by using network infrastructure elements, such as routers and switches, to maintain multicast distribution path information and to replicate/forward multicast packets as needed. These network devices can also use layer 2 protocols over multipoint media, such as Ethernet, to deliver multicast packets.

IP multicast networks have a number of characteristics:

- There can be one or more senders and any number of receivers. The most frequent implementations are a small number of senders and a very large number of receivers.
- Receivers explicitly join or leave a specific multicast group (normally notated as "G"). They will not receive multicast traffic unless they join a group. The group is identified by an explicit multicast address.

- No list of group members is maintained. Unless upper level protocols provide feedback, the sender has no information on the number or location of the receivers.
- IP multicast is a "best effort" service. Delivery of messages to the receiver is not guaranteed.

The most important elements of multicast are:

Addressing: Special IPv4 and IPv6 addresses are used to denote multicast traffic. See Section 4.4.2.

Group management: As mentioned earlier, receivers are sent traffic based on the multicast group or groups they join. The management of these groups requires both client and network infrastructure support. See Section 4.4.4.

Delivery infrastructure: This includes the architecture or type of multicast delivery (Section 4.4.3) and multicast routing protocols (Section 4.4.5).

4.4.2   Multicast Addressing

The multicast address is used to identify a specific multicast session. It is not specifically associated with any single host, but is used as a session identifier by all devices that participate (including the network devices). There are three major address types used for multicast, depending on what IP addressing scheme is used for transport:

IPv4: IPv4 address Class D is used for multicast. This means the first 4 bits of the address are "1110" and the remaining 28 bits are used to specify a multicast address. Specifically, this is the range of addresses from 224.0.0.0 to 239.255.255.255. These addresses are administered by the IANA.[5] To permit the use of a "private" multicast address space (e.g., within a company's network), the address block 239.0.0.0 to 239.255.255.255 is administratively (organizational or local) scoped.

IPv6: The IPv6 addresses prefixed with ff00::/8 (or otherwise shown as ffxx:) are defined as multicast. These addresses are also administered by the IANA.[6]

Ethernet MAC to IPv4: Within certain limits due to address sizes, the Ethernet MAC multicast addresses and IPv4 addresses can be mapped. As illustrated in Figure 4.14, the lower 23 bits of the IPv4 address are mapped to the corresponding bits in the Ethernet address. It is important to note that because IPv4 has a larger address range, there are 5 bits that are not mapped to an Ethernet address. This will not cause any ambiguity as long as IPv4 multicast addresses are carefully chosen. There are also a number of reserved, non-IPv4 Ethernet multicast addresses.[7] Ethernet multicast packets are flooded to all connected stations. Modern Ethernet controllers normally reduce end-station CPU load by discarding packets not defined in a table initialized by system software.

---

[5] http://www.iana.org/assignments/multicast-addresses/multicast-addresses.xhtml.

[6] http://www.iana.org/assignments/ipv6-multicast-addresses/ipv6-multicast-addresses.xhtml.

[7] http://www.cavebear.com/archive/cavebear/Ethernet/multicast.html.

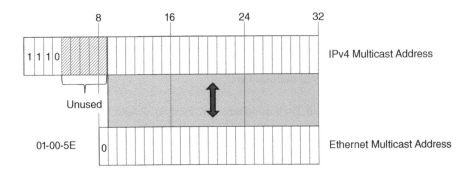

Figure 4.14   Mapping Ethernet and IPv4 multicast addresses

4.4.3   Types of Multicast

There are three fundamental models that can be used to deliver multicast traffic. In all cases, the receiver joins a multicast group "G" by specifying the multicast address. The difference in the three types is based on how the traffic is sourced.

AnySource Multicast (ASM):   The underlying network accepts traffic from any source addressed to the multicast address. The network is responsible for accepting traffic from any source and routing it to any receiver that has joined the multicast group "G." This was the first multicast model defined [158]. It is best suited to many-to-many applications such as audio/video conferences and groupware.

Source-Filtered Multicast (SFM):   In this case, the receiver, in addition to joining the multicast group, can specify one or more source addresses from which it will accept multicast traffic. This request is received by the last-hop router and forwards only traffic from sources requested by the registered receivers. As with ASM, it is most often used for many-to-many applications, especially when primary and backup or geographically dispersed sources are used.

Source-Specific Multicast (SSM):   In this case, the concept of "channel" is added. A receiver will join a multicast channel by specifying both the source address and the multicast address (S,G). By limiting the source of the traffic, the load on the network is reduced and security is improved. Specific support for SSM is required only in the receiver and the last-hop router using the IGMPv3 and MLD protocols [161]. IANA defined the address range 232.0.0.0–232.255.255.255 for SSM multicast traffic. From the receiver viewpoint, SFM and SSM are identical.

It is important to note that ASM has a number of issues. Since there are no restrictions on who can source traffic to the multicast group, an undesired source could inject packets into the stream of traffic sent to the clients. A second related issue is the potential for address collision. In an uncontrolled environment where multicast address assignments are not managed, it is possible for two separate multicast streams to use the same multicast address. If this happens, the network will view both streams as a single multicast session and merge traffic from all the sources and deliver it all clients. The use of Time-To-Live (TTL) in the IP packets can be used to restrict the scope, but it is not always possible to predict the maximum hop count and guarantee no overlap with the competing multicast group. Since company and service provider networks frequently manage address assignment, this normally only occurs in public networks. The use of SSM also eliminates this issue as the combination of group and source address (G,S) uniquely identifies multicast sessions and prevents unwanted source traffic.

4.4.4 Multicast End Delivery

The protocol used by routers and clients in a local, bridged environment to manage membership in multicast groups is called *Internet Group Management Protocol* (*IGMP*). The first version of IGMP was defined in 1989 [158]. This was followed in 1997 by IGMPv2 [162]. The current version, IGMPv3, was defined in 2002 [163]. There is also a lightweight version of IGMPv3 defined [164]. Although modern implementations include support for IGMPv1 and IGMPv2, they are not commonly used. The rest of the information presented here will focus on IGMPv3. IPv6 uses the Multicast Listener Discovery (MLD) protocol to support multicast. MLDv1 [165] was derived from IGMPv2 in 1999. In 2004, MLDv2 [166] was updated using the IGMPv3 protocol, translated for IPv6 semantics, and therefore includes support for SSM. Most of the IGMPv3 concepts discussed here also apply to IPv6 MLDv2.

IGMP messages over IPv4 are sent as IP datagrams with an IP protocol number of 2. The multicast routers use the local multicast address 224.0.0.1 (All Connected Systems Multicast) to send message to hosts. The hosts use the address 224.0.0.22 (All Multicast Routers) to send packets to the routers. The following are the main functions provided by IGMPv3:

Join: A host joins a multicast group by transmitting an *Membership Report* message on its network interface. The connected multicast router will receive this message and forward any multicast traffic for that group to the network segment connected to the host.

Leave: The *Leave Group* message instructs the multicast router to stop forwarding multicast traffic for the specified group.[8] If this host was the last receiver for this group on the connected network segment, it will cease forwarding traffic.

Query: The *Membership Report* message is sent by the multicast router on a network segment to determine the multicast reception state of connected hosts. The query either can be general (all groups) or may contain a list of specific groups. This means that if a host silently drops out of a multicast group (e.g., system crash or other unexpected termination), the router will eventually discover that they are no longer a member and drop them from the group.

The following is a typical exchange between a host and the multicast router for the LAN segment:

- An application on the host identifies a multicast stream the host wishes to receive. This consists of a multicast address and possibly a source IP address.
- The host broadcasts a Membership Report Message containing the multicast and source IP addresses to all multicast routers on the LAN.
- The multicast router receives the join request and now immediately forwards any traffic for that multicast address from the identified source IP address to the LAN segment connected to the host.
- When the host no longer wishes to receive the multicast stream, it issues a Membership Report message with a leave request for that group.

---

[8] Leave message first introduced in IGMPv2.

- When the multicast router receives the leave request, it broadcasts a Membership Query message for that group to determine if any other hosts wish to continue receiving the stream.
- Any remaining interested hosts will send a Membership Report Message to the router.
- If there are no interested hosts, the router will cease forwarding that group's multicast traffic to that LAN segment.

In Figure 4.15, details of the format of the Membership Query packet are illustrated. The most important fields are:

Group address: This is the multicast address of the group being queried. This value is 0 for a general query (all groups).
Source address: If this query is restricted to a specific Group and Source(s), then these fields contain the IP address of the source(s).
QRV, QQIC: Query's Robustness Variable (QRV) and Querier's Query Interval Code (QQIC) are fields are used to control the timers used in IGMP.

The packet sent by the hosts in response to a Membership Query is detailed in Figure 4.16. It is a set (one or more) of Group Records that describe the host's multicast reception state. As can be seen in Figure 4.17, each group record describes the host's membership with the multicast address and any source addresses associated with the group.

In the Membership Report message, there are a list of Group Records that describe the membership of the host in various multicast groups. The format for the Group Record is shown in Figure 4.17.

When the state of a multicast group changes with joins and leaves, there is the potential for a significant amount of broadcast/flooding. This can easily occur at the end of a broadcast when many receivers leave the group. To mitigate this impact or provide a *shock absorber* to a group membership change, timers were implemented to slow down the reaction to the change. The IGMP state model shown in Figure 4.18 shows how timers are used to generate a waiting period before the group change is processed.

However, there are situations where a more immediate reaction is needed to minimize user impact. For example, if a user wishes to change the channel in IPTV, we do not want to make

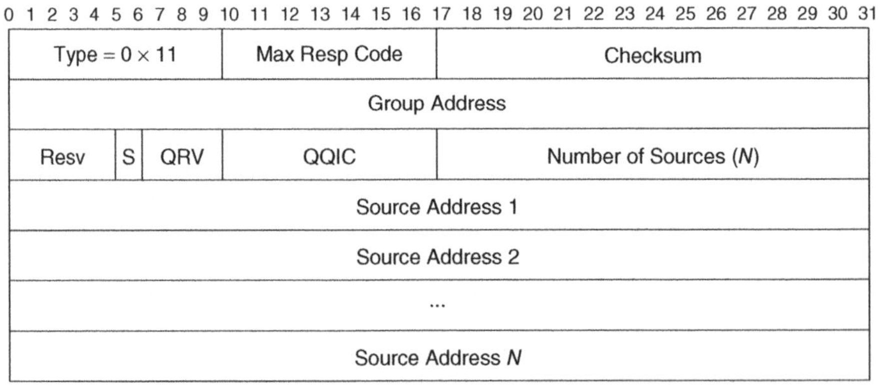

0 1 2 3 4 5 6 7 8 9 10 11 12 13 14 15 16 17 18 19 20 21 22 23 24 25 26 27 28 29 30 31

Type = 0 × 11			Max Resp Code		Checksum		
Group Address							
Resv	S	QRV	QQIC		Number of Sources (N)		
Source Address 1							
Source Address 2							
...							
Source Address N							

Figure 4.15 IGMP membership query packet

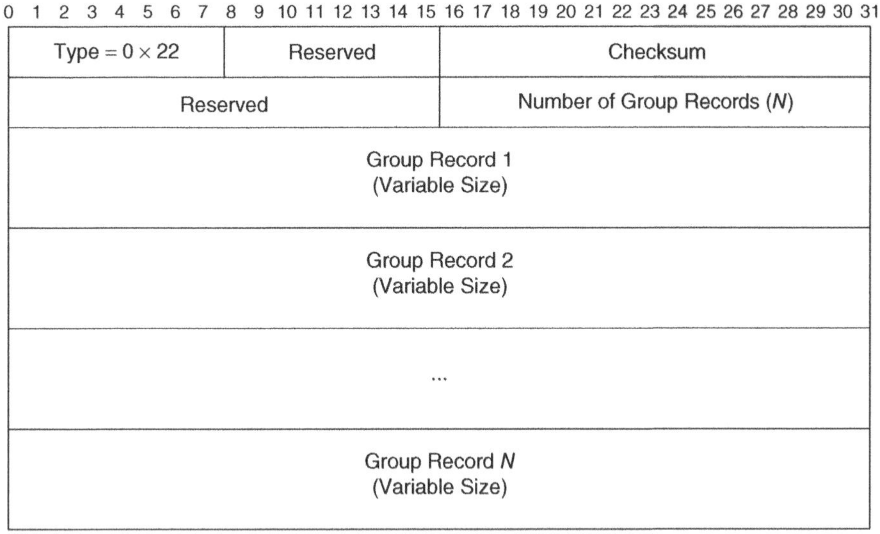

```
0  1  2  3  4  5  6  7  8  9  10 11 12 13 14 15 16 17 18 19 20 21 22 23 24 25 26 27 28 29 30 31
```

Type = 0 × 22	Reserved	Checksum
Reserved		Number of Group Records (N)
Group Record 1 (Variable Size)		
Group Record 2 (Variable Size)		
...		
Group Record N (Variable Size)		

Figure 4.16 IGMP Membership Report packet (Note: IGMPv1: 0x12, IGMPv2: 0x16)

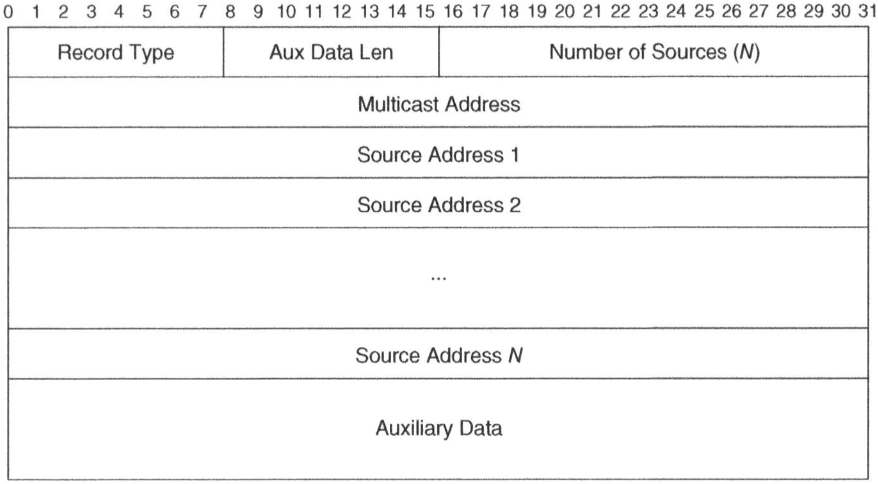

```
0  1  2  3  4  5  6  7  8  9  10 11 12 13 14 15 16 17 18 19 20 21 22 23 24 25 26 27 28 29 30 31
```

Record Type	Aux Data Len	Number of Sources (N)
Multicast Address		
Source Address 1		
Source Address 2		
...		
Source Address N		
Auxiliary Data		

Figure 4.17 IGMP group record

them wait until an IGMP timer expires. Therefore, a different technique is used to make a fast switch. With appropriate Group Report messages in IGMPv3, a host can *leave* and *join* a new group with a single Membership Report message. As this is important for the smooth operation of IPTV, home networks must support IGMPv3. See Chapter 11.

Layer 2 switches often use a technique called *IGMP Snooping*. With this technique, the switch listens to the IGMP traffic between the host and the multicast router, maintains a state table of active multicast groups and their members, and then filters traffic for those attached LAN segments with no interested hosts.

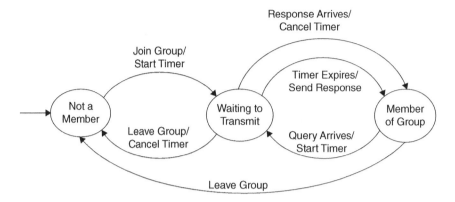

Figure 4.18 IGMP state diagram

4.4.5 Multicast Routing Protocols

Until now, we have discussed the communication between the host systems and the edge routers. A different mechanism is required to provide routing in a distributed router-based network. As with normal IP traffic, the network must be able to properly route the traffic from the source to the destination hosts. In addition, the network must control the forwarding of multicast traffic to minimize bandwidth use. For multicast, there are two fundamental methods to create the distribution or spanning tree[9]:

Reverse Path Forwarding (RPF): This method "floods" multicast traffic to the entire network. This variant is frequently called *Source-based Tree* because the creation of the tree begins when a source begins transmitting multicast traffic. When this traffic is received by a router, it then retransmits or "floods" the traffic to all other interfaces. When a router determines that it has no interested receivers, it sends a "prune" message back to the upstream router. The upstream router then stops forwarding traffic over that interface.

    While this method is very easy to implement, it does not scale well to large networks, unless the density of interested receivers is very high. This may be practical for applications such as IPTV providers in a relatively closed network or a company that broadcasts traffic internally to a very large number of users.

Center-Based Trees (CBT): Here, a fixed point in the network acts as a meeting or rendezvous point for building a spanning tree for a specific multicast stream. Since all of the multicast trees pass through this central point, these are often called *Shared Trees*. Join requests from receivers are forwarded by routers attached to the hosts to the central rendezvous point. The intermediate routers record the path of the request to create the distribution tree.

    The advantage of CBT is that the network can scale far more than RPF flooding with less overall load on the network. However, shared trees (through a central point) are generally not as optimized as source-based trees (RPF) and the central point can become a bottleneck. This can lead to increased delays both for registration and for delivery of real-time content.

---

[9] Multicast requires setup and maintenance. For small groups, direct unicast connections to a server are more efficient.

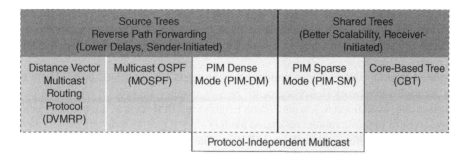

Source Trees Reverse Path Forwarding (Lower Delays, Sender-Initiated)			Shared Trees (Better Scalability, Receiver-Initiated)	
Distance Vector Multicast Routing Protocol (DVMRP)	Multicast OSPF (MOSPF)	PIM Dense Mode (PIM-DM)	PIM Sparse Mode (PIM-SM)	Core-Based Tree (CBT)
		Protocol-Independent Multicast		

Figure 4.19 Multicast routing protocols

The various implementations of these routing protocols are illustrated in Figure 4.19.

Distance Vector Multicast Routing Protocol (DVMRP): This protocol is an extension of the unicast routing protocol RIP and is defined in Ref. [167]. It formed the basis of the Internet's Mbone multicast backbone.

Multicast over OSPF (MOSPF): MOSPF was developed as an extension of the Open Shortest Path First (OSPF) routing protocol. It was defined in 1994 by Moy [168].

Protocol Independent Multicast – Dense Mode (PIM-DM): This is the first of the two routing protocols that do not depend on any underlying unicast routing protocol. It is similar to DVMRP and works best in a dense LAN environment. There is an experimental RFC that proposes the PIM-DM specification [169]. Although PIM-DM shares the same packet formats as PIM-SM, its characteristics and operation are completely different.

Protocol Independent Multicast – Sparse Mode (PIM-SM): This is another routing protocol that is independent of the unicast routing protocol. It uses shared trees and rendezvous points to establish the distribution paths from the source(s) to the receivers. It is commonly used in environments with a sparse population of receivers such as a WAN. It was first introduced in 1996 [170] and is now defined in Ref. [171].

Core-Based Tree (CBT): This experimental protocol was defined in 1997 [172]. It is also shared tree routing protocol with a defined central point. CBT is not widely deployed.

Border Gateway Multicast Protocol (BGMP): As an attempt to create interdomain multicast, BGMP was defined [173]. Due to commercial and addressing issues, it was never deployed. Note: This protocol does not appear in Figure 4.19.

4.4.6 Protocol Independent Multicast – Sparse Mode

The most often used multicast routing protocol is PIM-SM. There are three key entities in any PIM-SM deployment:

Rendezvous Point (RP): A router is configured to act as the root of the non-source-specific distribution tree for (normally) a range of multicast addresses/groups. Implementations include the ability to define one or more backups to prevent a single point of failure. The router should be in a central location in the network hierarchy close to the most frequent sources of multicast traffic. The location (IP address) of the RP is either configured directly into each multicast router or discovered through the use of a bootstrap protocol [174]. In large implementations, there are normally a number of potential RPs defined in the network.

Designated Router (DR): These routers act as an agent for connected hosts. There is one DR for each subnetwork (LAN segment). *Hello* messages are sent on every PIM-enabled interface and, in addition to learning about neighboring PIM routers, are used to elect a DR in the event of multiple routers on shared media.

Multicast Routing Information Base (MRIB): The MRIB is the multicast topology table. This is normally derived from the unicast routing table.

The establishment of a multicast session consists of three major phases. These phases are depicted in Figure 4.20:

Phase 1–RP tree: When a receiver wishes to join the multicast stream, it sends an IGMP or MLD join message to the multicast address. The DR recognizes the join request, makes an entry into its MRIB for that group (*,G), and sends a PIM Join message toward the RP.

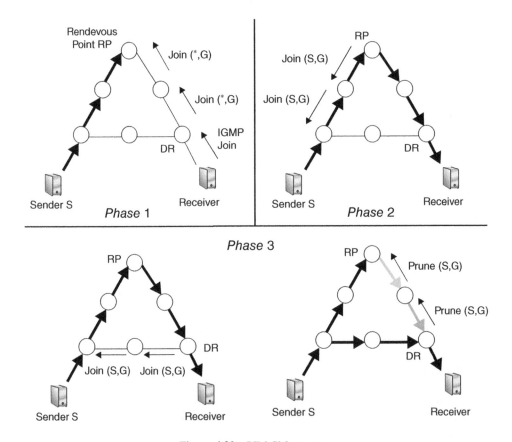

Figure 4.20 PIM-SM sequence

This process continues hop by hop until the message arrives either at the RP or at a router that already has a multicast routing entry for (*,G). This creates a structure called the RP Tree (RPT). This is a shared tree since all sources in this group use this tree.

A multicast sender simply starts to send multicast traffic. The router (DR) connected to the source recognizes the traffic as multicast and makes entries for both (*,G) and (S,G) into its MRIB. It encapsulates the traffic into a unicast message called a PIM Register message and sends it to the RP. The RP then takes the multicast traffic (unencapsulated) and forwards it to the DRs for that group. At this point, multicast traffic is flowing from the source to all joined receivers.

Phase 2 – Register-stop: Encapsulating the source multicast traffic and having all traffic pass through the RP is inefficient. Therefore, PIM-SM utilizes both Shared Trees through the RP (*,G) to establish the initial "connection" and Source Trees originating at the source (S,G). Shared trees are very efficient at establishing a link between the sender and the receiver, but the path is not always optimal. Therefore, after the initial join messages are sent, the receiver is linked to the source tree to utilize the shortest path. In Phase 2, after the first Phase established the traffic flow, the RP will send a source-specific join request (S,G) upstream toward the source. Multicast traffic continues to flow during this process. When the join request arrives at the source's DR, it will begin to send multicast traffic natively (unencapsulated) toward the RP. At some point, the RP will be receiving two copies of the packets. The RP will then send a PIM Register-Stop message back to the source's DR to shut down the encapsulated traffic flow. At this point, multicast traffic will flow from the source to the RP using the source tree (S,G). The traffic will then be forwarded by the RP to the RPT.

Phase 3 – Shortest-path tree: The next step to optimize the path is for the receiver's DR to find the optimal path to the source. The DR will examine the source IP address of the multicast traffic and determine the next hop to the source by examining its unicast routing table. The router will then send a join request for that source tree (S,G) over that interface. When it begins to receive duplicate packets (similar to the aforementioned RP example), it will prune the original path back to the RP (*,G) and receive traffic only via the source tree (S,G).

The combined PIM Join/Prune packet format is described in Figure 4.21. After the initial header, the message consists of one or more blocks with each block representing a multicast group. The key fields in this packet are:

Upstream neighbor address: This is the IP address of the recipient of this message. Upstream means a router closer to the root of the tree, either the RP or a source. This router will then either add or remove entries in its MRIB for joins and prunes, respectively.

Multicast source address: This is the multicast group, G. All following addresses in this specific block apply to G.

Joined source address: These are specific source addresses that should be forwarded when sending traffic to group G.

Pruned source address: Conversely, the traffic for group G from these sources should not be forwarded. This is sent when there are no remaining downstream group members for this source (S,G).

The join/prune messages are sent periodically to maintain the state of the tree and, if needed, drop receivers that have silently left the group.

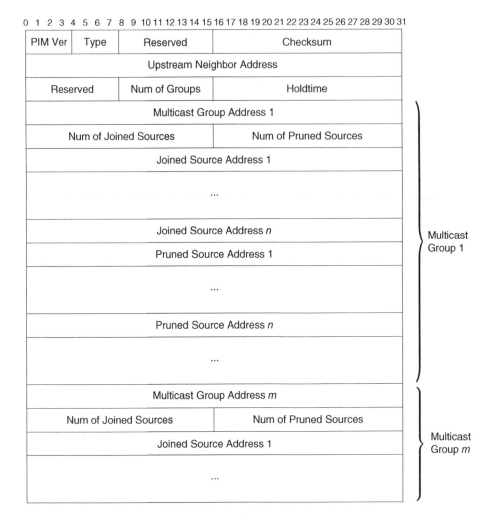

Figure 4.21   PIM-SM join/prune message

Multiple PIM routers on a shared LAN can sometime cause more than one upstream router to have a valid forwarding state for a packet. This can cause packet duplication. When this is detected, a single forwarder from the upstream routers is elected using *assert* messages. These messages are also received by the downstream routers and cause any further join/prune messages to the elected upstream router.

It is possible to implement a Source Specific Multicast service model using a subset of the normal PIM-SM protocol mechanisms. There is a range of multicast addresses reserved for SSM: 232.0.0.0/8 for IPv4 and FF3x::/32 for IPv6. If an SSM model is implemented, the restrictions are detailed in Ref. [171].

There are a number of different environments where PIM-SM has be implemented.

- The experimental IP multicast backbone, *Mbone*, was originally implemented based on DVMRP, but later migrated to PIM-SM for better scaling. The Mbone is no longer relevant.

- Implementing and maintaining PIM-SM on routers takes significant planning and effort. For this reason, virtually no ISPs have implemented PIM-SM.
- For providers that stream IP TV based on DVB, PIM-SSM/SFM has been implemented. This requires that home routers support IGMPv3, which is part of the DVB specification [175]. As stated in Section 2.3, there are approximately 100 million subscribers in 30 countries, all of whom use multicast.
- For companies and universities that stream Webcasts, A/V-conferences with large numbers of users around the world, PIM-SM is a good solution and is often implemented.

It is important to note that although there are some in the Internet community that believe multicast is dead, but this is actually true only for the Internet.

4.4.7   Application Layer Multicast

As discussed earlier, the implementation of multicast in an IP network requires planning, coordination, and maintenance. In spite of experiments such as Mbone and wide-area protocols such as BGMP, infrastructure-level multicast essentially does not exist on the Internet. With rapidly increasing demand for multiuser audio/video conferencing, massively multiplayer games, and content sharing, methods for multicasting with minimal or no infrastructure support were developed.

There have been attempts to overcome the limitations of network layer multicast, such as the experimental protocols Pragmatics General Multicast (PGM) [176] and XCast Protocol [177]. In the case of XCast, the design goal is large number of groups each with few members and minimal router overhead. The packet generated by the source includes all destination IP addresses in an XCast header. Intermediate routers then replicate the packet only when paths to the destinations diverged. While this reduces the router overhead, it requires multiple unicast routing table lookups per packet and can only scale to a limited number of receivers.

However, the most common method is Application Layer Multicast (ALM), which is also called by a number of different names: Overlay, End System, or Peer-to-Peer Multicasting. It is normally implemented as an end-system application using unicast links in a normally meshed overlay network. It is relatively easy to implement as it requires only an application running on an end system. However, it is inherently less stable, less secure, and less efficient.

There have been many alternatives proposed [178] with very limited deployment. The most well-known variants are: NICE [179], ZIGZAG [180], and Omni [181].

An early attempt to provide this capability is IETF Standard Internet Relay Chat (IRC) [182, 183]. This early chat protocol, which is still in use, links multiple servers together with common channels and state. Clients can connect to any server and communicate with any other client on any server. This requires the maintenance of a consistent global state in all servers, which limits the scalability of this architecture.

The most common (and successful) ALM implementations have been based on Peer-to-Peer (P2P) networking. These implementations have the following characteristics:

- Some of the more popular P2P Multicast systems are based on proprietary software, such as Zattoo and Pando.
- There is, as of now, no IETF standards involvement [184].
- The overlay is either meshed-based, such as BitTorrent [185], or tree-based, such as NICE.

- Registration is normally also made using a mesh overlay and is generally tree-based with the tree being built parallel to the data transfer network. In the case of NICE, the tree is built on Round-Trip Time (RTT) between the hosts.
- BitTorrent uses a meshed-system registration. The client contacts a central host and, for a specific file, downloads a list of one or more trackers. The trackers maintain a list of hosts connected to this file's meshed network. The client then downloads this list and connects directly to these peers to exchange chunks of the file.
- P2P client software applications frequently maintain a list of peers locally using Distributed Hash Tables (DHT) [186, 187]. Clients can exchange lists of peers without the use of a central repository (trackers).
- The proprietary Chinese service PPLive uses P2P to distribute IPTV streams to a huge audience. An analysis was done to show that PPLive is mesh-based [188]. A key criterion to the success of any P2P IPTV streaming service is maintaining the play-out deadline to prevent stuttering or breaks in the playback. See Section 5.4 for more details.

4.5   Quality of Service

Providing a service over a shared network infrastructure poses a number of challenges. When that service requires a guarantee of performance, such as the delivery of voice and video, the challenges multiply. Generally, the ultimate goal is a satisfied end user. The user is interested in good response times or seamless operation for their most important applications. This is called Quality of Experience (QoE).

One tool to achieve this is network QoS. QoS is a "guarantee" or promise by the provider to deliver network services within defined quality parameters such as minimum bandwidth and maximum delay or jitter. Translating the user's QoE into definable QoS parameters is a nontrivial exercise that requires significant analysis using network delay/packet loss simulation and monitoring server and client application characteristics.

An important point to note is that many applications' network performances vary widely and depend on the applications' "awareness" of network characteristics. A good example is the use of database query applications in the mid-1990s. These applications were designed to run in a LAN environment. As the use of WANs increased and these applications were rolled out to remote locations, the performance of these applications rendered them unusable in many cases. Changes in the application architecture were required to make these applications usable in a WAN environment with longer delays and limited bandwidth.

For those applications that cannot be changed, the use of a WAN accelerator device may prove useful. These devices intercept traffic destined for a remote location and apply a variety of optimizations, including compression, protocol modification, caching, and proxy functions.

For pure voice and video applications, extensive testing has been done to define the relationship between the user's perception of quality (QoE) given a certain set of network characteristics. Figure 4.22 is a graph showing the user's satisfaction with voice quality based on a certain delay from the mouth to the ear (encoding, network delay, buffering, and decoding). Based on this information, any delay <150 milliseconds cannot be discerned by the user.

In Table 2.1 in Section 2.6, we summarized the user requirements and stated in multiple places in the row "Delay within session" that it should be "Not recognizable." Now, we can replace in this table "Not recognizable" by "< 150 milliseconds" since we now know the measurable unit for this user requirement.

There are two fundamental approaches to provide QoS in modern packet networks: *Integrated Services* and *Differentiated Services*.

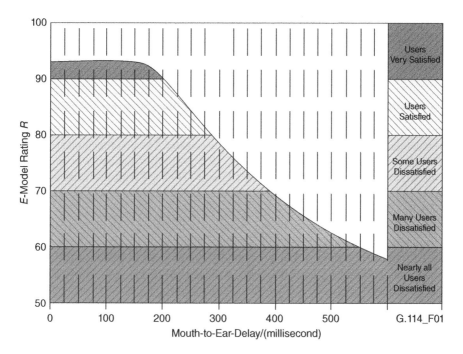

Figure 4.22   Acceptable voice delay (Source: Courtesy of ITU. © ITU)

4.5.1   Integrated Services (Intserv)

Intserv is a QoS model for packet networks that reserves a specified level of service from source to destination. It was created in the early 1990s by the Integrated Service Group and uses the Resource reSerVation Protocol (RSVP) [189] to reserve necessary resources to support the service requirements requested by the data flow. There are three Classes of Service (CoS):

Guaranteed service [190]:   This CoS provides strict bounds on the permitted delay and bandwidth.

Controlled load [191]:   This class provides the same level of service to the application as if the network were completely unloaded. This includes packet loss and delay. Even if the network becomes overloaded, this service class would not be affected.

Best effort: Traffic is delivered without any guarantee.

The implementation of Intserv has the following consequences:

- Every router must be configured to support RSVP. See Section 4.5.2.
- Connection admission control must also be configured throughout the network. This enables the network to reject a connection if sufficient resources are not available.
- Routers must maintain the state and packets per flow of any RSVP connection that passes through it and prioritize traffic accordingly.
- Connection states must be periodically refreshed.
- Packets of a connection flow follow a predetermined path. If a change to a link or router causes rerouting, the flow must re-reserve resources over the new path.

- This method is not commonly implemented on large networks due to the significant overhead to support RSVP. It is found in some company networks that provide services such as high-definition Telepresence – see Section 12.4.

4.5.2 Resource Reservation Protocol (RSVP)

The basic concepts of RSVP were first proposed in 1993 [192]. This was the first reservation protocol that supported multicast with multiple senders. The first functional specification (RSVP Version 1) was published in 1997 [189, 193, 194]. This standard was augmented by follow-on RFCs to define how to deliver the Class of Services Controlled Load and Guaranteed QoS [195]. RSVP admission control was extended to include Policy-based Admission Control in Ref. [196].

The use of RSVP was broadened by the extension of the protocol to support Multiprotocol Label Switching (MPLS) networks. MPLS permits multiple protocols for the distribution of labels used to forward traffic. Awduche et al. [197] introduced extensions to RSVP to support establishing Label-Switched Paths (LSPs) within an MPLS network. This is called RSVP Traffic Engineering (RSVP-TE). With this protocol, reservations are used to instantiate unicast paths and reserve appropriate resources.

A large part of the RSVP specification deals with multicast flows and how reservations from multiple branches can be merged. However, there are very few multicast deployments using RSVP and the remainder of this RSVP discussion will focus on unicast reservations.

To limit the impact of RSVP on high traffic core routers, proposals have been made to create a hybrid using Intserv in the periphery and DiffServ in the core routers.

RSVP has the following important characteristics:

- An RSVP reservation is for a simplex flow. That is, it only reserves resources in a single direction.
- The host systems fully participate and initiate the RSVP process.
- The native routing protocol is used to send RSVP messages.
- The sequence begins by the source host sending an RSVP PATH message in the direction of the receiver (or down the multicast tree). As the message passes through each router, the characteristics of the flow are stored but not reserved. See Figure 4.23 to know how this works.
- The receiver then issues an RESV message back in the direction of the source. As each router receives the message, it reserves the resources, if available, and passes the message to the next upstream router. If the resources are not available, a REJECT message is sent back to the receiver. Figure 4.24 demonstrates this process.
- The most important fields of RSVP messages are:

  Session ID: Contains the IP destination address, IP Protocol ID, and port.

  Flowspec: Defines the requested QoS. Specifies elements such as peak data rate, maximum packet size, expected traffic volume [198] (only resv message).

  Previous hop: IP Address and interface of router that sent this message.

  Filter spec: Defines the subset of packets that should receive desired QoS (only resv message).

  Refresh time: The refresh period used by the message creator.

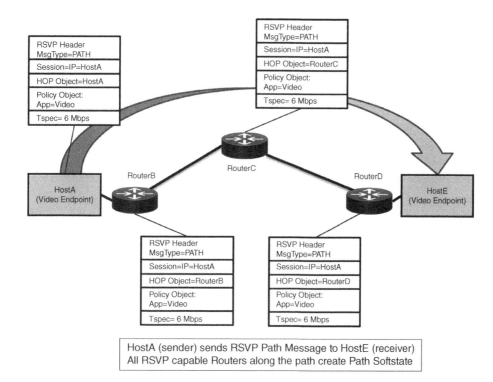

Figure 4.23   RSVP PATH messages

- There are three basic reservations types: Fixed Filter, Wild-Card Filter, and Shared Explicit. The difference between these styles is which senders are permitted to use the reservation: single host, any host, or explicit list of hosts, respectively. The Fixed Filter type with a single sender is used for unicast flows.
- Other messages include *path/resv tear-down* and *path/resv error*.
- The reverse method for creating reservations was chosen mostly to support multicast environments.
- The RSVP state information stored in the routers is "soft." This means that the source host must periodically refresh the state information with *path* messages.

4.5.3   Differentiated Services (DiffServ)

DiffServ approaches QoS in a completely different way. Instead of an end-to-end fine-grained reservation of resources, DiffServ specifies a simple, scalable, course-grained method to classify traffic and act on it hop by hop.

DiffServ was initially developed in the late 1990s with the first specification published in 1998 [199], with clarifications and new definitions of terms in Ref. [200]. This standard defines the differentiated services (DS) field in both IPv4 and IPv6 headers. This field is used to define different service levels or priorities for packet traffic. As mentioned earlier, each

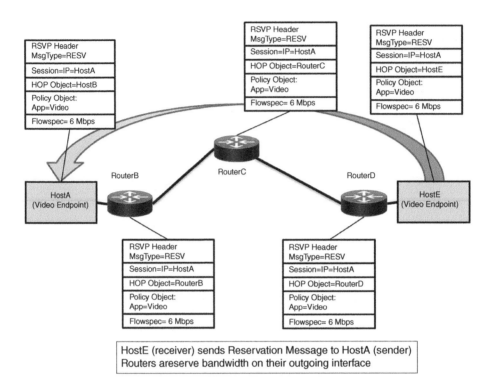

Figure 4.24 RSVP RESV messages

routers act independently (per hop) on the traffic, based on the DS field and its rules for each service level.

The first step for traffic in a DiffServ network is to mark the packets with a value in the DS field, In IPv4, this is the Type of Service (ToS) field, and in IPv6, this is the Traffic Class (TC) field. In any well-designed network, these values are explicitly defined during the design of the QoS implementation. The traffic is normally marked as it arrives at the first router, which is frequently the LAN router connected to the source host. Although it is possible for the host to set the DS field during packet creation, it is often ignored and overwritten by the network to prevent inappropriate use of QoS by unauthorized traffic. The marking by the router can be made based on a number of different parameters such as source address, destination address, port numbers, or traffic type. Ingress routers can also apply QoS policies in the form of rate limiters, traffic policers, or shapers.

Figure 4.25 shows the DS field's location in an IPv4 packet. The six most significant bits of this 8-bit field belong to the *Differentiated Services Code Point* (*DSCP*). The two least-significant bits belong to the *Explicit Congestion Notification* (*ECN*). Theoretically, a network could implement 64 different traffic classes. The various DiffServ RFCs recommend certain encodings, but they are not required. While most network service providers offer recommended encodings, they generally offer a subset of them, normally between 3 and 6 classes.

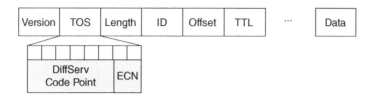

Figure 4.25 TOS marking

Table 4.5 AF behavior group

	Class 1 (lowest)	Class 2	Class 3	Class 4 (highest)
Low Drop Prio	001010 (AF11)	010010 (AF21)	011010 (AF31)	100010 (AF41)
Med Drop Prio	001100 (AF12)	010100 (AF22)	011100 (AF32)	100100 (AF42)
High Drop Prio	001110 (AF13)	010110 (AF23)	011110 (AF33)	100110 (AF43)

Once the packet has been marked, the router (and each subsequent router) then handles the traffic based on its configured *Per Hop Behavior* (*PHB*). As with the packet marking, there are recommended PHB configurations [201], but implementations are not required to follow them. The rest of the discussion will cover recommended PHB implementations. There are four commonly defined PHB classes:

Default PHB: This is the only required PHB. This is typically best-effort traffic with no guarantees. Any traffic that is not classified in one of the other classes is placed here. The DSCP for default PHB is 000000_B.

Expedited Forwarding (EF) PHB: This class is dedicated to low latency, low loss, and strictly controlled jitter [202]. This is the class that is normally used for voice and video traffic. This traffic is often placed ahead of all other traffic. To avoid an overload of EF traffic, ingress is controlled by rate-limiting, policing, and other admission control mechanisms. In general, networks will restrict EF traffic to between 25% and 35% of link capacity. The DSCP for EF PHB is 101110_B.

The IETF also defined a variation of EF explicitly for voice traffic called Voice Admit (VA) PHB [203]. This class has the same characteristics as EF PHB, but adds a Call Admission Control (CAC) procedure. The DSCP for AV PHB is 101100_B.

Assured Forwarding (AF) PHB: This class provides a level of service that is less time-critical than EF, but more important than Best Effort [204]. Typically, traffic for mission-critical applications would be placed in this class. To further subdivide the AF class, 12 different subclasses have been defined. See Table 4.5. This permits more granular control over the service levels offered in AF. Within AF, Class 4 traffic has the highest priority and will take precedence over any lower class. If congestion occurs within a class, the high drop priority packets with be discarded first.

Table 4.6   Class selector values

DSCP	Binary	Decimal
CS0	000 000	0
CS1	001 000	8
CS2	010 000	16
CS3	011 000	24
CS4	100 000	32
CS5	101 000	40
CS6	110 000	48
CS7	111 000	56

Class Selector (CS) PHB: Before DiffServ was defined, IPv4 networks could use the *Precedence* field in the TOS byte to define the priority of the traffic. At that time, the TOS field was not widely used. However, to maintain backward compatibility with existing implementations, DiffServ uses the most significant 3 bits of the DSCP to map to the original Precedence bits [199]. See Table 4.6.

Table 4.7 lists a recommended mapping of PHB to application.

Explicit congestion notification

In 2001, an extension to IP and TCP, ECN, was defined in Ref. [205] to allow end-to-end notification of network congestion without dropping packets. TCP/IP without ECN detects congestion based on packet loss. With ECN-aware routers and hosts, congestion can be signaled to the transmitter and the traffic flow adjusted accordingly. Note that this does not apply to RTP, but only TCP-based streams.

To optimize both the QoS provided within a router and maximize the throughput on a given interface, routers implement *shapers* and *droppers* on their input queues.

- **Shapers** place packets into different queues in order to affect the PHB of specific traffic. This is also called policing.
- **Droppers** discard packets under certain conditions [206]. In some cases, packets are dropped even before the queues are full, that is, Random Early Detection (RED) [207]. There are many variations of packet dropping mechanisms, such as Weighted RED [208]. These processes can also work together with congestion avoidance methods, such as ECN, to avoid dropping packets.

Table 4.7   QoS baseline recommendations

PHB	DSCP	DSCP Bin.	Reference	Protocols
EF	EF	101110	RFC 3246	Interactive voice
AF1	AF11 AF12 AF13	001010 001100 001110	RFC 2597	Bulk transfers, Web, general data
AF2	AF21 AF22 AF23	010010 010100 010110	RFC 2597	Database, transactions, interactive, preferred data
AF3	AF31 AF32 AF33	011010 011100 011110	RFC 2597	Locally defined, mission-critical apps
AF4	AF41 AF42 AF43	100010 100100 100110	RFC 2597	Interactive video and associated voice
IP routing	Class 6	110000	RFC 2474	BGP, OSPF, etc.
Streaming video	Class 4	100000	RFC 2474	Often proprietary
Telephony signaling	Class 3	011000	RFC 2474	SIP, H.323, etc.
Network management	Class 2	010000	RFC 2474	SNMP
Scavenger	Class 1	001000	Internet 2	User-selected service
Other	Default or Class 0	000000	RFC 2474	Unspecified traffic

There is a new QoS variant that detects traffic levels and informs network routers before congestion actually occurs. This is called Pre-Congestion Notification [209, 210]. This architecture uses a combination of instrumentation, DiffServ, and admission control to prevent traffic from overloading links.

In practical terms, the implementation of DiffServ has the following consequences:

- Advanced planning is required to define the required QoS classes and the appropriate DSCP marking. This also includes identification and classification of the applications requiring elevated QoS and prioritization of these applications.
- Network analysis is required to determine appropriate PHB to provide required QoS service levels. This includes intelligent capacity management. The use of externally managed networks (Multiprotocol Label Switching – MPLS networks or IP Network Service Providers) will require comparison and mapping of QoS offerings.

- Procedures and tools are required for configuring the ingress routers (setting appropriate header bits – marking, access control) and internal routers (PHB). Auto-marking may be possible in some cases, but requires a level of trust with end systems.
- Although the up-front work is significant, the actual operation of DiffServ is straightforward and with low overhead. However, monitoring and some maintenance are required to adapt to changing conditions.
- Interoperability and performance are normally good.
- There is no end-to-end assurance inherent in the architecture of DiffServ.
- The use of DiffServ is common in company networks, especially those that implement in-house transport of voice and video.

4.5.4   QoS on the LAN

Until now, we have discussed how to provide QoS primarily on the layer 3 WANs, where bandwidth is somewhat restricted and cost is higher than in a LAN. However, even with the much higher capacity of modern LANs, there can still be choke points in local networks. Since layer 2 LAN switching is commonly used in LANs, there are also QoS mechanisms that work at that level.

The most common method of providing QoS on a LAN is the use of VLANs. The VLAN originated with a number of proprietary versions, but was standardized in 1998 with IEEE 802.1Q [211]. It is common to separate time-sensitive traffic, such as voice and video, and place it in separate VLANs. The packets from these VLANs can then be easily identified and appropriately prioritized. The standard system for tagging 802.1Q packets is the addition of a header in the Ethernet frame. See Figure 4.26. The 802.1Q standard is also valid for other layer 2 IEEE protocols. VLANs are also part of the DVB IPTV specification for delivery over the "last mile" to the home gateway. Providers can then prioritize the traffic destined for the port carrying the video stream.

Since the standard also permits double tagging of packets, service providers can implement their own VLAN structure while passing the customer's VLAN traffic unmodified. Although this is possible, it is not often used. VLANs can be also transported over serial links or MPLS networks that are based on the Virtual Private LAN Service (VPLS) [212].

In many cases, traffic is segregated into VLANs through manual configuration of either physical ports or IP addresses. Some automatic registration protocols exist, such as Multiple VLAN Registration Protocol (MVRP), defined in the IEEE 802.1ak [213] amendment to IEEE 802.1Q. There are also proprietary implementations to assign VLAN membership based on a variety of port criteria.

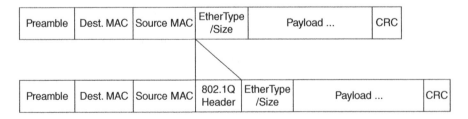

Figure 4.26   Ethernet frames with 802.1Q header

VLANs are also commonly used to provide an enhanced level of security. For example, many large corporate networks require the separation of mission-critical VoIP traffic from the normal data network, often with some form of firewall between them. This isolates the VoIP traffic, to a certain extent, from the possibility of a data network infection or other security incidents.

One challenge with the use of VLANs is the high level of maintenance to design, implement, and maintain them. This is especially true when the use of VLANs is expanded to support activities that were not in the original scope of VLANs. These topics include resource isolation, access control, decentralized management, and host mobility [214].

4.5.5   QoS in the Real World

For   all intents and purposes, QoS does not exist on the Internet at present. QoS will often exist within a service provider, especially when the provider (such as a cable company) offers a range of services such as Internet access, streaming TV, and VoIP. However, the QoS will not extend beyond the provider's network. For a broader analysis, see [215]. There are a number of issues related to QoS on the Internet:

- Who would pay the extra fees for QoS above best-effort? Few end users would be willing to pay for enhanced QoS and only for a limited number of services. Some companies would offer better access to their services, but the end user would pay the difference in the end.
- There is an extremely huge debate about *Network Neutrality*. This is the principle that Internet Service Providers (ISPs) must treat all data on the Internet equally, without discriminating or charging based on user, content, platform, and so on.
  - In 2005, the FCC (US Federal Communications Commission) issued an Internet Policy Statement that states that consumers are entitled to: "any lawful content, any lawful application, any lawful device, and any provider."
  - In 2009, the FCC added "ISPs must not discriminate against any content or applications, and ISPs need to disclose all their policies to customers." This triggered a huge debate of what Net Neutrality is and whether there are any possible compromises [216].
- Big companies already enhance access to their services by purchasing higher bandwidth connections and replicating/caching data regionally. This avoids the politics around Net Neutrality.

Company networks will continue to implement more QoS as the tools continue to get simpler. This increase will also be driven by mission-critical systems and multimedia (voice and video). This extends down to the LAN to support applications such as videoconferencing. Most of the implementations are based on DiffServ with occasional Intserv in networks that can support it.

Provider networks generally have plenty of bandwidth and often do not use all of it. Those providers that also offer multimedia services such as streaming TV will use QoS to maintain their service levels. QoS is also used in specialty services such as intercompany High-Definition videoconferencing connections.

Home networks have inconsistent QoS implementations. Unless the service provider owns and operates (or outsources) the equipment, QoS is not possible [217]. For implementations with multimedia, traffic is separated and routed to different VLAN ports, then prioritized on the LAN ports. Access to the router and the multimedia VLAN is strictly controlled by the provider.

4.6 NTP

Several times in this chapter, we have discussed the use of timestamps for synchronization and calculation of delays and jitter. Timestamps can also be used for error analysis. These actions and calculations are most effective when the clocks of the system are well synchronized. To achieve this, the systems use NTP [218, 219] to set their clocks.[10] NTP Time servers are arranged in a hierarchy.[11] Stratum 1 NTP servers are directly connected to reference clocks and have an accuracy normally within 10 microseconds of UTC. They also communicate with the other stratum 1 time servers for sanity and backup. Stratum 2 servers use NTP over the network to synchronize their clocks to the stratum 1 servers. These clocks are normally within 0.5–100 milliseconds of the stratum 1 systems. Each subsequent stratum adds between 0.5 and 100 milliseconds inaccuracy to the clock.

Conventional wisdom recommends that systems configure five upstream (lower stratum number) servers. The NTP protocol will compare the results from these 5 servers, discard any "falsetickers," and choose the best remaining sources. More servers can be configured, but NTP, when configured properly, does not utilize more than 10 servers.

The format of the NTP packet, shown in Figure 4.27, includes, as you might expect, a number of timestamps and reference information. Many of these will be used in the calculations to determine the current time. The variables used for the calculation are listed in the descriptions.

LI: The Leap Indicator field is a 2-bit integer that warns of an impending leap second that will be inserted or deleted in the last minute of the current month. "1" indicates that the last minute will have 61 seconds, and "2" indicates 59 seconds.

VN: NTP Protocol version, currently 4.

Mode: NTP can operate in a number of different modes including client, server, and broadcast. This integer specifies the mode that is being used.

Stratum: This defined the stratum, as described earlier, of the information in this packet.

Poll: This is the requested maximum interval between successive messages.

Precision: This defines the precision of the local clock. That is, what is the value of one "tick" of the system's clock.

Root delay and dispersion: These fields define the total round-trip delay and dispersion (variability) to the reference clock.

Reference ID: The interpretation of this field depends on the stratum of the server. For primary servers (stratum 1), this will contain a four-character string identifying the reference clock connected to the server. The authoritative list of identifiers is maintained by IANA.[12] If the stratum is 2 or greater, this ID is the IPv4 or hashed IPv6 address.

Reference timestamp: This field is the time when the system clock was last set or corrected.

---

[10] A Simplified NTP (SNTP) also exists in RFC 5905 for low-end devices.

[11] http://support.ntp.org/bin/view/Servers/WebHome.

[12] http://www.iana.org/assignments/ntp-parameters/ntp-parameters.xhtml.

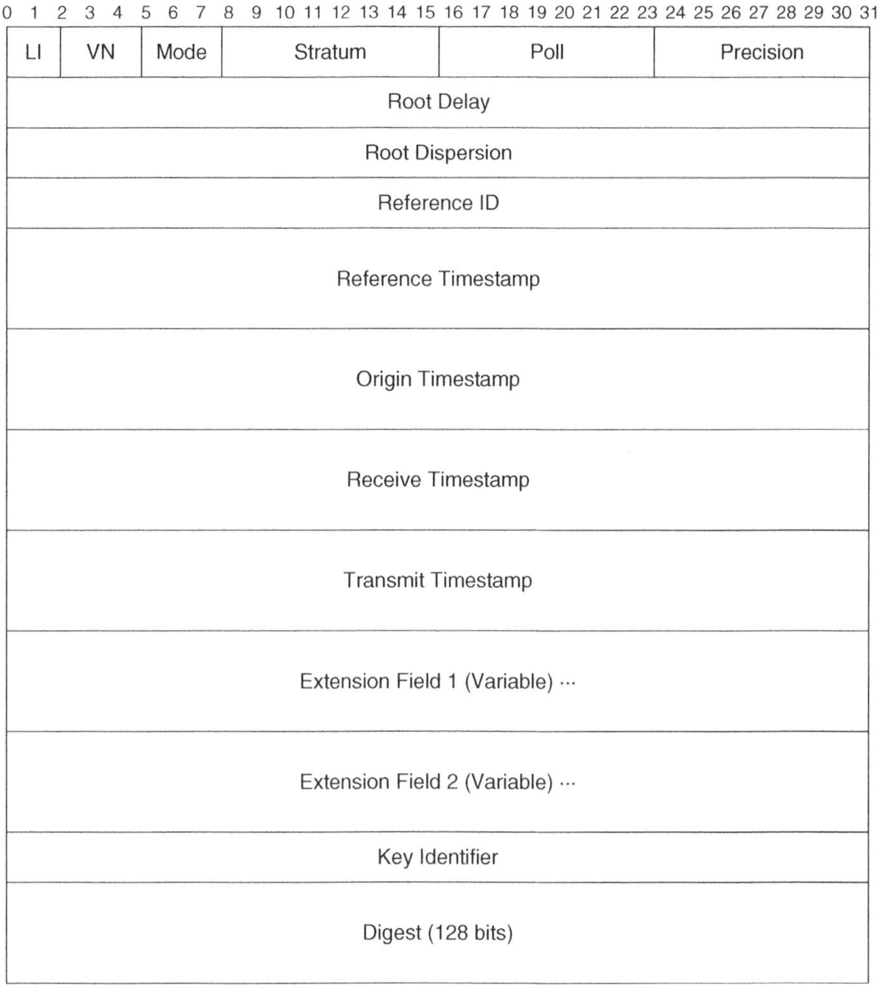

0 1 2 3 4 5 6 7 8 9 10 11 12 13 14 15 16 17 18 19 20 21 22 23 24 25 26 27 28 29 30 31

LI	VN	Mode	Stratum	Poll	Precision

Root Delay

Root Dispersion

Reference ID

Reference Timestamp

Origin Timestamp

Receive Timestamp

Transmit Timestamp

Extension Field 1 (Variable) ...

Extension Field 2 (Variable) ...

Key Identifier

Digest (128 bits)

Figure 4.27   NTP packet format

Origin timestamp: The timestamp T_1 is the time at the client when the request departed for the server.

Receive timestamp: Timestamp T_2 is the time at the server when the request arrived from the client.

Transmit timestamp: T_3 is the time at the server when the response was sent to the client.

Destination timestamp (not included in packet): T_4 is the time at the client when the response from the server was received.

Extension fields: These can be used for public encryption mechanisms.

Key identifier: This field is used by both the server and the client to designate a secret MD5 key.

Digest: This is the MD5 hash of the NTP header and extension fields.

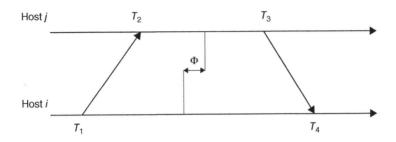

Figure 4.28   NTP offset calculation

Similar to the RTP calculation shown in Figure 4.7, the clock offsets between two hosts are calculated as shown in Figure 4.28. The following are the formulas used to determine the RTT and clock offset:

- Assumed RTT $\delta = (T_4 - T_1) - (T_3 - T_2)$.
- Assumed offset $\Phi = \frac{(T_2 - T_1) + (T_3 - T_4)}{2}$.

By assuming there is a difference between the two clocks ($\Phi \neq 0$), NTP will adjust the clock of the client gradually to match the server clock.

These formulas are correct if delay in each direction is equal. If that is not the case, then the maximum possible error is $\frac{\delta}{2}$. The NTP servers maintain a table of values per NTP peer, i. These values include ongoing δ_i and Φ_i and are used to identify "falseticker" with demonstrably bad times and "truechimers" that can be used to supply correct times and adjust clock drifts.

For maintaining correct timing in large bridged networks, IEEE has standardized Precision Time Protocol (PTP) in standards 802.1AS and 1588 [220, 221].

4.7   Caching

As has been mentioned many times so far, there has been a huge increase in the transport of multimedia content on the Internet. The relatively large data streams, coupled with the need for timely delivery, limit scalability when using a single origin for content. Distributed caching has proven to be critical for the successful implementation of these applications. Although caching applies to the delivery of any content to a client, in this section we discuss only those caching mechanisms that apply to multimedia content delivery. Until now, all of these successful multimedia caching techniques have been proprietary [222].

4.7.1   Caching Elements

To begin with, here are some key elements that might exist in various caching implementations:

- **Content distribution/maintenance** – The creation, distribution, and maintenance of content are not discussed in this section.

- **Content structuring** – As discussed in Section 4.1.3, video content can be stored in different resolutions and bandwidths to provide different levels of service over different or varying network conditions. The cache takes an active role, often providing the lower quality streams first and adjusting to better quality if supported by the network [223].
- **Cache location** – A cache can potentially be placed anywhere from the server up to and including the client. Some of the more common locations include:
  - Proxy – HTTP proxies are usually deployed at the border of a company, institution, or service provider. They provide caching services for that network and normally require the browser to be configured to use that proxy.
  - Reverse proxy – In this case, the proxy is located close to the content provider. The two proxy methods are independent and can be used simultaneously.
  - Dedicated cache/content servers – These systems are often placed in a variety of geographical locations and provide content to clients closest to them.
  - Client – For completeness, we should mention the cache that exists in each client and is used for previously viewed content. In addition, some delivery mechanisms, such as P2P, can use individual clients to deliver content to other clients.
- **Location identification methods** – In those cases where multiple caches exist, there must be a method to determine which cache should deliver the content. Although there are a number of purely scientific implementations [224–226], we will concentrate on commercially available options. The most common methods will be discussed in detail later.
  - Adapted DNS – The client's DNS query is redirected to the appropriate server.
  - Router interception and cache forwarding – An active network device, such as router or switch, monitors traffic and forwards appropriate requests to a cache. This uses the Web Cache Communications Protocol (WCCP). This protocol was developed by Cisco Systems, is supported many other vendors, and is now an IETF draft [227].
  - Intercache communications – Caches can communicate between themselves to determine if a requested object exists. Examples include the Internet Cache Protocol (ICP) [228] and the Cache Array Routing Protocol (CARP) [229].
  - Pear-to-peer – With P2P networks, the location of content is dynamically determined by querying the end-user systems.
- **Timed caching** – Although it is technically possible to store most deliverable media on a cache, in many cases, it is simply not practical or cost-effective. There are several methods that can be used to cache portions of media stream that are detailed in Ref. [224]. To provide this service, the cache must be able to receive, interpret, and forward RTSP, RTCP, and other messages. Figure 4.29 shows the architecture of this system.

  The following are two methods that can be used to stream media without storing the entire broadcast. Note that these methods are intended to save on storage and memory. With the significant reduction in cost of these items, the need for these methods is reduced.
  - Pre-loaded – The initial portion of the media is always loaded on the cache. Since the media is accessed sequentially, the cache can request the next portion from the server while the first block is being transferred to the client. The first block can be a limited quality stream, with a higher quality downloaded from the server, if the client requests. Another example of preloading content is Web or network acceleration devices. These systems will anticipate content that will be accessed, including audio and video, and preload some or all of that content.
  - Sliding interval – This method stores a sequence of media parts in the cache, which is large enough to service clients playing different points in the media stream.

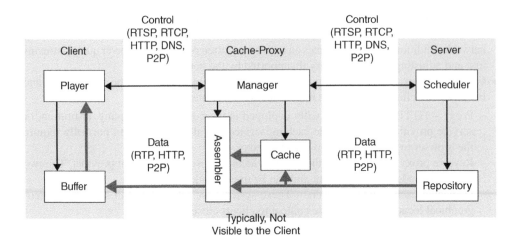

Figure 4.29   Architecture of streaming media cache

There are two caching methods that have met with significant commercial success, WCCP and Content Delivery Networks (CDN).

4.7.2   Web Cache Communications Protocol (WCCP)

The WCCP method of caching involves monitoring of traffic passing through the WCCP server (switch, router, or appliance) and, based on a set of rules (usually IP address and port and normally only HTTP traffic), forwarding that traffic to one or more caching servers [230, 231]. The protocol describes the communication between the interception point and the cache. It does not cover cache-to-server or cache-to-cache communication. It is designed primarily for use in a LAN environment. A common environment is to install a set of WCCP servers and clients (caches) at the Internet connection points of a company or institution. Outbound traffic is monitored by the WCCP server and forwarded to the cache through an IP tunnel that connects the two systems. If the object is available in the cache, it delivers the object directly back to the user. If not, then the cache connects directly with the destination, obtains the object, caches it, then delivers it to the user. The entire process is transparent to the user.

4.7.3   Content Delivery Networks

A CDN is a collection of geographically dispersed servers implemented to provide high performance and reliable delivery of content to end users. This content can include any downloadable object such as Web pages, images, applications, social networking content, and both on-demand and live streaming media. Every major content provider uses some form of CDN. While some are internal, most are purchased from CDN providers such as Akamai. CDN is treated extensively in Section 9.6.

4.7.4 Use of Cache Servers in Private Networks

One common implementation of caching servers in private networks is worth noting. Many larger companies and institutions use Webcasting to stream video to a large number of geographically dispersed internal users. These companies have found that implementing and maintaining full end-to-end multicast is quite expensive. Therefore, many have implemented caching servers in key remote locations. The end users are normally redirected to a caching server close to them. Depending on the remote infrastructure, the transport of content to the user from the caching server may be either unicast or local multicast.

5

Synchronization and Adaptation

This chapter treats the influences of time in the communication and multimedia application. A special chapter on this topic makes sense since timely behavior is so critical in multimedia as in no other networked application. Also, timing issues were touched upon in many topics as side remarks but have never had the concentrated treatment that this subject requires.

We lay the groundwork by providing an End-to-End (EtE) model in Section 5.1 that starts with decoding and ends with the play-out. The model allows us to work on the aspects in the different sections homogeneously. Section 5.2 discusses the variations of how traffic arrives. Section 5.3 handles the characteristics of packet loss and their treatment. Video and audio can only begin to play out when sufficient data has come in, and Section 5.4 discusses how to find out what sufficient means. Congestion is a well-known phenomenon that we analyze in Section 5.5. Delay is so common that we might think it is not worthwhile to handle it, but Section 5.6 probably gives the reader new insights. Queues are a primary reason for delays, and Section 5.7 helps us understand it better. The term player is used in this chapter and in the whole book, and Section 5.8 defines what we mean by it. Multimedia data needs to be stored in a proper way that we describe in Section 5.9. Finally, various elements of a multimedia application need a synchronized play-out. The synchronized presentation can be programmed by a multimedia synchronization language as shown in Section 5.10. After we have learned everything that makes it is so hard to serve an application right, we learn about possible optimizations in Section 5.11.

5.1 End-to-End Model

The model shown in Figure 5.1 is based upon the MPEG Media Transport (MMT) model [232], which is still a preliminary draft with ITU.

The time variables used in Figure 5.1 are:

T_{Sam}: This is the sampling time when media frames are ready to be encoded. After encoding, frames are waiting in an encoder buffer.

T_{Enc}: At the encoding time, the data is ready for media packetization – using Network Abstraction Layer Unit (NALU) or MPEG/PES/TS as treated in Sections 3.4.4.2 and 3.4.1. Note that the data can alternatively come from a precoded file if the video is not live.

Multimedia Networks: Protocols, Design, and Applications, First Edition. Hans W. Barz and Gregory A. Bassett.
© 2016 John Wiley & Sons, Ltd. Published 2016 by John Wiley & Sons, Ltd.

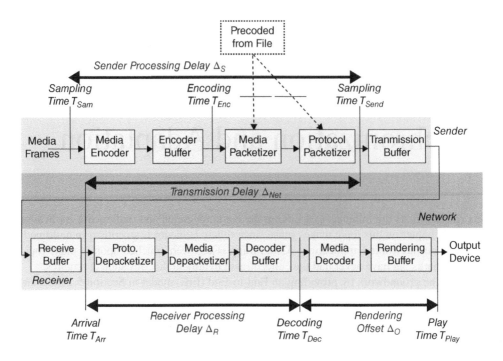

Figure 5.1   End-to-End (EtE) timing model

T_{Send}: At sending time, the information just goes in the transmission buffer and is ready to be sent out. The overall sender processing delay is Δ_S.

T_{Arr}: When the packet arrives, it has been transmitted with a delay of Δ_{Net} and sits in the receive buffer for further handling.

T_{Dec}: At the decoding time, the data is ready for decoding and the protocol and media packet information has been removed. Since leaving the receive buffer, this procedure needs Δ_R. Because H.264/5 packets are not arriving in the order of decoding, packets need to wait temporarily in the decoder buffer.

T_{Play}: After a further rendering offset Δ_O, the data is displayed at the play time. The rendering buffer might be unclear here, but we will explain this in detail later.

Note,

$$\Delta_{EtE} = \Delta_S + \Delta_{Net} + \Delta_R + \Delta_O \qquad \text{is the End-to-End delay}$$

$$T_{Arr} = T_{Send} + \Delta_{Net} + \Theta \qquad \text{with } \Theta \text{ timing offset}$$

The timing offset Θ results from the difference in the system time of the two clocks at the sender and the receiver – see Section 4.6.

If we need to refer to the timing of the ith sample, we add i to a time value.

With this EtE model, we can understand the transmission graph in Figure 5.2. The shown communication uses RTP since the sending intervals are regular and a send-out every 20 milliseconds is quite a normal value for audio – see Section 3.1. Also, the send data volume is equal per time slot since the transmission occurs with a Constant Bit Rate (CBR). In the case of a video, we would typically see a Variable Bit Rate (VBR).

Volume (kB/second)

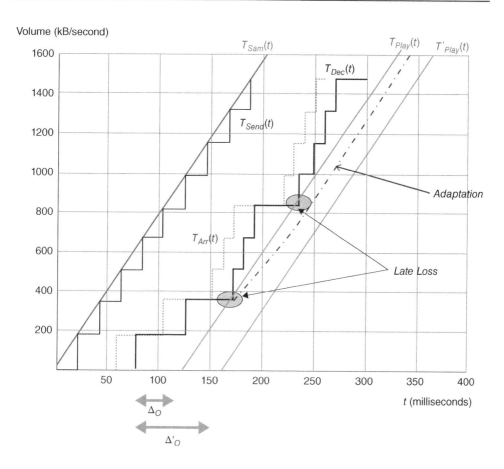

Figure 5.2   Transmission between sender and receiver using RTP for voice

The sampling curve $T_{Sam}(t)$ and the $T_{Play}(t)$ are parallel lines since sampling and playing multimedia should occur with the same timing. The sending curve $T_{Send}(t)$ and the $T_{Arr}(t)$ are definitively not parallel since delays in the network occasionally hinder an analog arrival. The dotted arrival times $T_{Arr}(t)$ lead to later decoding times $T_{Dec}(t)$, and in this example, on two occasions, the line to $T_{Play}(t)$ is touched. The late arrival or often named "late loss" causes an interruption of the play-out. Initially, the system chose a small rendering offset Δ_O so that this late loss occurred. With an alternative offset Δ'_O and a $T'_{Play}(t)$, the late loss would not have occurred. However, in Section 4.5, the acceptable voice delay of maximal 150 milliseconds is explained. With the offset Δ'_O, this maximal voice delay is violated as we can read on the time axis.

The additional curve marked Adaptation is treated in Section 5.4.

We illustrate another transmission in Figure 5.3. This communication does not use RTP because regular sends are missing. The higher volume makes video most probable. It is also not a live stream since the initially sent packets come in faster than T_{Play} will play the information out. Consequently, the sampling rate is hypothetical since it occurred when the video was

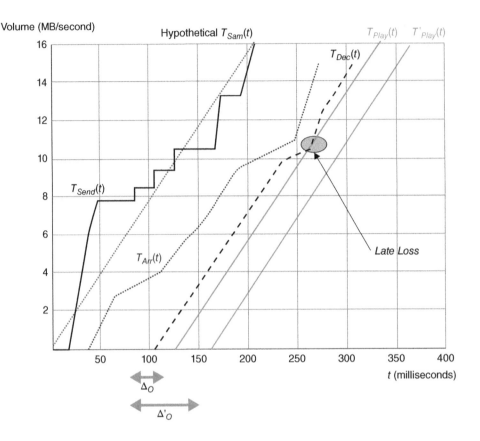

Figure 5.3   Transmission between sender and receiver using HTTP for video

created a while ago. Due to a late arrival, we also see in this case a late loss. However, choosing a rendering offset Δ'_O would not have caused any problem since it is not interactive. The only consequence would have been a later start of the play-out.

5.2   Jitter

Jitter is the standard deviation μ of the transmission delay Δ_{Net}. Typically, the jitter μ is a Gaussian distribution.

Within RTP [109], receivers continuously calculate the jitter for each source. So, when the jth packet from a source k arrives, the receiver recalculates μ^j by

$$\mu^j = \mu^{j-1} + \frac{\left| \left(T^j_{Arr} - T^j_{Enc} \right) - \left(T^{j-1}_{Arr} - T^{j-1}_{Enc} \right) \right| - \mu^{j-1}}{16}$$

We need to adjust to RTP terms by stating that $T_{Enc} = Timestamp - Timestamp_{Ini}$ with $Timestamp_{Ini}$, the first packet in the RTP session – see Section 4.1.1. In each Receiver Report of RTCP to a source k, the current value of μ^j is contained in the field *Interarrival Jitter* – see

Section 4.1.4.1. Note that RTP does not try to keep track of the drifts of the sender and receiver clocks. Another name for this drift is skew.

Note that there also exists another definition of jitter. The aforementioned definition is also named "Interarrival Jitter" or "Inter-Packet Delay Variation" (IPDV). Another definition is the "Packet or Cell Delay Variation" $\mu_j^{PDV} = \left(T_{Arr}^j - T_{Enc}^j\right) - \min_{i=1,\cdots,j}\left(|T_{Arr}^i - T_{Enc}^i|\right)$ used by the ITU [233]. For a comparison of such measures, see Ref. [234]. In the following, we stick with the definition provided by RTP.

If the jitter increases on a connection, this shows a typical temporal congestion on the link and a sender might proactively adjust to this situation – see Section 5.5.

Jitter is also added by queuing on "slow" links. Sending a 1.5 kB frame on 128/512/768 kbit/second link introduces 93/23/15 milliseconds serialization delay and finally jitter for the connection. Sending big video frames of 64 kB over an Ethernet also results in well-measurable jitter increases. The connection-line-induced jitter is sometimes named frame jitter in contrast to the more variable packet jitter.

To cope with the jitter, the receiver has to have enough buffer or, in other words, the rendering offset Δ_O must be large enough. The rendering buffer size is controlled by the application, and at the start of an application, there is typically no information available on the jitter for a connection. Of course, for an interactive application such as voice or videoconferencing, the rendering offset must still fulfill the overall EtE delay $\Delta_{EtE} < 150$ milliseconds. When addressing the different applications later, we will mention the allowed jitter and the relevant buffer sizes.

5.3   Packet Loss

When the receiver does not receive a packet in time, the packet is considered as lost – or late loss as stated in Section 5.1. In Section 3.7, we have seen what the video application can do in such cases to cope with the loss. On the network side, a retransmission is only possible if there is enough time – that is:

$$2\Delta_{Net} + Gap < \Delta_O - \Delta_R \qquad (5.1)$$

In Equation 5.1, *Gap* is the time to recognize the lost packet. The left side of the comparison is defined by the fact that the missing packet has to be requested first and then transmitted again. The right side approximates the time that is available before playing starts – that is, time for buffering minus processing time. Retransmissions have their challenges since drops occur when there exists an overload in a network path element. Figure 5.4 shows the probability of consecutive lost packets summarizing the data measured by Yajnik et al. [235]. The provided data shows that the next packet will also be lost with a probability of 10%. During overload times, the average transmission time Δ_{Net} is also typically increased. So, in reality, a retransmission is only feasible if we have "≪" instead of "<" in Equation 5.1. So, the repeated sending is only possible for noninteractive applications or when sender and receiver sit in the same LAN.

RTCP will report back to an originator in the field *Fraction Lost* of the RTCP Sender Report on the number of lost packets – see Section 4.1.4.1. This report is only a summary. However, there exists the possibility for immediate feedback by an RTCP-based feedback

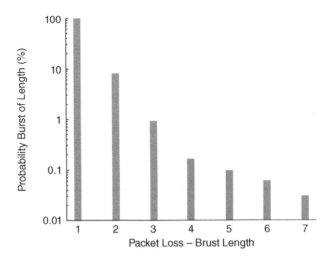

Figure 5.4   Consecutive Error Probability

mechanism [146]. This feedback allows positive and negative acknowledgments on a transport or application layer. RTP itself has no means to act on this accordingly, but the application could use this information.

Since it is probably too late to retransmit in the case of an error, the sender application could send a new I-frame when it receives a NACK. This method would at least minimize the consequences of a loss. The send-out of an I-frame always generates a lot of traffic, and another option would be that the sender adapts the incremental frames to a known, good previously received I-frame at the receiver. This technology exists seldom, but we will see one example in Section 12.4.2.

In case of a reported error, the sender can also increase the redundancy. Other options are provided in Section 5.5.

Also, the adaptive play-out techniques can help to restrict the consequences of an error that we treat in Section 5.4.

Proactive methods such as Forward Error Coding (FEC) were mentioned already in Section 3.7 and are also handled in Sections 5.5 and 5.7.

5.4   Play-Out Time

The play-out is the presentation to the user at the play time T_{Play}. The initial question is at what time the rendering software or player should start the play-out. Once the play-out has started, the goal is always to continue the playing uninterruptedly until the end or until a command from the user. Note that different players are treated in Section 5.8.

In principle, it is good to have more buffering to increase the stability of the play-out to cope with variations in packet delivery times. However, for interactive multimedia applications, the buffering is restricted by the rule $\Delta_{EtE} < 150$ milliseconds. Moreover, for noninteractive applications, the tolerance of the user for the initial waiting time is critical.

If jitter values were known at the beginning of a session, the buffer could be sized correctly. For example, using the Gaussian distribution, the probability of a late loss is 0.2% when the buffer holds packets for $3 \times \mu$. However, jitter values are commonly not known before a session starts, and so, a receiver just chooses typically a buffer to hold packets for 40 milliseconds.

5.4.1 Hypothetical Decoder

The calculation of the rendering buffer size $Size_{Ren}$ has, however, also follows another rule. As we know, video encodings have a VBR and the load on the decoder varies. Remember that the MPEG coding is based on the assumptions of dumb decoder – introduced with Audio MPEG in Section 3.1.3.1. With a dumb decoder – for example, a less powerful smartphone – the encoder must be sure that the decoder can handle the compressed video in time until play-out. So, the encoder also has to be sure that the receiver has enough buffer available for the variances in the amount of data. For these calculation tasks within an encoder, a particular unit is responsible: the Hypothetical Reference Decoder (HRD). HRD is the name in H.263, Virtual Buffer Verifier (VBV) in H.262, and Generalized Hypothetical Reference Decoder (GHRD) in H.264 [236, 237].

We have already seen the HRD in Figure 3.24 for H.261 but named as admission control. The admission control adapts the resolution quality of the quantizers by adjusting the step size. Each video standard includes performance assumptions for the decoder. By using these assumptions, the HRD can also calculate if the decoder can handle the current degree of resolution and if the queue is not overfull. Consequently, the HRD also knows at what time each frame needs to be decoded and presented. We have seen this time information in MPEG packets as *Presentation Timestamp* and *Decoding Timestamp* in Section 3.4.1 and named it here T_{Play} and T_{Dec}.

It is now obvious that the HRD knows the size of the decoder buffer at each point of time during the video. Since the HRD does not include any jitter influences, this calculated size is, however, a theoretical one. For the correct calculation of the initial play-out time and the buffer size, the HRD needs the transmission speed V. The higher the transmission rate, the smaller the buffer size can be. Also, with a higher transmission rate, the initial play-out time is shorter because the buffer is initially filled faster to start the video.

Figure 5.5 is based on a calculation in Ref. [236], and we see an example of the sizing of $Size_{Ren}$ for a video with an average bit rate of 0.6 kbit. For the two marked points **A** and **B** in this figure, the following holds:

A: With a transmission speed V of 0.6 kbit/second, a minimum overall $Size_{Ren} = 16.5$ kbit is necessary. When the minimum size is filled initially, it takes 27.5 seconds. Consequently, play-out can start at the earliest after 27.5 seconds after the first packet of this video arrives.

B: With a transmission speed V of 2.4 kbit/second, a minimum overall $Size_{Ren} = 0.37$ kbit is necessary. When the minimum size is filled initially, it takes 0.15 seconds. Consequently, play-out can start at the earliest after 0.15 seconds after the first packet of this video arrives.

For any other speed V, the $Size_{Ren}$ can be read from the curve in Figure 5.5.

In the example shown in Figure 5.5, we assumed an initial filling of the buffer equal to the minimum overall decoder size. However, the HRD can also calculate a smaller initial filling ratio of the buffer Ini_{Ren}.

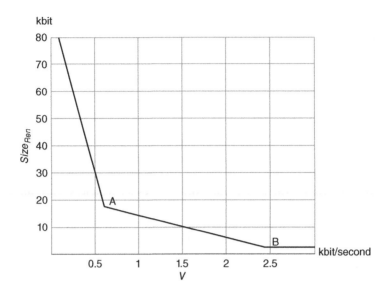

Figure 5.5   Minimum Rendering Buffer Size $Size_{Ren}$

During encoding, the later transmission speed is not known to the HRD. So, H.264 calculates for 32 assumed transmission speeds V the values $Size_{Ren}$ and Ini_{Ren}. The Video Usability Information (VUI) contains all these 32 value triples. The SEI of an NALU contains the VUI – see Section 3.4.4.2. With this initially transported information, the decoder can choose the $Size_{Ren}$ and Ini_{Ren} based on a transmission speed by using earlier experience or by other methods.

So far, there are no IETF standardized parameters to transfer the hypothetical decoder information to the receiver. However, additional standards [238] introduced SDP parameters such as *Min-buffer-time="a=min-buffer-time:* and *Average-Media-Bitrate="a=avg-br:.*

5.4.2   Multiple Streams

For all the calculations of the play-out, we assumed a single stream of information. However, we can also obtain multiple streams such as separately encoded audio and video streams. Typically, the first media from each stream $A, B, \cdots$ should play out together at a time T_{ini}. As discussed in the preceding section, each stream $A, B, \cdots$ has an additional specification, which determines the first play-out time for this stream $T_{Play}(A), T_{Play}(B), \cdots$. The overall initial play-out time T_{ini} is then $\max(T_{Play}(A), T_{Play}(B), \cdots)$.

After starting the combined media play of the streams $A, B, \cdots$, all media has to play in sync. The synchronization during the play of the streams needs to be based on either the timing information of the streams or the timing information of RTP. For simplification, we use the field timestamp from RTP – see Section 4.1.2. Let us describe the received timestamp of the ith packet of a stream A as $timestamp_i(A)$. Media within a packet i of stream A should then play at $T_{ini} + (timestamp_i(A) - timestamp_1(A))$.

RTP creates the time-stamps for different senders based upon different computer clocks. Since clocks may drift over time, the synchronization between the streams can also drift. If the streams contain audio and video, a human recognizes a bad lip-synchronization for differences of > 80 milliseconds [239]. The senders can avoid this drift at the receiver side by using the RTCP reports. RTCP reports to each media sender its NTP-timestamp plus information to calculate the round-trip time – see Section 4.1.4.

For a complete overview on how to synchronize streams from different senders for one receiver, see Ref. [240].

There also exists the requirement to play associated streams at different receivers in a synchronized form – for example, for certain multimedia multiuser games. Inter-Destination Media Synchronization (IDMS) solves this by a new extension of the RTCP protocol described in Ref. [241]. More general background on this topic is contained in Ref. [242].

5.4.3 Adaptive Play-Out

So far, we assumed that the play-out has occurred as originally specified by the sender. However, there exists the possibility to speed up or slow down the play-out within small boundaries. This adaptation is called Adaptive Media Play-out (AMP). In the case of dropped packets, it is obvious why a slowdown can help. We have already shown an adaptation in Figure 5.2 marked "Adaptation" for the curve $T_{Play}(t)$.

Adaptation of video play-out is performed by repeating or dropping a certain percentage of complete frames in the rendering buffer. The adjustment of audio signals requires more complicated interpolation methods as summarized in Ref. [103]. Silence periods in audio can easily be shortened or expanded. So, the receiver can set up an algorithm, which starts to slow down the output when the rendering buffer is below a critical value. If the rendering buffer goes back to normal, the receiver speeds up the play-out to adjust to the original timing. Since the probability of successive drops is high, more elaborated algorithms are possible. Those advanced algorithms look at the latest arrival times of packets to determine when to start a slowdown. This is discussed in much more depth in Ref. [243].

With such adaptation techniques, a receiver also can start the initial play-out earlier than calculated and adapt the play-out until the rendering buffer is at the preplanned stage.

5.5 Congestion Control

Congestion of a network node happens when the amount of traffic has filled its buffers completely so that packets need to be dropped. We list here the most common measures to avoid congestion:

Explicit Congestion Notification (ECN): As explained in Section 4.5.3, routers can inform hosts of existing congestion for reducing the traffic load. The router does not drop any packets as part of ECN.

Droppers: Section 4.5.3 introduced different drop methods – such as RED or WRED – which drop packets in routers based on various criteria. Additional newer techniques are included in Section 5.7.

TCP congestion control: All TCP control methods are described in Ref. [244]. When a router drops packets out of a TCP stream, the corresponding acknowledgment does not reach the sender. The sender will, of course, repeat the send-out but also decrease the sending rate and only slowly increase it again. This method reduces the load on this network path and avoids congestion. Also, the slow start behavior helps to avoid congestion.

NALU: Each NALU packet – see Sections 3.4.4.2 and 4.1.3 – contains information on the relative importance of this packet. A router can drop those packets with less criticality so that the client can still maintain a decent quality for the user.

RTP congestion control: In the original RTP specification, there exists no built-in mechanism to avoid congestion in the network. Currently, new Internet drafts exist, which address this topic:

> **Circuit breakers:** This proposal [245] acts from the sender side, and the sender breaks an RTP connection if there is a media timeout, an RTCP timeout, or a congestion. A media timeout situation occurs if consecutive RTCP packets do not show increasingly higher sequence numbers. An RTCP timeout is signaled by missing RTCP reports. A congestion is diagnosed in case of nonzero loss rates, and the sending rate is 10 times higher than a calculated sending rate based on RTCP values.

> **Application interaction:** This draft [246] describes a conceptual model of how RTP could interact with a video application to control the media flow. The mentioned interface groups are configuration, codec, state interchanges, and RTP exceptions such as the circuit breaker as mentioned earlier. With such interfaces, RTP and the application may decide to submit another scale – see Section 3.4.5 – depending on the network conditions.

> **Network-Assisted Dynamic Adaption (NADA):** NADA [247] identifies the congestion situation also in the sender via a variant of the aforementioned circuit breakers. It calculates an adapted video rate and sending rate. A forthcoming video standard could use these calculations to react on the network situation.

> The IETF RTP Media Congestion Avoidance Techniques (RMCAT) working group governs the last two proposed methods.

> The overall impression is that the standard bodies for video and networking are not working together. So, time will show if any of the work of RMCAT substantiates.

Queue control: Section 5.7 contains additional measures to manage the congestion.

For a broader list of measures, check the RFC overview on congestion [248, 249].

If a congestion with considerable user impact has happened in well-operated environments, the operation team has to find out what caused the congestion and where it occurred. For identifying the root cause, one needs routers or probes in the network that monitor the connections. Many routers have this functionality – for example, Netflow [250] – but this feature is often not activated due to high resource usage. Network probes could offload this function to separate boxes. Also, an analysis tool on top of the captured data is required to work efficiently.

To analyze the cause, further additional information on the quality of this connection is required, such as bit rate, drop rate, errors, and timing information. For a multimedia congestion analysis, the following data elements are helpful:

Media delivery index: RFC 4455 [251] defines the Media Delivery Index (MDI), which is composed of the Delay Factor (DF) and the Media Loss Rate (MLR). DF is the maximum difference, observed at the end of each media stream packet, between the arrival of media data and the drain of media data. This calculation is similar to the jitter calculation μ, but in this case, it is calculated for multiple connections. MLR is the count of lost or out-of-order media packets.

V-Factor: This is a complex measurement system, which uses NALU information in routers. As part of the analysis, it delivers at each router an overview on I/P/B-slice losses. For more details, see Ref. [74].

Multicast tree: Different tools [252] can monitor the proper functioning of multicast trees but are not commercially available.

Proprietary tools: Many network equipment providers have special monitoring tools available – such as Cisco VoIP Monitor Server, Netscout NGenius for VoIP. The major problem with all monitoring solutions is an integration or combination into an overall monitoring solution. Often, network operation personnel has to combine a set of different tools manually.

5.6 Delay

Network elements including their software, end systems, and the length of the transmission lines create delay. We will see that guaranteeing an overall delay of less than 150 milliseconds is sometimes challenging.

Table 5.1 gives an impression of the typical delay created by devices and software. We have not listed in this table the delay introduced by buffers in routers and computers because of their high variability. We treat buffer latency separately in Section 5.7. All the following values are One-Way Delay (OWD); two-way delay is Round-Trip Time (RTT).

The transmission lines introduce delay, and Table 5.2 lists the minimal latency values for some connections. This data comes from rental lines, which were ordered from a Telco with minimum delay. Of course, these values include network elements on the path.

The Network Dashboard [253] provides recent Internet data on global transmission delay, and Table 5.3 gives a summary of data from a router in Munich to diverse locations.

Table 5.3 shows high variations for India and Nairobi. The best values to reach California via the Internet are comparable to the minimum latency. The variations in delay get clearer if one checks Figure 5.6 for the global submarine cables.

Figure 5.6 shows that there are fewer sea cables from Europe to Asia in comparison to the number of sea cables between Europe and the United States. Note that there are currently no land lines connecting Europe to Asia. In the past years, the capacity from Europe to Asia was substantially increased but still seems not always to work out.

Table 5.1   Typical delay

Device or software	Created delay in milliseconds
De-Jitter buffer	Typically 40
En-/Decoder	2–5
Multipoint Control Unit (MCU) – see also Section 3.8	5–20

Table 5.2   Minimum line transmission delay

Connection between	Minimum delay in milliseconds
Basel (CH) to Paris	10
Basel (CH) to New York	67
Basel (CH) to San Francisco	92
Basel (CH) to Singapore via Egypt	95
Basel (CH) to Sydney via the United States	175

Table 5.3   Typical Internet transmission delay in January 2015

Connection between	Delay in milliseconds
Munich to Amsterdam	19–22
Munich to Los Angeles	91–125
Munich to Sydney via Suez Channel	150–192
Munich to India	218–800
Munich to Nairobi	89–302

There are only a few parallel cables available in the Suez channel from/to Asia, and if one cable is cut, the overall capacity drops remarkably. Sea cables break much more often than it is broadly known. Currently, around 50 breaks per year occur only between the United States and Europe. Most of the cables break because of fishing or anchors. However, undersea earthquakes can also be very harmful. On 26 December, 2006, the Hengchun earthquake with an epicenter off the southwest coast of Taiwan cut nine sea cables at once, and it took 49 days until all were repaired. The traffic was rerouted, but the overall capacity was reduced, and delay increased considerably. If cables in Egypt break, the use of other cables to Asia – via South Africa or the United States – increases the latency minimally by 40 milliseconds.

When designing a reliable global infrastructure for interactive multimedia in company networks, all of the aforementioned information needs to be combined to achieve the required quality and delay.

Example: Planning a high-end videoconference network topology Assume that we need to plan a company network with the regions of Europe, Asia, and North America for a videoconference network supporting multisite conferences. There exist not only global conferences between all the regions but also regional conferences.

We concentrate only on the topology and do neither plan the capacity nor backup links. Planning the capacity and the details of the links would need much more information on the

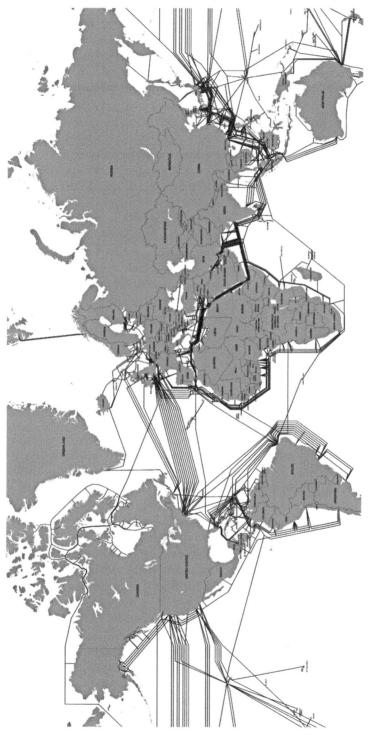

Figure 5.6   Submarine Cable Map. (Source: TeleGeography, www.submarinecablemap.com)

number of expected parallel conferences, number of sites participating in a typical conference, quality expectations per site, and required availability per site.

For the overall topology, the first decision is on the locations of the MCUs. MCUs are required because we have to serve multiple endpoints and need to switch between active and passive participants. First of all, we know that the delay in displaying at each multisite endpoint must be less than 150 milliseconds.

For High-End Videoconferencing, we need less jitter and minimum latency. These requirements lead to rented lines – see Table 5.2 – instead of building Virtual Private Network (VPN) tunnels over the Internet – see Table 5.3. However, we see that we cannot serve three continents at once because the delay becomes too high, but we can serve two continents at once. Some videoconference suppliers state that working with 250 milliseconds delay is still fine to overcome the restriction by the circumference of the world. However, the 250 milliseconds should be the exception but not the rule. For the calculation, one needs to add 10 milliseconds for an MCU.

For North America and Europe together, one MCU in New York would be sufficient and also regional European conferences could use the MCU in New York. Of course, the regional conferences now require bandwidth on the links to the United States, and this may trigger an additional MCU in Europe. Also, videoconferences in Asia could use a European MCU, but the latency is under risk due to possible cable cuts. Consequently, an MCU in Asia should be put in place.

5.7 Queuing

Each queue in a computer system or a router introduces delay and jitter. Also, congestion means overfull queues, and consequently, there are interdependencies between the topics presented in the preceding sections and the theme in this section.

Queues in routers have been investigated for decades, mostly assuming a Poisson distribution for the arrival times of packages. This distribution may be correct if looking at the arrival times of all packets from different sources and connections. But most of us know that traffic patterns have a lot of intrinsic dependencies.

TCP packets leave the sender after an initial stabilizing time at fixed intervals. The regular arrivals of the acknowledgment packets from the receiver define the intervals. During congestion, the confirmation does not arrive, and no packets are sent until with the first acknowledgment the initial stabilizing time begins again. Figure 5.7 shows the steady state, and the reason for the fixed timing is the capacity of the bottleneck, which the sender fills with the packets. The acknowledgment packets are shown to be slimmer than the data packets since they do not carry so much information.

We have shown in Figure 5.7 only one queue from the sender to the receiver. With multiple routers on the path, queues before and after the bottleneck hold packets with equal time distance. Only the bottleneck routers have the additional waiting packets – in this example, three packets. During steady state, there are always three packets waiting as shown here.

The TCP sender knows how many packets are in transit, but the sender has no information that his sliding window is too big – in this example, the sliding windows can decrease by 3. The insightful presentation by Jacobson [254] provided most of the information mentioned so far on queues.

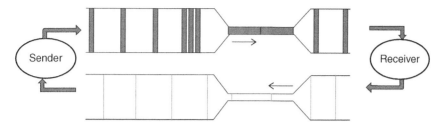

Figure 5.7   Steady-state TCP flows

Figure 5.8   Typical lumpy traffic

RTP packets are sent as defined by their clock rate without waiting for any acknowledgment and without any initial phase or intermediate halting/restarting behavior.

Measured traffic in the network is often lumpy as shown in Figure 5.8 with packets with the same letter belonging to the same connection. Jacobson [254] argues that two effects cause this structure. The first effect is that once packets have come close to each other in a burst, they will probably stay in this configuration until they reach their common destination. The burst occurs when free capacity is available at once, and a host sends all waiting packets immediately out. This behavior has intensified with varying quality connections over wireless LANs [255]. With a bad connection, the computer fills the outgoing queue with packets. When the connection improves, the aforementioned behavior occurs.

With bigger queues, the lumps will get bigger since more packets can wait until they leave together. As a matter of fact, hosts have rather big queues, which trigger this effect. If the lumps are bigger, the chance for a drop at a router due to an overfull router queue increases. TCP will reduce its sending rate in case of drops and restart slowly. So, it is advantageous to have only small outgoing queues on all devices in a WLAN instead of large queues. With big queues and varying quality connections, the TCP traffic will often be in a stop-and-go situation. Oversized and overfull queues were named "bufferbloat" by Gettys and Nicols [255].

This argumentation has shown that a queue should contain only a minimum of buffered TCP packets. The minimality is true in the steady state and also when a host is waiting for free capacity on his data link.

The following proposed methods should solve the bufferbloat problem:

Adaptive Random Early Detection (ARED): ARED [256] is an old adaptive extension of RED – see Section 4.5.3. RED calculates the drop probability from a configured average queue length. ARED's drop probability increases linearly above a minimum queue length threshold until a maximum queue length. The minimum and maximum queue lengths are calculated based on the link bandwidth and a reference delay value.

Controlled Delay (CoDel): CoDel [257] checks the sojourn time of packets in the queue. If the stay time is more than 100 milliseconds, a packet is dropped from the tail of the queue. The next check time for possible drops is calculated based on the number of executed drops.

Proportional Integral controller Enhanced (PIE): PIE [258] measures the queue draining
rate. Incoming packets are dropped with a probability depending on the draining rate and
the queue size.

None of those algorithms is currently available on commercial systems. Since PIE was codeveloped by Cisco, the chances that PIE gets implemented are the highest. Simulation comparisons exist and do show that there is no technique that is always superior –see, for example,
Ref. [259].

The bufferbloat problem also affects HTTP-based streaming such as DASH [260]. There
exists a proposal by Mansy et al. [261] to calculate the send-out of HTTP requests based on
the rendering buffer size. This proposal leads to a more uniform request rate, which fights
bufferbloat for DASH in the steady state.

5.8 Media Player

The media player is the software that takes compressed content and plays it on the output
devices (screen, loudspeaker). A simplified version of the operation of the media player is
seen in Figure 5.9. This figure is just another view of the basic Figure 5.1, which we have used
throughout this chapter. A media player can get its input from a file or a live event. In addition
to combining and synchronizing different input streams such as audio and video, the player
also integrates metadata. Such metadata can be subtitles for videos or just information such as
album and title for an audio file.

The media player is not just a receiver that takes what it gets but requests explicit content
parts. This request can be reading a complex structure from the file system but can also request
explicit data over the network. In Chapter 9, we see how players generate requests within
HTTP-streaming.

There are many media players on the market – RealPlayer (first built in 1995), VLC, Apple's
QuickTime, or Microsoft's Windows Media Player. The players are differentiated by the supported environments (Operating System, Browser), user interfaces, supported codecs, supported input stream types, and possible output types. Some of those are not only monolithic
products but also media development environments such as Microsoft Silverlight [262] or a

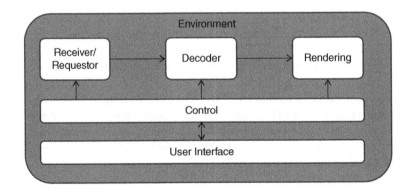

Figure 5.9   Schema of a media player

JavaScript toolset combining HTML5 and Media Source Extensions (MSE) [263]. Individual media players can be built using the aforementioned media development tools. For example, the Netflix client today is a Silverlight-based media player in browsers and will probably be a "HTML5-MSE" media player in the future since Silverlight will get out of support in 2021 [264].

Different players can be combined into an overall media output within a browser by multimedia integration scripting languages – see Section 5.10.

5.9   Storage and Retrieval

Movies need to be stored in some way and can contain video, audio, and additional information. Such a file can be used in different situations. As shown in Figure 5.1, a file is read and sent over the network. In this scenario, it would be perfect when RTP packets would have been prebuilt in the file. When the file is read from storage and sent to a decoder, an MPEG-TS format would be the best – see Section 3.4.1. In HTTP-streaming, files for the transfer get separated into chunks. Moreover, even another structure of the file might be perfect. The base file format for storage that can support all of these scenarios is the ISO Base Media File Format (ISOBMFF) ISO 14496-12 [265].

The MBFF format is an object format and everything is based on boxes. Boxes can include other boxes in a hierarchical way. Figure 5.10 shows a part of the main elements of the structure. All boxes have a four-letter name that is already partly commented in the figure. The *ftyp*

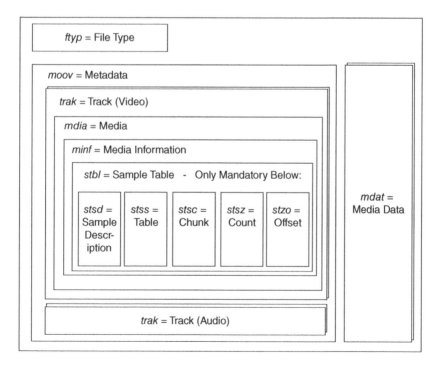

Figure 5.10   Object structure of ISO BMFF

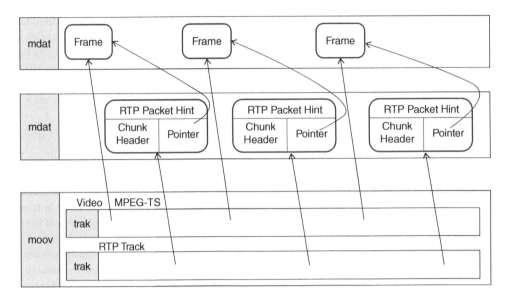

Figure 5.11   BMFF example for storage of RTP packets together with MPEG-TS

specifies the file type that consists of the *moov* with the metadata and *mdat* with all the video and audio samples. The *moov* contains at least one *trak* for videos and one *trak* for audio. The *trak* contains a media box *mdia* with an included *minf* as media information box and again an included *stbl* sample table box. The boxes in the last substructure maintain the information about the coding type and initialization (*stsd*), an index from decoding time to samples (*stts*), and a count of the samples plus size per sample (*stsz*). The media data in *mdat* is separated into chunks. The box *stsc* contains an index to find a sample in a chunk, and *stco* contains the offset of each chunk in the file.

Each *trak* delivers one structured view on the media data. Figure 5.11 provides an example of how different views of the data can be accomplished in the same structure serving different purposes. The figure shows that the pointers from the first *trak* of *moov* deliver a sequence of frames. The second *trak* of *moov* provides a list of ready-to-send RTP packets when the pointer is replaced by the resulting frame. This type of additional structure is often named "Packet hints." With that example, we see how different views are represented in the same file. Now the two optional arcs in Figure 5.1 for inserting a file are clear. It depends on the provided file structure.

On top of ISO BMFF, other formats have been developed over time. The ISO Standard MP4 or ISO 14496-14 [266] was the first extension. MP4 additionally includes pictures (JPEG, PNG) and a Binary Format for Scenes (BIFS) [267]. BIFS is a description language for content or scenes and is typically used for subtitles.

When H.264 became available, the NALU structure (Section 3.4.4.2) required integration, and this happened in the Advanced Video Codec (AVC) file format ISO 14496-15 [268]. AVC stands for H.264.

A recent expansion was defined by W3C by Google, Netflix, and Microsoft to include JavaScript in the file for a later generation of media streams [269].

Registered extensions of ISOBMFF are published on the official registration authority [270].

When playing information on a device, the ISOBMFF is not used directly but is embedded in a container. Containers wrap different types of media (audio, video) together. There are many container formats in use today. The common ones that include the ISO format are:

3GP: This is common in mobile phones, and the format is defined by the 3GPP [271].

MP4: This was listed earlier and is also a container format.

Flash: There are two flash formats, FLV and F4V. The F4V format is based on ISO BMFF and was introduced with the Flash Player 9 update [272].

Quick time: The Quick Time File Format (QTFF) has been used when defining ISO BMFF, but it is a little bit different and was never aligned [273].

The given list is not exhaustive since there exist many more.

5.10   Integration Scripting Languages

The World Wide Web Consortium (W3C) standardized SMIL (Synchronized Multimedia Integration Language) [274] in 2008. It is the most prominent example for player scripting languages. This XML language allows the specification of how to order the execution of different multimedia output sources – audio, video, images, text – in time. There are many examples of the use of SMIL:

- Adding subtitles in various languages for videos or presentations for a defined time [275].
- Show a list of pictures as thumbnails in a browser and display a selected picture enlarged in another subwindow [276].
- Running a digital signage system using multiple displays and each display gets an appropriate sector of the picture [277].
- Prepare an on-demand Webcast after a live Webcast in such a way that the presentation slides change at the correct point in time – see Sections 2.5 and 12.1.
- Define server manifest files for download in progressive streaming as in Microsoft Smooth Streaming. Microsoft's Silverlight supports Smooth Streaming [149, 278].

Algorithm 5.1 shows a valid SMIL code.[1] For many browsers (Chrome, Firefox, IE, etc.), SMIL plug-ins exist, and, for example, the RealPlayer already has built-in SMIL. This example SMIL code shows the basic fundamental features of SMIL.

In the *<head>* part from line 2 to 8, the display regions are defined. The *<root layout>* defines the overall window for the later code, and the *<region>* defines the two subareas within the overall window. The *<body>* contains a *<par>* block, which executes the included code parts in lines 11–19 in parallel. Here, three code parts are executed in parallel *<audio>* and two *<seq>* statements. An *<seq>* statement specifies that the included code is executed sequentially. The first *<seq>* statement in lines 13 and 14 displays two images one after the other in region *reg1*. The first image is exposed for 7 seconds and the second for 10 seconds. In parallel to the picture displays, the texts, *First text* and *Second text*, are shown in the region *reg2* with the same timing as the images – see lines 17 and 18. Music, as specified in line 11,

---

[1] For an execution, two images (image1.jpg,image2.jpg) and an audio file (music.rm) need to be in the same directory.

Algorithm 5.1: Simple SMIL program

```
1  <smil>
2     <head>
3        <layout>
4           <root-layout width="350" height="300" />
5           <region id="reg1" width="350" height="280" left="0" top="0" />
6           <region id="reg2" width="80" height="15" />
7        </layout>
8     </head>
9     <body>
10       <par>
11          <audio src="music.rm" />
12          <seq>
13             <img region="reg1" src="image1.jpg" dur="7s"/>
14             <img region="reg1" src="image2.jpg" dur="10s"/>
15          </seq>
16          <seq>
17             <text region="reg2" src="data:,First text" dur="7s" />
18             <text region="reg2" src="data:,Second Text" dur="10s" />
19          </seq>
20       </par>
21    </body>
22 </smil>
```

---

accompanies the image and text displays. This example is a building block to prepare a slide presentation with music and subtitles.

Of course, there are all kinds of additional operations available to define positions, synchronization, transitions, and animations – see the complete standard [274].

SMIL is integrated in the W3C SVG (Scalable Vector Graphics) [279], which describes two-dimensional objects. SMIL is then able to animate SVG objects.

5.11   Optimization

In this section, we provide an overview of methods to run applications faster over wide area networks. Since we know that multimedia applications are most critical in terms of delay, these techniques should be used even more for multimedia applications.

Computer queues are often not optimally programmed – for example, parts of a packet are copied during the packeting or depacketing process instead of only using pointers to restructure a packet. Specialized software packages are known to boost performance by a factor of 10. Rizzo [280] showed this improvement between a new package and the standard Free BSD package. However, in many cases, one cannot change the delivered packet stack within a computer.

Application developers can optimize their code in many ways to run much faster over the network – see also Refs [281, 282]:

Reduce: The developer should decrease the number of requests and the contained payload as much as possible. Each request requires at least an RTT to complete, which might be negligible in a LAN but is definitively not negligible over the WAN. Combining multiple packets in one can substantially increase the performance.

   The programmer should enable the client to execute as much as possible locally. A typical fault executes all validations per input mask field separately at the server.

   Also, packets should contain as much application payload as possible, minimizing the per-packet overhead.

Compress: Images need to be compressed or resized to shrink the payload. SPDY – an enhanced HTTP protocol – compresses automatically all data [283, 284].

Cache: Browsers cache resources when the programmer has included the information on caching time.

Adapt: The program should detect the type of connection and adapt as much as possible to the expected throughput.

Tolerate: A program should try as much as possible to tolerate different latency behaviors. Instead of sending a single request to one server, it should send parallel requests to multiple servers if this is possible. The program should monitor the latency per server and eliminate slow ones from further requests.

If delivered applications cannot be changed, the last resort is the deployment of the so-called network accelerators – for example, Riverbed. Network accelerators remove redundancies, cache content, and improve protocol behavior [285]. Typically, devices are placed at the egress and ingress of the network. On the egress side, client requests are analyzed and communicated in a more intelligent way to the egress accelerator device. The egress device sends the recalculated original request to the server. One finds these devices very often in company networks.

6

Session Initiation Protocol

The Session Initiation Protocol (SIP) was originally specified in 1999, and the latest version dates from 2004 as RFC 3261 [286]. In addition, 16 updates to this main RFC 3261 exist with the latest one from March 2015 [287]. When building a complete telephone system with SIP, more than 240 additional RFCs are available, which specify detailed features. There are three main reasons for the large number of supplementary RFCs. Initially, telephone systems can be quite complex, and this requires a comprehensive specification. Secondly, the SIP document only describes a regular session setup and nothing else. Thirdly, SIP can also be used to initiate other nontelephony sessions such as videoconferencing. It would not have been possible to specify a complete telephone system as a single RFC because of the tradition of the Internet Engineering Taskforce (IETF) to start simple. This tradition of the IETF is great for new functionality but is problematic when specifying complex existing functionality for use over Internet protocols.

The number of updates and the additional RFCs show, on the one hand, the interest of the Internet community in SIP and, on the other hand, that the topic is hard to conquer. With so many RFCs, the risk of overlaps and contradictions exists. The IETF has addressed this by a guideline for SIP extensions [288] and a description for the change process [289]. The RFC5411 [290] categorized 160 SIP standards until 2009, but many have been added since. In the following overview of this chapter, we use the categories of the RFC5411 to describe the contents of the sections.

SIP is a Voice-over-IP (VoIP) standard. Other VoIP standards are treated in Chapter 7. VoIP software is implemented on regular phone sets, on every type of computer including smartphones and on servers. Softphone describes a VoIP client implementation on a computer.

Section 6.1 treats the basic elements for a telephone system as used in company environments – see also Section 2.1. The basic elements cover core specifications utilized in each phone conversation and a few primitives for manipulating SIP dialogs. Each telephone system needs to interoperate with the Public-Switched Telephone Network (PSTN),which we handle in Section 6.2. Section 6.3 explains how conferences are established using SIP. We described in Section 2.1, the requirements for presence and show in Section 6.4 the implementation methods within the SIP event framework. SIP protocols have problems to pass through firewalls, and Section 6.5 presents the mechanisms for a traversal. Section 6.6 provides an overview of

Multimedia Networks: Protocols, Design, and Applications, First Edition. Hans W. Barz and Gregory A. Bassett.
© 2016 John Wiley & Sons, Ltd. Published 2016 by John Wiley & Sons, Ltd.

the most common APplication Interfaces (APIs) for SIP. APIs are traditionally not specified in RFCs but are proposed by other bodies. Telephone systems should be very reliable, and conversations should stay private. Consequently, security and safety of the overall system are a concern. Section 6.7 handles SIP and VoIP security solutions, while we look at the safety in Section 6.8.3. The final section of this chapter is devoted to planning of a VoIP company telephone system. This section uses knowledge of audio coding (Section 3.1) and statistics to set up such a system correctly. Of course, this part is not covered in the SIP standards but is essential to understand the challenges of deployment. In the final section, we also explain the setup for emergency calls.

This chapter does not treat a few categories mentioned in RFC5411 [290]: Minor extensions, Operations and Management, and Quality of services. We did not include the SIP RFCs on QoS since implementations with underlying QoS measures do not use these standards.

In this book, SIP is treated in one chapter, but there are good books solely devoted to SIP [291, 292]. Note that these two books do not explain planning or deployment issues for telephones. Also, the interconnection to the public telephone network is only rudimentary covered. For the latter two aspects, [293] is a good source.

6.1 SIP Basics

6.1.1 First Steps with SIP

The two main actors of SIP [286] are the User Agent Client (UAC) and the User Agent Server (UAS). A User Agent (UA) is either a UAC or a UAS. As the name describes, these two actors work on behalf of users. An additional element is the proxy that will forward messages and may induce minor changes to the message content. Messages between UAC and UAS can be a Request or a Response. The messages are sent in clear text – UTF-8 [294]. The format is very similar to RTSP – see Section 4.3.1.

A Request message has three fields:

- Method – This is essentially the Request to the UAC/UAS.

INVITE	Starts the build-up for a session
ACK	Acknowledges a request
OPTIONS	Queries capabilities
BYE	End a created session
CANCEL	End a session setup before it is finally created
REGISTER	Registers a UA at a registrar – see Section 6.1.3.

   Many more methods will be explained later in this chapter on top of the basic one.
- SIP-URI – This is the Uniform Resource Identifier (URI) of the partner for a request with an initial *sip:*. The SIP-URI appears in different forms, but the most typical one is a usual E-mail address *sip:user@hostname* or *sip:user@IP-address*.
- *SIP/2.0* – This last field identifies the protocol and the version.

Example 6.1 The initial part of a Request message could be

INVITE sip:hans.barz@ethz.ch SIP/2.0

After the first part of the Request message, a list of CR/LF separated message headers follows. Note that we concentrated on the common elements; the complete syntax is provided in Ref. [286]. Each of those message headers can have one of these forms:

To: name SIP-URI	This specifies the destination of the message. The name is a clear-text description for the SIP-URI. Here, also other URIs are possible, such as the telephone URI *tel* [295] and the secure SIP *sips*.
From: name SIP-URI	A specification where the message originates.
"; tag=" digits	The digits of the *tag* are a unique key from the sender helping identify the later session.
Call-ID: word	The requester creates a word of letters and digits for the *Call-ID*, which identifies the session or the dialog together with sender and receiver *tags*.
Cseq: digits method	The digits count the number of requests for a listed method.
Max-Forwards: digits	Specifies the maximum number of hops over proxies until reaching the UAS.
Via: SIP/2.0/protocol hostname:port; *branch* = word	The *Via* elements record the path over proxy servers until reaching the UAS. Each proxy adds a *Via* statement with his own hostname:port. The used protocol for the transfer is also recorded with a possible value *UDP* or *TCP* or a few others.
	The *branch* is a unique word per dialog and host. It helps to avoid loops. The first Request always starts with the magic cookie *z9hG4bK* as a word. The element *received:* together with an IP address can be added at the end of the expression to indicate a host.

All of the aforementioned message headers are mandatory.

Example 6.2 Example 6.1 needs an expansion for mandatory fields:

INVITE sip:hans.barz@ethz.ch SIP/2.0
Via:SIP/2.0/UDP hub.ethz.ch:5060; branch=z9hG4bK64378
Max-Forwards: 50
To: Hans <sip:hans.barz@ethz.ch>
From: Greg <sip:greg.bassett@ucla.edu>; tag=78342
Call-ID: j6qrtwe14568
Cseq: 1 INVITE

Even if this example shows all mandatory fields, we still need a few more additional message headers for an *INVITE* statement:

Contact: SIP-URI	The meaning varies according the context it is used. Commonly, it is the device of the originator of the message, in contrast to the SIP-URI in *To/From*, which specifies a person. The later use has the name Address of Record (AOR).
Content Type: media type	The media type is the specification for the format of the media session. The media types [296] are managed by the IATA[1]. Typically, *application/sdp* is used and specifies the use of SDP – see Section 4.2.1.
Content Length: digits	The length in octets of the content description. The CR/LF counts for 2.

Each Request is answered by one or multiple responses, which have the following initial part and are followed by one or more CR/LF separated message headers:

SIP/2.0: Status-Code Reason-Phrase	The status codes can be positive (2xx), provisional (1xx), or negative (3xx–6xx). The reason-phrase is a clear-text explanation of the status code.

Figure 6.1 shows a simple SIP call sending first the INVITE, receiving a provisional response "*180 Ringing*" and a final positive response "*200 OK.*" The requestor finalized the call setup by an *ACK* request. The UA begins the call by building a media stream between the partners using the information provided in the *Contact* headers and the SDPs. The media handling is not part of SIP. One of the participants ends the session by a *BYE*, and the session is closed after the positive response "*200 OK.*"

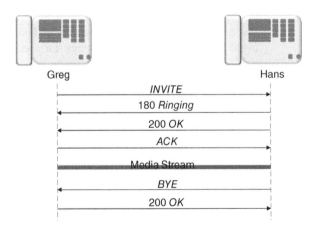

Figure 6.1   A simple SIP call

Example 6.3 We can now finalize the INVITE – as initially shown in Example 6.2. See Section 4.2.1 for the interpretation of the SDP part starting with *v=0*. Note that PCMU stands for Pulse Code Modulation μ-law [16] – see Section 3.1.1.

---

[1] http://www.iana.org/assignments/media-types/media-types.

INVITE sip:hans.barz@ethz.ch SIP/2.0
Via:SIP/2.0/UDP hub.ethz.ch:5060; branch=z9hG4bK64378
Max-Forwards: 50
To: Hans <sip:hans.barz@ethz.ch>
From: Greg <sip:greg.bassett@ucla.edu>; tag=78342
Call-ID: j6qrtwe14568
Cseq: 1 INVITE
Contact: <sip:greg.bassett@ucla.edu>
Content Type: application/sdp
Content Length: 142
v=0
o=Greg 2890844526 2890844526 IN IP4 ucla.edu
s=Phone Call
c=IN IP4 128.3.61.231
t=0 0
m=audio 49172 RTP/AVP 0
a=rtpmap:0 PCMU/8000

With the *INVITE* as before, the first *200 OK* from Figure 6.1 could look similar to the following:

SIP/2.0 200 OK
Via:SIP/2.0/UDP hub.ethz.ch:5060; branch=z9hG4bK64378; received=195.1.2.3
To: Hans <sip:hans.barz@ethz.ch>
From: Greg <sip:greg.bassett@ucla.edu>; tag=78342
Call-ID: j6qrtwe14568
Cseq: 1 INVITE
Contact: <sip:hans.barz@195.1.2.3>
Content Type: application/sdp
Content Length: 142
. . .

The BYE request by Hans would look similar to:

BYE sip:greg.bassett@ucla.edu SIP/2.0
Via:SIP/2.0/UDP lab.ucla.edu:5060; branch=z9hG4bK7324f
Max-Forwards: 40
From: Hans <sip:hans.barz@ethz.ch>
To: Greg <sip:greg.bassett@ucla.edu>; tag=77457
Call-ID: h34560gh23w
Cseq: 145 BYE
Content Length: 0
. . .

We show here the SDP parts only once for completeness but will not do this in future examples. The UAS answers a request with the same SDP or a subset SDP. The UAS replies with a subset if the UAC sent a number of options in his SDP – for example, multiple audio encodings. If there is no possible choice for the UAS available, the UAS will deny the session. This offer/answer model is described in Ref. [136]. In addition, [297] specifies alternative ways of how to negotiate in this model. An expansion of this negotiation model is by first

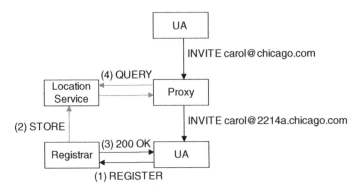

Figure 6.2   Registration in SIP

exchanging capabilities before deciding on the used variant as described in Ref. [298]. This capability model is usually not available in commercial applications.

6.1.2   SIP Servers

The previous section briefly mentioned the method REGISTER. Each client needs first a registration with a registrar before it is reachable. The *REGISTER* method requires the message headers *To, From, Call-ID, Cseq, Contact, Expires*. *To* specifies the name that shall get registered. *From* contains the person who is responsible for the registration. Each request from a client to its registrar has the same value for *Call-ID*. *Cseq* is incremented for each registration. The *Contact* contains the SIP-URI and denotes where the name is reachable with an IP or domain address.

The previously not mentioned message header *Expires* specifies in seconds how long the registration should be valid. The addresses of registrars are preconfigured in each client or a multicast to 224.0.1.75 will reach a registrar. In implementations, preconfiguration is often performed with the Trivial File Transfer Protocol (TFTP) [299]. The registrar confirms a successful registration with a *200 OK*. A registration is removed by sending a registration message with *Expires=0*.

Queries for registered clients go, however, to a location server. The standard specifies neither how the registrar stores his information with the location server nor how the query process looks. Typically, the functions Registrar, Location Service, and Proxy are combined in one system, and no communication specification is necessary. Figure 6.2 exemplifies the concepts of registration whereby the actions STORE and QUERY are performed internally. The numbers for the activities highlight the required order. After registration of the user carol@chicago.com, the proxy can resolve this to the correct server 2214a.chicago.com.

Usually, all initial requests and responses pass through a proxy – as it is seen in Figure 6.3 revised from Figure 6.1. Note that there can be many more proxy servers on the path from UAC to UAS and not only two proxies as shown here.

Apart from resolving SIP-URIs, a Proxy has many more tasks. A proxy acts as an intelligent application router by writing a new *VIA* line and decrementing *Max-Forwards*. Mostly, proxys are stateful by maintaining previous requests and act in case the proxy does not see a response.

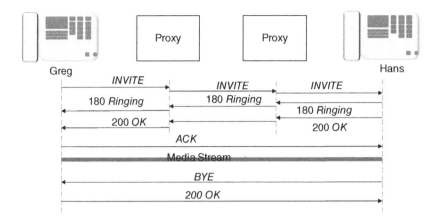

Figure 6.3   A more typical SIP call

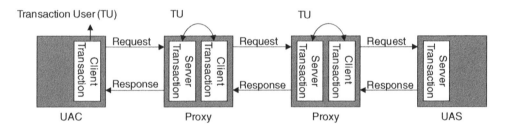

Figure 6.4   Client- and server-transaction interworking

Also, a proxy can fork requests to multiple recipients. Since a proxy can relate requests and corresponding responses, it is built very similar to a UAC or a UAS. It is seen in Figure 6.4 that a proxy combines a client and a server transaction as the UAC and the UAS. For special purposes, there can also be two clients in one box: the so-called Back-to-Back UA (B2BUA). Later, we see usages of the B2BUA.

Now, let us look at the client transaction and the state information, which SIP maintains as shown in Figure 6.5. After an *INVITE*, the client state is "Calling." If no response comes back for some time, a resend is done. If a provisional response arrives (A), the state changes to "Proceeding" and the Transaction User (TU) also gets this provisional response. In case of a client transaction within a proxy, the server transaction is the TU and receives the response for further handling. Otherwise, the user receives the response.

If a provisional response arrives – for example, *180 Ringing* – in state "Proceeding," the answer to the TU is the same, and the state stays in "Proceeding." If in the states "Calling" or "Proceeding," a positive response (B) arrives – for example, *200 OK* – this is handed over to the TU and the state changes to "Terminate." When the client transaction reaches the state "Terminated," all data of the transaction is deleted. When the state is either "Calling" or "Proceeding" and a negative response (C) arrives – for example, *480 Unavailable* – the system changes to state "Completed." After that, it hands this response to the TU and sends an *ACK*

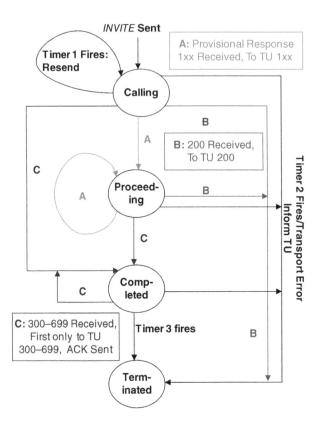

Figure 6.5   State diagram SIP client transaction

to the sender of the response. If the state is "Completed" and another negative response arrives, no further communication to the TU happens. The state "Completed" is only left if one of the two timers fires and it reaches the state "Terminated." The "Completed" state exists to hinder the additional error message causing any follow-up.

There exists, of course, a similar state diagram for the SIP server transaction, which we do not treat here. The server transaction diagram can be easily created by analogy and is part of the standard [286].

We still need to explain how the routing functions with a proxy. The numbers used in the following text refer to Figure 6.6. The UAC sends his *INVITE* (1) to the proxy. Either the address of the proxy is preconfigured with the Registrar or DNS can deliver this informa-tion using a SeRVice resource record (SRV) [300]. Proxies resolve the domain name of the recipient SIP-URI with a domain server (2,3). SIP can use multiple transport protocols, and DNS can store the preferred transport protocol for a destination. This transport protocol query uses the Naming Authority PoinTeR (NAPTR), and the proxy can use this information to for-ward correctly a message [301]. Before forwarding the *INVITE*, a proxy adds a new *VIA* line to the original request with its IP or domain address plus a new *branch* value. In addition, *Max-forwards* is reduced by 1. All other values stay unchanged during forwarding. When the message arrives at the proxy of the home domain (4) of the recipient, Proxy 2 queries the

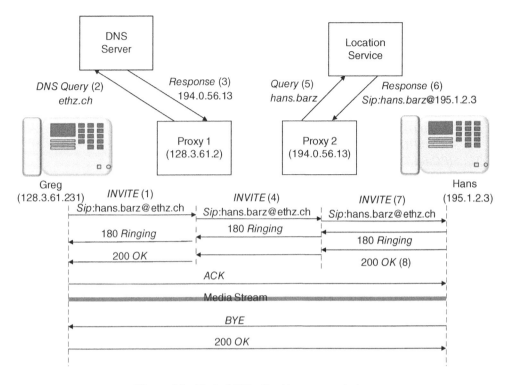

Figure 6.6   Typical SIP call with name resolutions

location service to look up the recipient IP address (5,6). Of course, the recipient has registered before with the registrar its name and IP address, and the registrar has added this to the location service. After adding its *VIA* line and decrementing *Max-forwards*, the proxy can finally deliver the *INVITE* (7) to the UAS. For further options on how to locate servers, refer [302].

The recipient responds to the request that passed over two proxies and includes all received *VIA* lines. With this information, the response can take exactly the path back that the request took. Each proxy in the path removes its own *VIA* information, forwards it to the next proxy or finally to the originator.

Example 6.4 This example uses the statements as in Example 6.3 and shows the adding and removing of information on the path over proxies.
The *INVITE* (7) in Figure 6.6 has the form:

INVITE sip:hans.barz@ethz.ch SIP/2.0
Via:SIP/2.0/UDP proxy2.ethz.ch:5060; branch=z9hG4bKA7834
Via:SIP/2.0/UDP 128.3.61.2:5060; branch=z9hG4bK67213
Via:SIP/2.0/UDP 128.3.61.231:5060; branch=z9hG4bK64378
Max-Forwards: 48
To: Hans <sip:hans.barz@ethz.ch>
From: Greg <sip:greg.bassett@ucla.edu>; tag=78342
Call-ID: j6qrtwe14568

Cseq: 1 INVITE
Contact: <sip:greg.bassett@ucla.edu>
Content Type: application/sdp
Content Length: 142
. . .

The confirmation response (8) from Hans has the form:

SIP/2.0 200 OK
Via:SIP/2.0/UDP proxy2.ethz.ch:5060; branch=z9hG4bKA7834; received=194.0.56.13
Via:SIP/2.0/UDP 128.3.61.2:5060; branch=z9hG4bK67213
Via:SIP/2.0/UDP 128.3.61.231:5060; branch=z9hG4bK64378
To: Hans <sip:hans.barz@ethz.ch>; tag=43568
From: Greg <sip:greg.bassett@ucla.edu>; tag=78342
Call-ID: j6qrtwe14568
Cseq: 1 INVITE
Contact: <sip:hans.barz@195.1.2.3>
Content Type: application/sdp
Content Length: 142
. . .

Note that the confirmation message of the last hop contains in the *received* field the IP address, as we know this already from Example 6.3. The *Contact* field holds the IP address of the recipient, and the originator can later send his *ACK* directly to his partner.

6.1.3   More SIP Methods

In Section 6.1, we already mentioned some methods without giving all the required information:

BYE: We have already seen the use of this method in Example 6.3 to terminate an existing session. The corresponding media session terminates earlier. The communication is directly between the UAS and the UAC without the participation of a proxy. When the clients have established a session after the response *200 OK*, the proxy has no information on ongoing connections between SIP clients. The *BYE* message gets acknowledged.

CANCEL: This method terminates a session that is not yet established. The message can be sent either by the server or by the client. A typical example is that a person has dialed a number, but during ringing, the person ends the call. This termination method passes through the proxies. Here, different scenarios are possible – for example, one could be that the UAS has already sent a response *200 OK* and at the same time the UAC sends the *CANCEL*. In any case, all states of involved proxies need to be cleaned properly.

OPTIONS: A UAC requests a list of its capabilities from another client or proxy. With this method, a later *INVITE* can take into account the capabilities of the other party. The other party answers as if it was an *INVITE* with the corresponding codes. In case of a positive response (*200*), additional message headers are used *Allow, Accept, Accept-Encoding, Accept-Language, Supported* to detail the capabilities. The first four message headers were defined previously in HTTP [151], and the last one, *Supported*, will deliver extensions available to the other party.

A number of additional methods are very familiar, but the original RFC3261 [286] does not contain them:

UPDATE: During the setup of a session, this method asks for a change of the used parameters [303]. If a UAC is already in a session, it can send a new *INVITE* – the so-called re-*INVITE* – with all message headers identical to the old invitation but only an updated SDP.

PRACK: The provisional responses – as *180, Ringing* – are not acknowledged. However, in interactions with the public telephony, it is required to mimic the PSTN protocols. If a UAC receives a provisional response with a *Rseq* message header, it answers with a *PRACK* (PRovisional ACKnowledge) [304].

INFO: During an existing session, additional information gets communicated. This information is not used to change any parameters of the current session, but the application can use it in diverse ways on top of the connection [305]. A widespread use for Dual-Tone Multifrequency (DTMF) signals are in call-center applications or voice-mail systems for deciding various options by pressing the buttons. Touch-Tone is another name for DTMF [306].

MESSAGE: This method was explicitly introduced to support Instant Messaging over SIP [307]. A *MESSAGE* may use the proxies for forwarding but does not set up a session. The receiver and intermediate proxies acknowledge the receipt by a *200 OK* positive response.

SUBSCRIBE/NOTIFY: The user agent can establish a subscription for an event by the *SUBSCRIBE*. When the event occurs, the user agent receives a *NOTIFY*. An unsubscribing method does not exist, but sending a *SUBSCRIBE* with *Expires=0* will do it.
These methods are specified in Ref. [308] that also contains a namespace for events managed by the IANA. The specific events for a function are contained in dedicated standards – that is, for conferencing, see Ref. [309] and, for presence, see Ref. [310]. We will show in Example 6.8 the functioning of *SUBSCRIBE* and *NOTIFY* in the presence event framework.

REFER: The requirements for telephony – see Section 2.1 – list various types of transferring calls. The SIP method to support these transfer requirements is the *REFER* request [311, 312]. A UAC can send this request any time – that is, within a call and outside a call. The UAC sends a *Refer-to* message header with a SIP-URI to a UAS. This message asks the UAS to start a session by an *INVITE* to the communicated SIP-URI. The UAS answers immediately to the request if it is willing to react on this – the positive message is *202 Accepted*. The outcome of the building of a new session is communicated back to the UAC with a *NOTIFY* message, which contains a partial copy of the latest exchange with the requested SIP-URI. The UAC can also tell the UAS initially to suppress the *NOTIFY* message by a *Refer-Sub:false* header.

Example 6.5
REFER sip:hans.barz@ethz.ch SIP/2.0
Via: SIP/2.0/UDP 128.3.61.231:5060; branch=z9hG4bK64378
To: < sip:hans.barz@ethz.ch>
From: Greg <sip:greg.bassett@ucla.edu>; tag=193402342
Call-ID: 898234234
CSeq: 451 REFER
Max-Forwards: 50
Refer-To: <sip:info@ritter.com>

Contact: sip:greg.bassett@ucla.edu
Content-Length: 0

NOTIFY sip:greg.bassett@ucla.edu SIP/2.0
Via:SIP/2.0/UDP hub.ethz.ch:5060; branch=z9hG4bK644278
To: Greg <sip:greg.bassett@ucla.edu>; tag=193402342
From:<sip:hans.barz@ethz.ch>; tag=49931234
Call-ID: 898234234
CSeq: 978503 NOTIFY
Max-Forwards: 50

It is obvious now how to implement transfers blindly or with consultation and even the more seldom option "Call Parking" – see Section 2.1.

6.2   PSTN Interconnection

Without the connection to the Public-Switched Telecommunication Network (PSTN), no SIP environment would be interesting to users. We treat this interconnection again in Section 7.3 with more details on mapping of VoIP protocols to public network protocols. Here, we concentrate on an overall architecture, addressing and support issues of Telcos for SIP.

PSTN networks use "regular" phone numbers, which we all use since our childhood. These "regular" numbers are ENUMs that stand for the ITU E.164 standard NUMbers or the "The international public telecommunication numbering plan" [313]. A DNS extension [314, 315] supports ENUMs. A phone number *+41 44 632 11 11* is translated to the DNS name *1.1.1.1.2.3.6.4.4.1.4.e164.arpa*. The reason for the reverse writing of the phone number lies in the fact that the refinements for ENUMs start from the left side – countries, region, individual. However, in DNS, the refinement starts from the right side – System, Domain, Top-Level Domain. The ending *e164.arpa* is necessary to separate the ENUM namespace from the usual DNS name space. An actual phone number does not need as many dots, but since countries structure their phone numbers differently, the dots were repeated throughout. For each ENUM, DNS will store the SIP-URI to make access possible over a gateway. Note that apart from the E-mail-based SIP URIs, there also exist ENUM-like ones as *tel:+41-44-6321111*, *tel:44-6321111;phone-context=+41*   or   *sip:+41-44-6321111@gateway.com;user=phone* [295]. However, these formats are rarely supported.

Figure 6.7 shows an infrastructure with one SIP phone, one old-fashioned PSTN-based telephone, a gateway between the two worlds, a proxy, and an ENUM DNS. Let the PSTN call to the SIP-recipient terminate at the gateway (1). The gateway will retrieve from ENUM a mapping to a SIP-URI (2), and the message arrives at the SIP client via the proxy (3,4).

When the SIP phone calls the PSTN phone, it sends a request to the SIP proxy (5), which interrogates the DNS. The answer of DNS (6) will point to the Default Outbound Proxy (DOP), and the proxy will forward the message to the DOP (7). When the originating call from the SIP phone used an SIP-URI with an included phone number, the gateway can map this and start the call to the PSTN-world phone (9). However, the DNS can also maintain a mapping between an ENUM and an SIP-URI for the PSTN phone. If such a mapping exists, the SIP phone can call the PSTN phone with an SIP-URI, and in that case, the gateway does a mapping lookup (8) before forwarding (9).

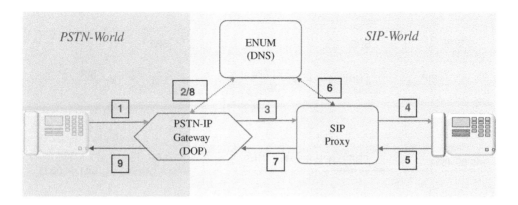

Figure 6.7 SIP-PSTN gateway with ENUM

The typical implementors for an ENUM DNS would be the Telcos that maintain the PSTN numbers in their environment in any case. If they do it, they use another form that is called the Infrastructure I-ENUM [316–318]. The I-ENUM is a DNS infrastructure within the sole control of providers. The information of this infrastructure is not accessible to clients and only selectively to peers.

The mentioned issues of the Telcos with the ENUM SIP are not the only problems for them. The SIP model is not a business model for Telcos because providers still live from the usage fees or the flat rate of their subscribers. Flat rate exists typically per country or subparts of countries but not globally. When SIP has created a call, there is a direct IP media connection with the partner and the proxy has no information on the duration of the call. With PSTN, the correct association from a number to a person was under the control of the provider and had to be accurate for charging reasons. Since SIP did not fit in the provider's business model, there was no interest for a global ENUM-DNS directory, and SIP providers always had to reach other countries or other providers via the PSTN network. However, each gateway on a path reduces the quality and creates potential mapping problems. In addition, the operation of gateways is costly. However, the major Telcos had to offer also SIP-based services because other smaller providers started this as a new business. So, another operating model that still uses SIP but allows proper charging was required.

The "Session PEERing for Multimedia INTernet" (SPEERMINT) architecture [319] achieves the aforementioned goals. Another RFC [320] contains several use cases. This design does not require any new protocol and is consequently easily implementable. Telcos have moved globally to this design or close variants of it. Figure 6.8 mentions all elements required to function. For security issues of SPEERMINT, see Refs [321, 322].

Calls in SPEERMINT between different service providers are set up by SIP with the Signaling Border Elements (SBE), which are composed of a Lookup Function (LUF), Locating Routing Function (LRF), and a Service Function (SF). Each call setup that leaves the service providers responsibility reaches the SBE, which queries the ENUM database with the LUF for the mapping to the DNS address. Each service provider may maintain his own ENUM database that other service providers may read.

The media data is transported via RTP using the Data Border Element (DBE) by the Media Functions (MF). Without the knowledge of the user agent client or the user agent server, the

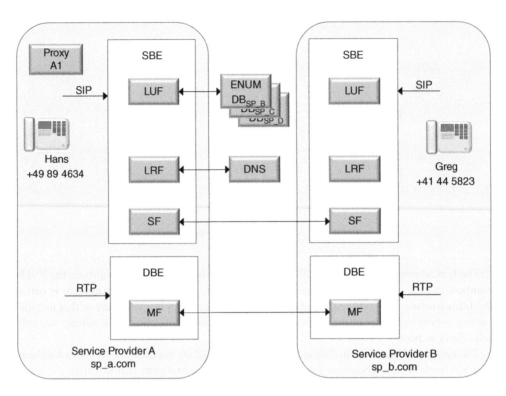

Figure 6.8   The SPEERMINT architecture

"direct" connection between them uses the MFs forwarding function during the whole call. At the end of the call, the MFs can calculate the charging for their corresponding agents.

Example 6.6 The following *INVITE* goes to the proxy A1 in the network of Service Provider A – see Figure 6.8.

INVITE sip:+41445823@sp_a.com; user=phone SIP/2.0
Via: SIP/2.0/TCP client.sp_a.com:5060; branch=z9hG4bK74bf9
Max-Forwards: 1
From: Hans <sip:+49894634@sp_a.com; user=phone>; tag=12345
To: Greg <sip:+41445823@sp_a.com; user=phone>
Call-ID: abcde
CSeq: 1 INVITE
Contact: <sip:+41445823@sp_a.com; user=phone; transport=tcp>

The client does not know the real domain for the recipient and just adds his own. If the client would specify instead *tel:44-5823;phone-context=+41*, it would be obvious that the domain name needs to be found. Since the proxy detects that the recipient address is a phone number, it queries the LUF. The LUF determines the correct database and returns the correct new domain *sp_b.com*. Proxy A1 directs the Invite now to SBE due to its routing policy and adds a new *VIA* plus an updated *INVITE*:

INVITE sip:+41445823@sp_b.com; user=phone SIP/2.0

. . .

Via: SIP/2.0/TCP A1.sp_a.com:5060; branch=z9hG4bKye8ad

. . .

The SBE receives the updated *INVITE*, queries DNS for the required transport protocol, and the Service Function forwards it to Provider B's SF. This SF can act as a proxy or as a Back-to-Back User Agent (B2BUA). As a proxy, it adds *VIA* entries, and as a B2BUA, it rewrites the *INVITE* as follows:

INVITE sip:+41445823@sp_b.com; user=phone SIP/2.0

. . .

Via: SIP/2.0/TCP sf.sp_b.com:5060; branch=z9hG4bKyeysx

. . .

The standards of SPEERMINT do no treat the RTP session setup. However, the path to the specified IP address of *Greg* will be forced over the MF, which records usage times.

Note that the function of a B2BUA is not specific to SPEERMINT. The original SIP standard [286] already introduced this function, which we will see being used further.

6.3   Conferencing

SIP can build conferences not only for voice but also for all types of multimedia. There exist different types of conference implementations, and all of them support voice. The four implementation variants for conferencing are:

Endpoint mixing: One dedicated client calls the participants one after the other and creates a conference. If the dedicated client does not reach a person, this participant can only join in later if he can ask the dedicated client by other means to dial him again. Client implementations only allow a few members – typically below 10 – due to restricted resources at the dedicated client. The dedicated client has to mix all input lines into one output signal. Another disadvantage of this solution is that the dedicated client has to stay in the conference even if it is only the other participants that want to continue the meeting. This solution solely exists for voice.

Server and distributed clients: A server establishes calls with each pair of participants. After the establishment phase, meshed sessions exist between all clients. Each participant mixes all media he receives from all other partners in the conference. Of course, the resources of the clients limit the number of participants. If a participant was not reachable during the establishment phase, the server might try again later and connect the missed participant in the call. In that setup, each participant can leave the conference at his discretion. This solution usually only exists for voice.

Conferencing bridge: A conferencing bridge is a dedicated device that is sometimes part of a voice switch – an umbrella word that also covers a proxy and a registrar. The client dials in with a number and possibly uses a code to join a particular conference. For commercial services, a call might first go to a support person, who then brings the caller into a conference. A bridge can also call out to the clients. The bridge mixes the media from all clients and transmits it back to the clients. Users can join and leave conferences at any time.

Note, a Multipoint Control Unit (MCU) is a conference bridge specifically for videoconferences. Often, MCUs can also integrate phone clients in a videoconference call. An important function of MCUs is the transcoding of different input formats – see Section 3.8. If all participants in a conference would use SVC-based coding, the MCU does not need to transcode, and the remaining functions can be performed by a Single Forwarding Unit (SFU).

Each conferencing bridge can run many but still a restricted number of connections. Since the client numbers are restricted, scheduling is required. Scheduling means that clients have first to reserve a slot of the existing resources of the bridge.

Multicast: A multicast can also be a conference. However, this is a rather unidirectional conversation since any questions and comments need to be communicated by means outside of the conferencing call. While a multicast is technically possible for audio only, it is never used in this way. Multicast is combined with video and presentation slides. We treat this in more detail in Section 12.1 on Webcast. Multicasts are joined with a URL from a browser – for example, *rtsp://example.com/myvideo.mpg*. SIP is not helping multicasts in any way. The usage of multicast requires that the complete network path is multicast-enabled, which is not commonly the case in networks and requires some preparation work.

For conferencing, two IETF framework standards exist. The SIP'ING conferencing framework [323] is based solely on SIP, and the XCON conferencing framework [324] is independent of the signaling protocol. However, XCON can use SIP as a protocol in the framework. XCON always has one central control point, while the SIP'ING framework does not require this. Since central control points are the rule in current implementations, the XCON model with its associated protocols is the more likely model to survive, and we concentrate on XCON here.

XCON has four main elements:

Signaling: Consists of a signaling client that manages with an associated protocol the focal point. SIP could do the signaling, and we explain this next.

Floor: The floor client manages the Floor Control Server (FCS) with the Binary Floor Control Protocol (BFCP). A floor is a shared resource, and this protocol controls the exclusive access to it.

Conference control: The conference/media control client manages a conference control server with the conference control protocol.

Notification: The client has a notification protocol to interact with the notification service. With SIP, the methods *SUBSCRIBE* and *NOTIFY* deliver this functionality together with a corresponding event package.

All four elements listed (Signaling, Floor, Conference Control, Notification) work together on conference objects. Figure 6.9 shows the interworking.

Within the SIP'ING conference framework, one additional protocol was developed. The Media Server Control Markup Language (MSCML) [325, 326] controls a media server in conjunction with a SIP server.

Three standards [309, 323, 327] cover SIP conferencing signaling whereby the first is solely the SIP'ING framework document; the second specifies the protocol, and the last is the event package for conferencing.

Each conference is identified by a conference URI. It is outside the scope of the SIP conferencing standards how this SIP conferencing URI came into existence. If a connection belongs

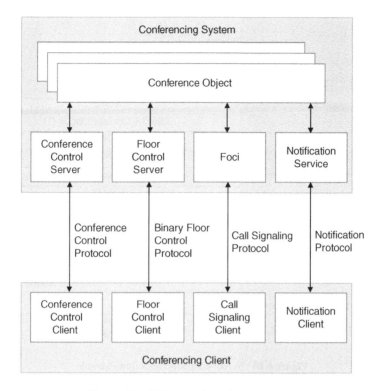

Figure 6.9 XCON conferencing framework

to a conference, the *isfocus* tag gets added, and the user agent for the conference is *focus*. SIP conferencing delivers the functions create, delete, and discover for a conference. For participants, the primary functions are join, leave, request another to join, and switch with another participant. The functions do not need new methods but are established by the existing methods *INVITE* and *REFER*. Figure 6.10 shows a typical signaling for a conference, and Example 6.10 augments some of the commands with the code. *NOTIFY* and *SUBSCRIBE* are used together with the event package as specified in Ref. [309].

Example 6.7 In Figure 6.10, we do not see an initial *INVITE* for Greg. If Greg was the first participant, his initial *Invite* created the conference in the focus.

Statement (1) in Figure 6.10 is detailed. In this statement, we find a specification for a supported message header *replaces* and an allowed description *application/conference-info+xml*. The *replaces* is an additional message header in SIP conferencing that describes how user agents switch in a conference. The *application/conference-info+xml* refers to the event package schema for conferencing as specified in the [309] and described in the W3C Extensible Markup Language (XML). Each *NOTIFY* uses this XML description. The XML description is registered with the IANA[2] as defined in Ref. [308]. In Example 6.8, we will see part of a similar XML description for presence.

---

[2] http://www.iana.org/assignments/sip-events/sip-events.xhtml.

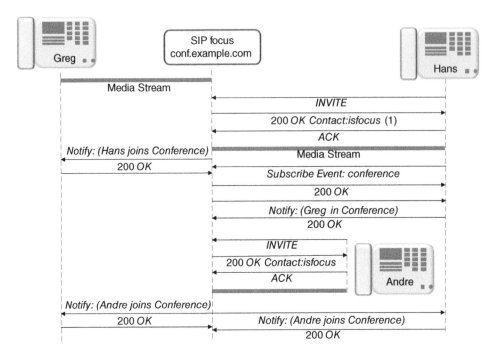

Figure 6.10 A SIP conference call flow example

SIP/2.0 200 OK
Via: SIP/2.0/UDP 128.3.61.231:5060; branch=z9hG4bK643; received=192.0.2.4
Max-Forwards: 50
To: <sip:3402934234@conf.example.com>
From: Hans <sip:hans.barz@ethz.ch>; tag=32331
Call-ID: d432fa84b4c76e66710
CSeq: 45 INVITE
Contact: <sip:3402934234@conf.example.com>; isfocus
Allow: INVITE, ACK, CANCEL, OPTIONS, BYE, REFER, SUBSCRIBE, NOTIFY
Allow-Events: dialog, conference
Accept: application/sdp, application/conference-info+xml
Supported: replaces, join
Content-Type: application/sdp
Content-Length: · · ·

In real deployments, the automatic creation of the technical side of a conference is nontrivial but essential for user acceptance. End users in companies agree to meetings by invites and acceptance within calendar applications. The end user is not willing to act explicitly on the technical side of a conference setup, and he would only like to book conference rooms at the diverse sites, and the rest works automatically. This automatism can only work smoothly with an integration of a videoconference scheduling in a calendar application.

Now, we come back to the XCON framework protocols. The most common one is the BFCP that allows to control common resources in a conference [328–331]. The literature also

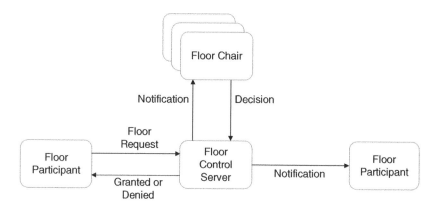

Figure 6.11   Binary flow control protocol functionality

mentions two Internet drafts since the original protocol undergoes an update for inclusion of UDP. Typical shared resources in conferences could be a shared presentation screen or the channel for the chosen outgoing video.

Figure 6.11 shows the main functionality provided by BFCP. The word binary in the protocol describes a binary rather compressed encoding – just the opposite of SIP or SDP. The floor participants are user agents in a conference that can request, release, or query a floor. A floor is a temporary permission to use a resource. The FCS knows the Floor Participants (FPs), the floors, the chairs plus the states of those elements. The FCS is the center of the communication of BFCP. Participants can ask for a combination of floors, but if one of the floors gets denied, the overall request is denied. Chairs queue requests. Floors and the associated resources are conference objects that are created or deleted by conference control.

Figure 6.12 shows the format of the BFCP packets. In the following, we mention only the important parts of the packet. The *Primitive* are the typical requests, releases, or queries. The *Conference ID* is the conference to which this message belongs. Transactions are logical units for identification of answers. Answers to the previous requests have the same *Transaction ID*. There exist a number of attributes of variable lengths determined by the corresponding *Length*

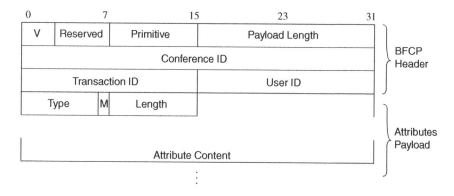

Figure 6.12   BFCP packet

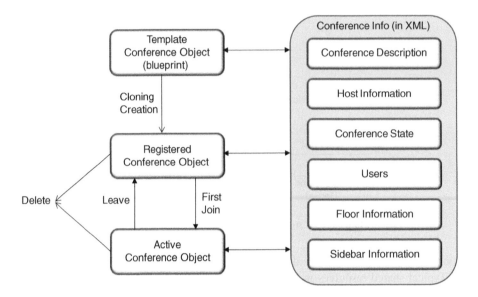

Figure 6.13   Centralized conference objects (XCON) lifecycle

field (multiples of 32). The attributes are, for example, *Floor-ID, Priority, Request-Status, Error-Code,* and *Information.* Note that there exists with H.239 [332] a comparable protocol within the ITU.

The last missing element of the XCON Conferencing Framework as shown in Figure 6.9 is Centralized Conference Manipulation Protocol (CCMP). The CCMP allows participants to create, retrieve, change, and delete conference objects [333, 334] as seen in Figure 6.13. All protocol actions are written in XML and HTTP transports the actions as payload. The details of the conference information elements – such as conference description, and conference state – are based on the XCON extensible data model as specified in Ref. [335]. All parts of the conference information from Figure 6.9 are easy to understand apart from sidebar information. Sidebars help to build side discussions and side conferences in addition to existing conferences. Sidebars may only be visible to certain participants of a conference – for example, a supervisor joins a customer discussion without being visible to the client [336]. An additional standard [337] helps SIP-registered conference users find the corresponding CCMP conference. From the XCON framework, the protocol BFCP is commercially implemented, and we touch on this again in Section 12.4. However, for CCMP, the authors are not aware of a commercial implementation.

There exists in addition an IETF Media Control Channel Framework (MCCF) [338, 339] that supports distributed applications and media. In Section 7.3, we treat in much more detail another Media Gateway (MG) protocol that in contrast is broadly implemented.

6.4   Presence

Presence is the information where a person is and how this person is reachable. For the understanding of reachability, one needs to understand the state of all communication links to this

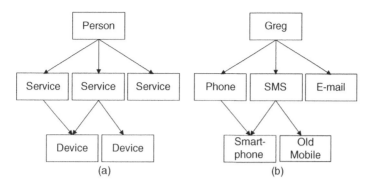

Figure 6.14   Presence: possible structure and example

person. The hope behind a system delivering this information is to contact a person at the right point of time with the optimal medium and no unsuccessful trials. Products supporting this are sold mainly to companies that can enforce that this information is available.

The presence model uses the element's person, service and device [340] – shown in Figure 6.14. Part (a) of the figure shows one instance of the elements of the presence model with a person having three communication services but only two of those services have a device associated. In part (b) of Figure 6.14, we see the same instance with actual values. The model uses an XML presentation named the Presence Information Data Format (PIDF).

A person can have many additional attributes such as "Is in a meeting," "In a plane," "Sleeping," or "Timezone." Services have characteristics such as "Last used," "Locations" for a "Postal," and very important also the URI of a service. The characteristics of a device may note, for example, "Low power," "In call," or "Switched off," which of course influences the reachability. A presentity is a combination of the set of devices and services associated with a person. A presentity describes the overall reachability of a person.

Of course, an event package exists [310], which allows waiting for conditions of a presentity. A Presence User Agent (PUA) manipulates presence information of a presentity. Multiple PUAs can exist per presentity since different devices can change the information. In Figure 6.14, the mobile phone and the smartphone may independently change the status of the presentity. The Presence Agent is a SIP user agent who receives *SUBSCRIBE* requests and creates *NOTIFY* in case of changes. A Presence Agent runs on a Presence Server. In addition, a watcher is a user agent who is interested in a presentity. Example 6.8 shows how a presentity is supervised with *SUBSCRIBE* and *NOTIFY*.

Example 6.8 Here, example.com is a presence server and the watcher *SUBSCRIBE* for an *Event: presence* and a time period *Expires:600.*

SUBSCRIBE sip:resource@example.com SIP/2.0
Via: SIP/2.0/TCP watcher.example.com; branch=z9hG4bKnas
To: <sip:resource@example.com>
From: <sip:user@example.com>; tag=xfg9
Call-ID: 2010@watcher.example.com
CSeq: 17766 SUBSCRIBE
Max-Forwards: 70

Event: presence
Accept: application/pidf+xml
Contact: <sip:user@watcher.example.com>
Expires: 600
Content-Length: 0

Now, example.com informs the watcher on a presence event that *basic* is now *open*. The *Subscription-State* is still *active* but *expires* in *323*.

NOTIFY sip: user@watcher.example.com SIP/2.0
Via: SIP/2.0/TCP server.example.com; branch=z9hG4bKna99
From: <sip:resource@example.com>; tag=ffd2
To: <sip:user@example.com>; tag=xfg9
Call-ID: 2010@watcher.example.com
CSeq: 8776 NOTIFY
Event: presence
Subscription-State: active; expires=323
Max-Forwards: 70
Contact: sip:server.example.com; transport=tcp
Content-Type: application/pidf+xml
Content-Length:253
<?xml version="1.0"?>
  <presence xmlns="urn:ietf:params:xml:ns:pidf"
  entity="sip:resource@example.com">
   <tuple id="77242676" >
     <status>
       <basic>open</basic>
     </status>
     <contact>sip:server.example.com; transport=tcp </contact>
   </tuple>
  </presence>

If several agents are interested in the same combination of events from different presence servers, each client needs to *SUBSCRIBE* for the combination of events. For better efficiency, an intermediate Event State Compositor (ESC) is placed between the watchers and the presentity servers. The presentity servers *PUBLISH* [341] the events to the ESC, and the ESC distributes corresponding NOTIFYs to the watchers.

Historically, presence and Instant Messaging (IM) were quite close since an IM showed the presence of a user. The IETF standardized two IM protocol sets. The SIP version was SIMPLE and stands for "SIP for Instant Message Leveraging Extension" [307, 342]. The other IM version is the eXtensible Messaging and Presence Protocol (XMPP) [343] that also covers presence and can interwork with SIP [344]. This protocol also has the name Jabber. Since IM is a text message system, we do not treat this in this book.

In Figure 6.14, we used mobile phones as an example for presence devices. This would require a SIP implementation on mobile phones and smartphones. Of course, there exist SIP clients for smartphones, but these are not the primary phone clients on these devices. For a presence system, these SIP smartphone clients cannot report on the relevant status of mobile phone calls. There are two options to solve this:

- Use a SIP 3G network as specified as the 3G IP Multimedia System (IMS). We introduce IMS in Section 7.4.1.
- One installs a Presence Network Agent (PNA) on the device or the path. Such a software analyzes the voice network traffic and acts on behalf of the nonexistent presence user agent [345].

Another unexpected device for the presence is a set-top box for IPTV – see Chapter 11. Since IPTV users register implicitly with their service providers, all type of IP-based services can be integrated. The activity "Viewing at the TV" may get registered within the presentity of a user. This use case of IPTV was proposed in Ref. [346].

The IETF standardized several extensions to PIDF since one can envision that there are many attributes that may be interesting to a partner – see Table 6.1, but note that even this table is not complete. In addition to those expansions of the presence XML model, the RFC [347] describes how to handle the access to the sensitive presentity information.

SIP presence is broadly implemented – for example, by Microsoft [355]. However, additional private XML schemas are common in those implementations.

6.5   Network Address Translation

A device performs a Network Address Translation (NAT) when it translates IP addresses and ports of a packet to other IP addresses and ports. Typically, the NAT device transforms an internal nonregistered address range to and from a registered address range. The NAT device keeps these address assignments for some time and applies it continuously in both directions until termination of connections. The systems pair IP address/socket behind the NAT carries the name "reflexive transport address."

SIP, RTP, RTSP, and other protocols have problems with the NAT function as described next – see, for more details, [356]:

1. Responses to SIP request go to port 5060 on the NAT device, but this port may not be open for traversal.
2. RTCP uses an incremented port from RTP. Again, this port might not be available for traversal.
3. Addresses in the SDP information or contained else internally in the payload – see, for example, Figure 4.12 – will not work.

Table 6.1   Standardized extension to presence information

Contact information (CIPID)	[348]
Description of presence in past and future time intervals	[349]
Partial presence	[350, 351]
Uncertainty and confidence	[352]
Better civic address information	[353]
Overall enlarged model – Rich Presence Information Data (RPID)	[354]

A comprehensive overview of the solutions is complicated by the fact that NATs show different behavior, that is [357]:

Mapping: A NAT initially creates the reflexive address for an internal IP address, port pair for the communication to an external host. In case the internal IP address is used again later to communicate with another external host, the NAT can react in two ways. It can reuse the same reflexive address or create a new one. In case of reuse, this is named as endpoint-independent mapping, else it is an address-dependent mapping. Additionally, address-port-dependent mapping exists if this behavior also depends on the used port.

Port assignment/parity/contiguity: A NAT may preserve or not preserve the internally assigned port number also for the reflexive ports. The same may hold for the parity or the contiguity of the reflexive ports.

Mapping refresh: This is the time until a new mapping of UDP packets is possible. Since UDP is connectionless, a NAT has to remember for some time a former association of addresses to guarantee a proper function.

Filtering: The filtering is endpoint-independent if any external host with any port reaches the internal client with the reflexive address. We have an address-dependent filtering if only the once contacted external host with any port can answer on the reflexive address. Finally, the filtering is port- and address-dependent if only the once contacted socket can answer to the reflexive address.

Hairpinning: Internal hosts can also communicate with their reflexive addresses. The name originates from "Hairpin Turn." This functionality may be a requirement for call transfers by an external partner to another internal partner.

ALG: Some NATs have Application Level Gateways (ALGs) active, which support the "seamless" transport of certain protocols.

Special NAT support: The NAT may have support for special protocols to avoid problems – see later in this section.

With this list of options, it is obvious that a general solution might not exist, and solutions only work in certain circumstances. Preferable for SIP and RTP is a NAT, which does endpoint-independent mapping, no port/address-dependent filtering, no assigned port preservation, port parity preservation, and UDP refresh after 5 minutes.

Another older terminology also exists [358], which, however, does not describe all potential features:

Full-cone NAT: This is endpoint-independent filtering and mapping.

Address-restricted-cone NAT: This is address-dependent filtering and endpoint-independent mapping.

Port-restricted-cone NAT: This is port-/address-dependent filtering and endpoint-independent mapping.

Symmetric NAT: This is port-/address-dependent filtering and endpoint-dependent mapping.

The Session Traversal Utilities for NAT (STUN) [359] provides a method for a system behind a NAT to determine its reflexive address of the last NAT on the path to the STUN server. The protocol starts with a STUN binding request from the client to the server. Each packet contains a client chosen transaction id to correlate answers on requests. The STUN

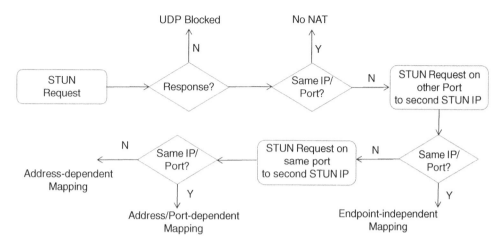

Figure 6.15 Discovery of NAT mapping based on Ref. [360]

server answers by sending the reflexive address in a packet back to the client. The STUN server converts the reflexive address by an exclusive OR to avoid any rewrites of a NAT. This answer helps to solve problem (3) when the client adds reflexive addresses in his SDP.

The client can also ask a STUN server to send the answer to another specified address. If two STUN servers are reachable for the client, this feature helps for identification of the type of NAT – that is, the mapping and filtering functions. To test this, the second STUN server is asked to test the reflexive address as seen by the first STUN server for reachability of the STUN client as well. Another option exists when the STUN server has two IP addresses and a message type to answer on the *Other-Address* as proposed in Ref. [360]. The second IP address gets returned after the binding request to the main address. In such a setup, an analysis of the mapping is shown in Figure 6.15. For other tests, see Ref. [360]. If UDP does not work at all, there is also a STUN TCP version [361], which might be used.

Solutions for the port issues (2) and (3) are possible with one of the following methods:

Port adjustment: SIP can change the port for responses by *rport* [362]. The same is true for RTP with *a=rtcp* [143] in the SDP.

Port reuse: SIP can reuse the initial port for responses [363].

Hole punching: This is a method to test for possible reflexive addresses as we have seen before with the two STUN servers. Hole punching is part of the overall ICE standards (Interactive Connectivity Establishment). We summarize ICE as follows.

PMP: The Port Mapping Protocol [364] interrogates the NAT for the reflexive address and proposes an external port. The client can reserve a reflexive address for a defined time. This protocol was introduced by Apple and is part of all their products since years. Consequently, all NAT products have this service implemented.

PCP: The Port Control Protocol [365] is an expansion of PMP, which has a compatible packet format. PMP can in addition to PCP work with IP6, create reflexive addresses for external devices, and has an improved keep-alive procedure.

SIP outbound: This technique requires that the user agent, the registrars, and the proxies include an extension [366]. The idea of this method is to reuse the path taken for the

REGISTER message for all future requests and responses. So, the proxies take over additional load by the forwarding and maintain a state for each client.

ALG: Application Level Gateways (ALGs) within firewalls or DSL routers understand SIP and act accordingly. Since no standard implementation exists, the effects are not determined.

ICE [367–369] is the overall solution to the NAT problem, which includes STUN and a media relay based on the Traversal Using Relays around NAT (TURN) [370]. The media relay takes all traffic that cannot go directly to the partner and forwards it afterward. It took the IETF 6 years to approve these standards, which shows the complexity. Here, hole probing is used and the results are communicated as candidates *a=candidates* in RTCP messages between the user agents and the STUN servers. We expand on ICE in Section 8.1.1.

6.6 APIs and Scripting

The IETF standardized the XML-based Call Processing Language (CPL) [371, 372] that has no loops and no calls of external programs. CPL allows incoming call handling dependent on given preferences, filters calls, and allows forwarding when busy or not available. The language is usable at proxies and clients and is widely available.

The W3C VoiceXML language specifies interactive voice dialogs and is broadly used in call-center applications – see Section 12.3.2 for details. In addition, W3C has created a Call Control eXtensible Markup Language (CCXML) [373] that augments VoiceXML by basic functions. The main focus of CCXML is the control of multiparty conferencing that includes advanced audio control and can place outgoing calls. Both protocols work within browsers.

The following APIs are worthwhile to be mentioned for SIP:

WebRTC: The API was primarily developed for browsers but can also be used in other environments – see Chapter 8.

SIP Java servlets: The current version 1.1 is also available within HTML5 [374] and is widely used.

JAIN: NIST has publicized years back Java APIs for Integrated Networks (JAIN), which is still maintained in variants [375].

TAPI: This TAPI was introduced by Microsoft 1993 and still exists in Version 3.0 with new variants such as the Media Service Provider (MSP) [376] and support for SIP.

6.7 Security and Safety

Security for SIP can fill a whole book – see Ref. [377]. For an overview on the existing security literature, see Ref. [378]. Here, we approach the security pragmatically and not dive into too many technology details. We highly emphasize on implementation aspects in major deployments.

Former telephone systems in companies operated in separate environments, which means their own cable infrastructure with voice servers. From the voice servers, trunks to the public PSTN existed and only these trunks could be attacked. Rare attacks occurred but typically were phone fraud. With an integration of the IP world server and clients share a common infrastructure, and all client–server traffic use the same cables as any other IP traffic. This reduction of necessary cabling is, on the one hand, a central element for a business case for VoIP but, on the other hand, any integration in a common infrastructure increases risks. For regular users in a corporation, the risk that someone looks at his E-mail or capture his phone call on the wire seems comparable. But fixed telephones are used for emergency purposes, and this means not only for personal needs but also for production facility issues. Sometimes, within a PBX, the whole alarming system for a production facility is programmed, and in the case of an alarm, all required personnel are called automatically. Since people in their daily life see how often computers and computer programs fail, the migration path to VoIP is often combined with extensive security and safety discussions.

The major possible threats are listed in Table 6.2, and we discuss the possible options for each threat.

The measures against call hijacking are authentication of the signaling and enhancing the identification of the identity. SIP authentication can use the HTTP authentication [379], which is based on a shared secret. While it is broadly available, security is known to be weaker than methods based upon certificates. SIP can use the Transport Layer Security (TLS) [380], which is in essence Secure Socket Layer (SSL) on the transport layer. Servers typically have certificates of a well-known certification authority, but clients very rarely have certificates. In companies, this would require a Public Key Infrastructure (PKI), which is very rare. Consequently, TLS will mostly identify only the server. As with SSL, the connection is henceforth encrypted with a computed key. Note that the used SIP-URI is in this case *sips: · · ·* [381].

To prevent registration hijacking, we need to protect the integrity of the registration. The options are again HTTP authentication or TLS. The same applies to the next two categories "Impersonate" and "Eavesdropping on signaling."

For protecting the media transfer, we have to secure RTP. The Secure Real-Time Protocol (SRTP) [121], which also covers the Secure Real-Time Control Protocol (SRTCP) provides this protection. SRTP uses symmetric keys that can be configured or need a secure way of creation. For the secure key creation, Zimmermann's method (ZRTP) [382] or the use of the

Table 6.2 Major SIP security and safety threats

Threat	Description
Call hijacking	User establishes session with other user but has not used signaling before
Registration hijacking	Incoming call to a user is diverted to a third party
Impersonate	Third party takes over call
Eavesdropping on signaling	Third party tracks who communicates with whom
Eavesdropping of media	Third party tracks and records sessions
Denial of service	Calls to and from a user are prevented
Tear down session	Calls to/from a user are ended by someone else
Viruses	Infection of voice servers
Unstable software–hardware	Software/hardware on the phones or on the server do not reach the availability goals of 99.999%.

Datagram Transport Layer Security (DTLS) [383] exists. DTLS is a UDP-variant of TLS. An additional RFC [384] specifies the key agreement on top of DTLS – the so-called DTLS-SRTP. The communication is established directly with the UA without proxies on the path since the hop-to-hop signaling path requires too much overhead. Certificates are sent within SDP *a=fingerprint*.

Proper traffic management for IP, SIP, and RTP can hinder possible Denial of Service (DoS) attacks. QoS methods such as Diffserv or Intserv – see Section 4.5 – partly implement the traffic management. Also, filtering the traffic in certain segments – that is, only allowing the "right" protocols per segment – reduces the risk of a DoS attack. We see in Figure 6.16 how filtering can, in addition, reduce other previously mentioned risks.

As stated before, the stability of the VoIP servers needs to achieve an availability close to the famous 99.999%. Viruses and any instability need to be avoided. Non-VoIP switches are very reliable systems using a number of principles that need to be reused for VoIP systems as well:

- Run only the Voice switch software on the system. Only use recommended additional security software.
- Apply updates – inclusive system, virus software, and data updates – only if tested thoroughly by the provider. With open-source software, one will hardly ever reach the availability goal.
- Use highly redundant hardware (CPU, Disks).

The problematic part often is software for additional functionality – for example, call centers or integration into applications. The combination of the add-on software and the primary software requires extensive testing before integration. In addition, all VoIP servers need to be hardened carefully in contrast to a non-VoIP system – for example, Ref. [385] – since the systems are exposed to network-based attacks.

For communication over the Internet, the described security means may be the right way forward. However, for VoIP systems within companies, the operation teams do not like encrypted traffic to each phone since the monitoring, and the analysis in case of problems is severely restricted. Companies should be secured well enough with firewalls to the Internet, and safeguarding each voice call seems to be an overkill. So, the main issue is by some method to separate data and voice as it was once the case with the non-VoIP systems. A typical approach within a company is to check the requirements and then build very few options for the security setup. The different categories have associated technical implementation guidelines with more or less separation from the data network. A typical outcome could be the following categories:

Low security: No separation between the data and the voice network
Medium security: Logical separation of voice and data as far as possible
High security: Physical separation of voice and data.

Figure 6.16 shows an example for medium security. The separation on the LAN for telephones and computers is achieved by VLANs (Section 4.5.4). The media gateways with connections to the PSTN sit on separate networks as the voice switches – media gateways are treated in Section 7.3. The only unrestricted traffic goes from the voice switch to the other two LANs. The firewall in case of dark arcs (**1**) restricts the traffic. When applications require integration into telephony, additional interconnections from the firewall need to be built. VLANs

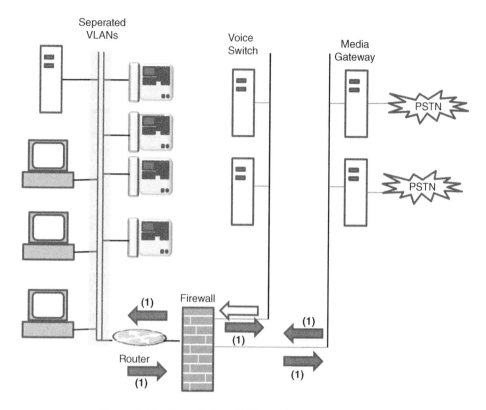

Figure 6.16 Example for a VoIP security zone in companies

offer the possibility to give telephony a higher priority on the network. On company WAN links, a higher priority is necessary, but this is not a requirement in normal situations on LANs. Also, service providers may consider such a medium security setup for deploying SIP to their customers in combination with priority adaptations.

The logical separation increases the costs for the deployment, but the increase is not as dramatical as with a physical separation. Each separation also creates usage restrictions. So, softphones will not work in a logically or physically separated network.

6.8 Planning a VoIP Company Telephony System

Each company telephone system needs careful planning since the systems with all their phone equipment are expensive investments that should last for many years. In the past, 12–15-year-old PBX systems were the rule and not the exception. This longevity contrasts highly with the comparably quickly changing client and server infrastructure within companies.

There are numerous elements required for the planning. We treat three elements in separate subsections: dial plan (Section 6.8.1), emergency (Section 6.8.2), and network planning (Section 6.8.3). Other planning elements are handled directly as follows.

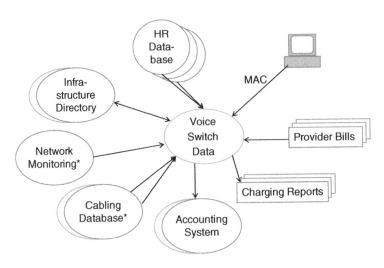

Figure 6.17   Possible data flows for a VoIP system in a company

Each telephone system needs a user directory to manage the association of phone numbers to users. In addition, user phone-call charges are billed, and the system has to cover the cost center information as well. Within companies, many directories already exist, such as Human Resources (HR) employee databases. Within the computer infrastructure, a global Active Directory (AD) may exist. The optimal solution would be the use of AD, but the question is whether an integration in the telephony directory is possible. In the past, telephone directories were synchronized per country with the HR databases. VoIP telephone systems allow a centralized installation for a whole continent and a single synchronization with HR in the region would be possible. However, global and regional HR systems are still rare within international companies, and numerous individual synchronizations may be the reality.

Linked to the directory setup is the billing planning. Historically, international telephony charges were quite expensive, and in many companies, admission systems for users were established. Also, management expected to see detailed overviews on calls per employee to control telephony costs. When introducing a new telephone system, it would be a good point in time to get rid of most of this overhead, taking into account the global lower charges. Today, overseeing mobile telephony costs is more critical for managers. In addition, companies are also billed by providers. These bills need cross-checking with the call records. One needs to decide on tools for the cross-checking in the planning process. Figure 6.17 gives a summary of the possible data flows for a telephone system. Note that items with * are treated in Section 6.8.2.

Also, the directory needs to support Unified Communication (UC) solutions. Voice mails may be communicated to the corresponding E-mail addresses, or at least an E-mail informs that a voice mail is waiting. Of course, the announcement of waiting voice mails can be communicated by SMS as well. Even E-mails might be checked with a phone call, and this is very attractive for sales personnel on the road. Mobile phones are part of a user communication environment, and the directory has to maintain this. Finally, the overall system might include a presence solution. This all shows the tight integration into the office infrastructure. If different manufacturers deliver the functionality, gateways are part of the infrastructure that could be a challenge to operate.

We already touched on the topic softphones in the section on security. Softphones are also a theme when options for the home user support need a decision. With VoIP telephony, it is straightforward that a remote user can use "his" phone in the same way as in the company, also from any other place when securely connected into the company network. So, the security department needs to get involved early. Remote phone users can also be part of a call-center infrastructure and may even work from outsourcer's offices. So, a supporting VPN solution should be able to cross firewalls.

Each telephone system needs a helpdesk and a support organization. In the past, the Move, Add, and Changes (MAC) in the telephony was an essential part of each move of an employee within a company. This support was typically different from the desktop customer support. However, with VoIP, all this can be integrated. So, the VoIP technical integration often leads to a support organization integration.

Each plan needs to include running cost optimization right from the beginning. We need to address mobile calls and international call routing as the two most important elements. Providers charge more for calls from fixed telephones to mobile than for mobile-to-mobile calls. As a consequence, mobile gateways are an option that include SIM cards from one or multiple mobile providers. The switch needs to route calls to mobile phones via those gateways. However, by negotiations with providers, one may be able to avoid this. If the planned system covers multiple countries, this needs to be addressed per country. International calls within a company are typically routed over the company WAN with VoIP. If a voice switch serves a whole region, even call signaling goes over the WAN. Also, international calls outside of the company can be routed through the own infrastructure to another country leaving at a country gateway. Of course, the caller then has a caller id from the company switchboard in the country and reverse calls could arrive there.

Business cases for VoIP often include the cost savings due to lower international calling costs. Other saving areas are cabling costs in case of new buildings and manpower reductions because of organizational alignment with infrastructure organizations. The international calling costs may be an argument for mid-sized and small companies. However, large international companies can typically negotiate good prices with their providers that the costs of international calls are lower than the additionally required bandwidth on the WAN.

6.8.1 Dial Plan

With SIP, phone numbers are not required any longer. However, even if phones have a good user interface for searching URIs, users want to use phone numbers [295]. People have learned to remember phone numbers, and they can use the number with different prefixes from any phone in the world. Within a company, phone numbers should follow a scheme similar to the public telecommunication numbering plan. So, the dial plan has a prefix for intrasite calls followed by a site code. Within a location, further differentiation is questionable. Differentiation based on organizational elements is a bad idea since no organization is stable. Sometimes, personnel require short numbers within the group or the department. In this case, a prefix for calls outside the department is required. This functionality can be programmed at the switch or the involved phones.

It is worthwhile to think about prefixes for voice mail, directory assistance, service codes, outside lines, and international outside calls. However, there are usually local preferences on those prefixes, and global and regional rules may not be acceptable. In the United States, 9 is typically used for an outside line, but in Germany, 0 is used. The same applies for the switchboards in different countries. Different providers and local authorities may enforce local emergency numbers. Even if the numbers are not enforced, users in a country know the usual emergency numbers, and using them in a unique way also helps internally. It is obvious that this needs clarification with all involved countries, and in the end, a site owns a set of local numbers that do not interfere with prefixes.

6.8.2 Emergency

Regulations in most countries require to deliver location information for phones used in an emergency call – for example, in the United States, the (enhanced) E911 obligation [5]. In old telephone systems, each fixed phone was cabled individually, and during the MAC process, the new location of the phone was entered in the voice switch database. In case of an emergency call, the switch could transfer this information to the alarming center. Alarming centers require the use of special phone trunks or a defined trunk out of many connected trunks of a company.

VoIP phones and computers are registering at the location where the device is currently located by a wired network, by a WLAN, or by another mobile network. For finding the location of a VoIP client, the following options exist – partly contained in Ref. [386]:

User provided data: During registration of the phone, the user also needs to enter the location information.

Cell information: The providers of mobile telephone networks know the cell where a mobile phone resides. The cell information can at best deliver a location with an accuracy of 50 m. If GPS works reliably at the location, better information is available. Both of those location methods work badly in a building.

WLAN: To determine actual information with WLANs, three Access Points (AP) need to be in sight in all locations of a site. The installation of so many APs is quite expensive. With less WLAN APs in sight, accuracy is worse than 100 m.

Switch/Room location: This option works with wired networks. The voice switch may, with DHCP, check the MAC address for the IP address of the client. Switches support the Link Layer Discovery Protocol Media Endpoint Discovery (LLDP-MED) as specified by IEEE in 2005 [387]. Similar provider specific protocols exist – for example, Cisco Discovery Protocol (CDP) [386]. Figure 6.18 shows an LLDP packet with EtherType 88CC representing the MED. The payload is organized in Type-Length-Value (TLV) format. There are four mandatory TLVs: Chassis (type=*1*), Port ID (type=*2*), Time to Live (type=*3*), and End (type=*0*). IEEE did not fix the type *127*. This type has been used afterward by the Telecommunications Industry Association (TIA) within the American National Standard Institute (ANSI) [388] to define additional location fields such as the US Emergency Location Identification Numbers (ELIN).

With LLDP or the provider-specific protocol, a switch regularly communicates with, for all attached devices, the MAC Service Access Points (MSAP) and the used MAC address. An MSAP is a chassis identifier plus a port identifier. This data is stored by an LLDP agent in a defined Management Information Base (MIB) of SNMP (Simple Network Management Protocol). A proxy can read the data from this source. In this way, the switch and the port used by a VoIP client are identified. If only the switch location is maintained in a central list, the building and the floor are normally clear. If a complete cabling database with cable endpoints per port on each switch exists, the room with the VoIP client gets derived. Since complete cabling database in companies is an exception from the rule, this complete information is seldom available.

In SIP, the emergency location communication uses a Uniform Resource Name (URN) with the top-level name *sos* [389, 390]. No RFC exists, which specifies the exact way to receive the location information via LLDP. The framework for emergency calling [391] covers LLDP but does not explain the use in detail. The UA initiates the emergency call with the *sos*-URN and sends it to his proxy. The proxy recognizes the emergency and directs the call to the Public Safety Answering Points (PSAP) [391], which is a call center for emergency calls. When the call arrives at the PSAP, location information has to accompany the call. This information can be a presence location object – Presence Information Data Format Location Object (PIDF-LO) as specified in Ref. [392]. The location information can come from the UA or the proxy.

6.8.3 VoIP Network Planning

Before designing the network in detail, the location of the VoIP servers needs a decision. VoIP servers can be up to 100 milliseconds round-trip delay away from the client devices. When looking at Section 5.6, we see that only a single location with servers for Europe or North America is needed. For high resilience, a cluster of VoIP servers is possible, but the intercluster delay must be below 50 milliseconds.

A central installation of VoIP servers does not deliver sufficient availability for remote sites. Even with redundant access lines to a site, an availability of 99.999% is hardly achievable. In such cases, small surveillance systems need to be installed per site. These surveillance systems can also route voice traffic to the national provider – see also Section 7.3. Routers can integrate surveillance functions, but these fall-back systems offer much less functionality than the central system. With such a surveillance system, one can often avoid the redundant site access.

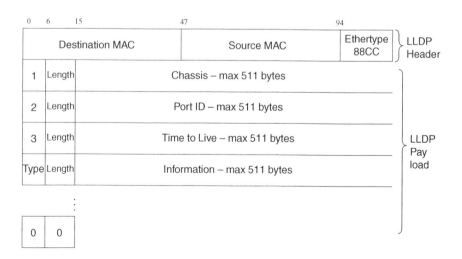

Figure 6.18   LLPD packet

When the overall system includes unified communication elements, the localization of the involved other servers too plays a crucial role. It is quite favorable if all other servers – such as Fax, Voice mail, E-mail, and applications – are on the same central site. However, centralizing all E-mail servers in a region creates a measurable delay for E-mail client synchronization at remote sites. Especially in the morning when many E-mail client synchronizations run parallel, the WAN to the center can get overloaded, and the users are impacted. So, a centralization possibly creates a whole avalanche of other required changes.

We touched earlier on user expectations. Voice is an application with high end-user visibility, and all planning needs significant end-user involvement. Due to the integration of voice into the network, the former voice operation teams sometimes may not be so much involved in the rollout planning. However, these voice personnel have the insight in the end-user perspective and need to be part of the preparation and the rollout.

In Section 3.1.3, we introduced numerous audio codecs, which were summarized in Table 3.3. One VoIP infrastructure typically uses a common codec. It is not advisable to transcode within the same installation due to quality losses.

The setup of phones requires defining the frame sizes for audio. The chosen codec determines the minimal frame rate. The bigger the frame size, the less the protocol overhead plays a role. However, with big frames, the delay for delivery potentially gets too large. So, one needs to choose a frame size that still fulfills the maximum allowed delay for the longest path.

When choosing a codec, some of the characteristics are important. Vocoders such as G.723.1 or G.729 are designed to optimally compress human voice but cannot work successfully on Fax, music, and DTMF. For many codecs, silence compression or Voice Activity Detection (VAD) can be used, which saves typically 40%. However, Fax cannot be sent when this option is active.

Example 6.9 We need to build a VoIP system for a company with headquarters in Zürich and sites in Geneva, Paris, and Moscow. The one-way delay between Zürich and Paris is 20 milliseconds and between Zürich and Moscow is 45 milliseconds. All VoIP servers will be installed in the headquarters. The system has an integrated Voice-mail system.

Of course, we would like to use the most efficient compression such as G.723.1 or G.729. However, DTMF is required in the user interface for a voice-mail system. We mentioned in Section 6.1.3 that the *INFO* signaling method can be used to transmit DTMF tones since the Vocoders cannot transport them. However, we can expect that Faxes are still sent over this line, and we have to think of Music-on-Hold (MOH). While holding a call, users might wish to hear music. If the music comes from the central voice switches, the codec has to support this too.

As a consequence, we may have to use the G.726 codec. The longest path on our VoIP system is between Paris and Moscow and takes 65 milliseconds. We should at most have 150 milliseconds delay overall, and we may choose a frame rate of 50 milliseconds with still some reserve for an additional delay. With 50 milliseconds delay, we will have 20 packets per second and 200 bytes VoIP payload per packet when choosing the 32 kbit/second option. For the calculation of the overall packet size, we need to add 40 bytes for IP, UDP, and RTP. For the transport over a serial link, HDLC (High-Level Data Link Control) may take another 6 bytes.

When we would switch silence compression on, the payload would shrink in average to 120 bytes. The overall packet size would then be 166 bytes versus 246 bytes. However, Fax would not work in this setup.

Over a WAN, the voice traffic needs prioritization to guarantee high quality. This is normally done with Diffserv – see Section 4.5.3 and also Table 4.7. In Table 4.7, the recommendation already was to have less priority for voice signaling in comparison to the very high priority for voice media. Regular phones sometimes have long call setup times, and end users also do accept non-instant VoIP setups. On the LAN, no priority for voice is required for a well-managed network.

In the next step, we have to calculate the required capacity on the WAN and the trunks to the PSTN. Trunks to the PSTN would be a T.x in the United States and an E.x in Europe. An E.1 connection has 2 Mbit with 32 channels, and two channels are used solely for signaling and synchronization. For each WAN, an analysis is required on how much capacity can be taken away for prioritized traffic. If the reserve is not sufficient, additional capacity needs to be installed.

For telephony, standard calculations for the traffic exist, and we will demonstrate a simple one named Erlang-B. Calculations for the number of channels c is based on Erlangs – named after a Danish mathematician. One Erlang a is a 1 hour traffic on the same channel. B denotes the probability that the system blocks a call due to insufficient bandwidth. We need the maximum simultaneous calls between sites and outside in busy hours – the so-called Busy Hour Traffic (BHT). The traffic during the busiest hour of the day is typically 20% of the overall daily traffic. The second required value is the Average Holding Time (AHT) – that is, the time a call takes in average during the day. Typical in business are 180–210 seconds for AHT. BHT and AHT should best be measured in the individual environment.

When we assume that AHT is distributed exponentially, traffic arrival is random, the number of callers is infinite, blocked calls get dropped and are not coming back, then the following Erlang-B formula holds:

$$B(c,a) = \frac{\frac{a^c}{c!}}{\sum_{k=0}^{c} \frac{a^k}{k!}} \tag{6.1}$$

When specifying the probability B that a call is blocked due to an insufficient bandwidth, we can derive the number of required channels. Before we elaborated that the capacity per channel depends on the codec and the packetization. Note that the packetization defines the overhead for transporting the payload. There are calculators available, which do all those calculations as exemplified in Example 6.10.

Example 6.10 We expand Example 6.9. The network drawing in Figure 6.19 contains more information, and additionally, we have:

- Twenty calls per user leave each site per working day with an AHT of 180 seconds. 50% of those calls go to the headquarters in Zürich, and the others go to partners outside in the corresponding country.
- The probability that a call is blocked has been agreed to 0.05%.

For each user, we then have 10 calls for 180 seconds to the headquarters. Using the guess that 20% of the traffic occurs in the busiest hour, we get 360 second per user and a BHT per user in the busiest hour as $\frac{360}{3600}$. Using the number of users per site, we have the links Geneva–Zürich and Moscow–Zürich with a BHT of 15 and the link Paris–Zürich with a

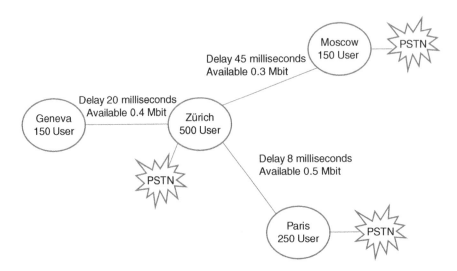

Figure 6.19   Example network map

BHT of 25. We insert this in a calculator[3] using the encoding and the packetization from Example 6.9. As a result, we get 20 channels or 788 kbit on the connections Geneva–Zürich and Moscow–Zürich and 31 channels and 1191 kbit on the link to Paris.

All network links need an upgrade to cope with the required bandwidth for Voice. For the PSTN connections in Paris, we need two E.1 links and in Moscow, one E.1. The Geneva users might use the PSTN connection in Zürich since the users are in the same country. However, an expanded reservation for the additional traffic of 788 kbit on the WAN link between Geneva and Zürich would be necessary. Probably, another PSTN link in Geneva is superior.

This calculation covered neither MOH nor any Fax traffic for the remote sites; nor did it include outside calls routed over Paris and Moscow. In this example, we only looked at Voice mail, but E-mail integration and application integration may change the required capacity again.

---

[3] http://www.erlang.com/calculator/eipb/, Last accessed 16 April 2015.

7

Other Standard VoIP Protocols

This chapter is mainly dedicated to standard ITU IP telephony protocols and mobile telephony SIP protocols.

Section 7.1 addresses the ITU VoIP standards within the H.323 family. At the end of the section, we compare the standard with SIP as presented in Chapter 6. Section 7.3 treats media gateways based upon H.248, which we already touched in Chapter 6. Those gateways base on an ITU specification that initially was codeveloped by the IETF. In Section 7.2, we treat multimedia ITU protocols that are not as well known as the two ITU families mentioned before. Mobile telephony, so far, has not used IP protocols, but this will change soon when the world moves to Voice over LTE (VoLTE) based upon SIP and IP Multimedia System (IMS). Section 7.4 addresses IMS and VoLTE. The last section looks at the most deployed VoIP protocol, Skype, which is, however, the least known protocol of all presented.

For the discussion in Sections 7.1 and 7.3, the book by Collins and Swale [293] is a well readable recommendation for diving deeper into the subject. For the discussion in Section 7.4, the books [393, 394] can be consulted.

In Chapter 4, a multimedia network stack is shown in Figure 4.1. With SIP+, we cover all SIP-based protocols from Chapter 6 and Section 7.4. As an alternative to "SIP+," the figure lists "H.323/H.248," which we treat in Section 7.3. Skype is not listed in the overview because there is no associated standard organization.

In addition, there exist other provider-specific VoIP protocols. The most known one is the Skinny Client Control Protocol (SCCP) from Cisco, which can coexist with H.323 environments [395]. We will not treat this further since we concentrate on standard protocols.

7.1 H.323 VoIP Family

H.323 [396] is an umbrella document that specifies a number of underlying standards. The first version was publicized in 1996 and initially foreseen for videoconferencing in a local area network environment. In 1998, a revised version became available, which did not focus solely on videoconferencing on LANs. This was one year before the approval of SIP.

In mid-1990s, it was still open whether the ISO protocols or the IP protocols would survive. All the historical ISDN (Integrated Services Digital Network) telephony protocols

Multimedia Networks: Protocols, Design, and Applications, First Edition. Hans W. Barz and Gregory A. Bassett.
© 2016 John Wiley & Sons, Ltd. Published 2016 by John Wiley & Sons, Ltd.

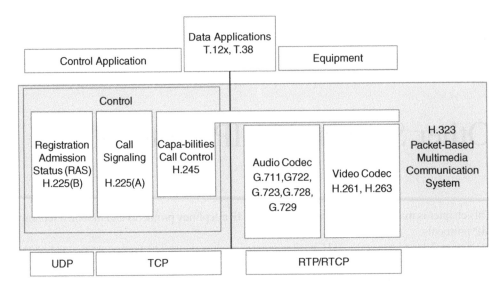

Figure 7.1   H.323 overview

used the ISO terminology. Since the H.323 protocol set incorporates many ISDN standards, the terminology for major parts is still ISO terminology, and the original protocol was only adapted to IP.

When using the H.323 family, today, there is always an ISO-to-IP adaptation layer installed [397, 398], named TPKT (TransPort PacKet), as a correspondence to the ISO Transport Protocol Data Unit (TPDU). Note that, as part of the ISO terminology, a Transport Service Access Point (TSAP) is a socket in IP terminology. Service Access Points (SAP) – that is, addresses – can occur in other layers as well, and then the SAP is preceded by additional letters.

Figure 7.1 provides an overview of the elements of H.323. The mentioned codecs apart from G.728 were already introduced earlier in Chapter 3.

The standard does not separate in H.225 (A) and (B), but we use it here to differentiate the functionality. H.323 was the first protocol to use RTP and RTCP. The protocols H.225 and H.245 are treated in the next two sections, while we handle the data applications T.12x in Section 7.2. The protocol T.38 [399] is mentioned as a standard in the figure because a voice switch is also responsible to transport Fax. T.38 is a fax relay standard, which transports faxes over IP between voice switches.

The standard uses the word "Terminal" for a UA in SIP. H.323 has a central control point for a zone, which is the "Gatekeeper" but often the names "Softswitch" and "Soft-PBX" are used. In the past, PBXs had dedicated hardware, so the term "Soft" exemplifies that the system can run on any hardware. The standard also mentions gateways, which are MCUs due to the foremost use in videoconferencing.

From Figure 7.1, the overall characteristics of the mentioned protocols should be obvious:

H.225: This protocol consists of the following two functional parts:

  A: The protocol initializes and terminate calls over TCP. It does more than the SIP signaling.

B: The protocol cares for the registration of terminals, controls the admission of terminals, and maintains the status of the terminals (RAS) on top of UDP.

H.245: This protocol controls calls during transmission and negotiates the capabilities. It selects codecs and issues start and end commands for the media. It runs on top of TCP.

Media: The possible audio and video codecs are explicitly stated in the standard. RTP and RTCP are responsible for the transport.

Addresses are URLs with IP addresses or DNS names. The system also supports alias names, which the gatekeeper translates. Alias names are useful for short code dialing.

In contrast to SIP, the functionality is not as visible in message headers or SDPs specifications, and we cannot show examples in the form of message headers and SDPs. The functionality is hidden in the Abstract Syntax Notation 1 (ASN.1). ASN.1 was created originally in the 1980s and had the standard number X.409 but has been re-created a few years back as X.680 – X.683 [400–403]. Figure 7.2 gives an idea of how ASN.1 works based on a rather simplified example based on Ref. [400].

Figure 7.2(A) shows a simple personnel record. This record is in line with the syntax provided in (B). The record (B) is an example of an ASN.1 syntax, and, of course, all types of records can be described in a syntax. Combined with ASN.1 are encoding rules that transfer instances of a syntax – for example, (A) – in packets. The most known one is the Basic Encoding Rules (BER) [404]. Other encoding rules are contained in the standards X.691–X.696. For each syntactical element, the associated encoder generates a unique label. Such a label is visible in the first field from (C) where the hexadecimal value *60* specifies the *Personnel Record* from (B). The elements are built as Type, Length, and Value. So, the first name "John" gets $\boxed{16}\boxed{04}\boxed{John}$. In the data packets, we find exactly the boxed parts from (C) starting from $\boxed{60}$ and ending with $\boxed{Smith}$.

ASN.1 has associated generators for programming languages that translate a syntax in a program data structure. When sending or receiving packets, the packet data gets filled in the generated data structure. As a consequence, ASN.1 is an effective tool for programmers and creates rather compact packets. However, the readability of the packets requires a corresponding ASN.1 syntax plus a converter. All packets and messages, of course, are defined in ASN.1 in the H.323 standard set.

In addition to the already mentioned elements of H.323 and the codecs, a few other standards are needed for a complete H.323 system:

- Q.931 [405] is the basic underlying ISDN control specification
- H.235 for the security and authentication
- H.450.1–H.450.12 include supplementary services such as "Call Hold," "Call Offer," and "Call Intrusion." All these additional standards share one common generic signaling protocol defined in H.450.1 [406].

7.1.1 H.225

This protocol [407] uses as a basis the ISDN Q.931 specification for basic call control and adds differences or adaptations.

Q.931 H.225 messages contain Information Elements (IE) and the most important ones are *Call Reference, Call State, Called and Calling Number, Cause*, and *User-to-User*. A *Call*

(A): Example Record

Name:	**John P Smith**
Title:	**Director**
Employee Number:	**51**
Name of Spouse:	**Mary T. Smith**

(B): ASN.1 Syntax

PersonnelRecord :: = [APPLICATION 0] IMPLICIT SET {
     Name,
     title [0] IA5String,
     EmployeeNumber,
     nameOfSpouse [2] Name,
Name :: = [APPLICATION1] IMPLICIT SEQUENCE{
     givenName IA5String,
     initial IA5String,
     familiyName IA5String}
EmployeeNumber :: = [APPLICATION 2] IMPLICIT INTEGER

(C): ASN.1 Encoded Example Record

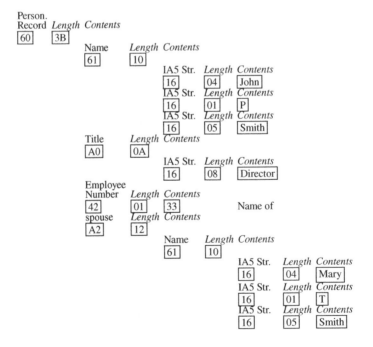

Figure 7.2   ASN.1 example

Reference is a unique value for the identification of a call and is chosen by the originator. The *Call State* represents the current state of a connection and can have 16 different values. The most important values of the *Call State* are *Call Initiated (1)*, *Call Delivered (4)*, *Active (A)*, and *Connect Request (8)*. The *Called Number* and *Calling Number* can be derived from standards such as an E.164 number or a Telex (F.69) or a private one. Also, a *Called Number* can be an alias name that is created dynamically and marked specifically. These alias names have a special marking. The *Cause* value is the return code provided for an earlier request. The *User-to-User* (USR) field was in the original Q.931, just a field to provide the user with any additional information. H.323 has reused this field to send information that the original protocol had not covered. One usage of the field is to send IP addresses – of course, they are encoded in ASN.1.

For the initialization and termination, the following main commands are used, which act on port 1720:

Setup: The Terminal issues a *Setup* for initialization of call. This message includes at least a call reference, the bearer capability, protocol discriminator, and USR field. The protocol discriminator originates from Q.931 and is always a code for the type of the message. Here, it is *00000101*.

Setup ongoing: A number of different optional messages can indicate an ongoing setup such as *Call Proceeding, Progress* sent by gatekeepers, *Alerting* when it is ringing or *Setup Acknowledge* from PSTN gateways.

Connect: The acceptance message *Connect* from the called party indicates that the call is now created.

End of call: The *Release Complete* shows the end of the call; a cause specifies the reason for the termination.

Status: The partner is asked for its status of the call by a *Status Inquiry*, which is answered by a *Status* message. *Notify* delivers information for the presentation to the user.

The other part of H.225 is the Registration, Admission, and Status Signaling and acts on port 1719. The most important commands are:

Association with a gatekeeper: The terminal has the first contact with a gatekeeper by sending a *Gatekeeper Request* (GRQ). This request can be answered positively by a *Gatekeeper Confirm* (GCF) or negatively by a *Gatekeeper Reject* (GRJ). The addresses of gatekeepers are either preconfigured or the central system can be reached by a multicast to 224.0.1.41. The GRQ may contain a nonzero gatekeeper identifier. Some other included fields with the gatekeeper are a sequence number, the type of the terminal, and alias names for this terminal.

Registering with a gatekeeper: After a terminal is associated with a gatekeeper, registration starts. The terminal sends a *Registration Request* (RRQ) to the gatekeeper. The gatekeeper allows registration by *Registration Confirm* (RCF) and disallows by *Registration Reject* (RRJ). A registration can contain a lifetime, addresses for signaling and transport, the type of the terminal plus vendor information, and alias names.

Unregistering with a gatekeeper: The commands are *Unregistration Request* (URQ), *Unregistration Confirm* (UCF), and *Unregistration Reject* (URJ).

Getting admission for a call: The terminal sends an *Admission Request* (ARQ) and specifies additional elements. These elements can be the type of call (multiparty, two-party),

party numbers with optionally alias numbers, bandwidth requirement and callModel. The callModel specifies if the terminals can signal directly with each other or with a gatekeeper in the middle. The gatekeeper answers with either *Admission Confirm* or *Admission Reject* (ACF, ARJ). Note that a gatekeeper can already issue with the registration the admission for a later call together, and in this case, this step is not required.

Figure 7.3 shows Terminal A, which has already associated with the shown gatekeeper. Next, it requests the admission to establish a call to Terminal B. After the confirmation message arrives, Terminal A does a *Setup* via the gatekeeper, which is forwarded to Terminal B. Terminal A did not get the permission to talk directly to other terminals during its admission with *ARQ*. So, the signaling has to pass via the gatekeeper. After Terminal B got the *Setup*, it also needs to get the admission to perform the call from the gatekeeper. After the confirmation message arrives, Terminal B can confirm the connection by a *Connect* via the gatekeeper to Terminal A.

We need to add a few more to the RAS functions of H.225:

Bandwidth: During an established call, any terminal can ask to change the associated bandwidth. Reductions do not need to get communicated, but a client can change it on its own. The *Bandwidth Request* (BRQ) can be answered by a *Bandwidth Confirm* (BCF) or a *Bandwidth Reject* (BRJ) message. Such a change requires additional actions in H.245. Note that a gatekeeper can also send a BRQ to a terminal, but, in this case, the only allowed answer is the BCF message.

Location: A terminal or a gatekeeper may request the resolution of an alias name to the real address by issuing a *Location Request* (LRQ). This is answered by a *Location Confirm* (LCF) or a *Location Reject* (LRJ) message.

Disengage: The terminals conclude a call by *Release Complete* to each other. After this communication, the gatekeeper as well gets informed by a *Disengage Request* (DRQ). The DRQ is usually answered by a *Disengage Confirm* (DCF) or abnormally by *Disengage Reject* (DRJ).

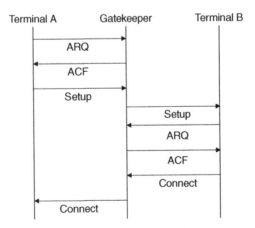

Figure 7.3 Registering with gatekeeper and call setup

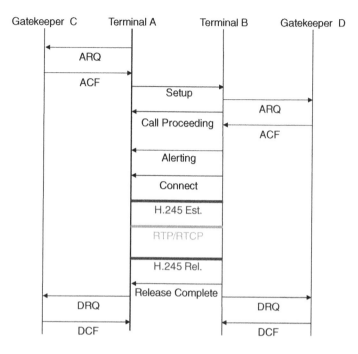

Figure 7.4   Registering with two gatekeepers, call setup and release

Figure 7.4 integrates the additional commands. In this example, the two terminals are permitted to establish the connection by themselves. After the *Connect*, the establishment of H.245 begins and soon the RTP session transmits. After the end of H.245, Terminal B issues a *Release Complete*, which is followed by the *Disengage* messages to the corresponding gatekeepers.

7.1.2   H.245

While many of the concepts in the previous sections have already been associated with SIP, H.245 [52] has some unique features. One of those features is the creation of a master – slave relationship for multiparty conferences. If there are many parties in a conference call, it is difficult to agree with each other on the capabilities of a call. Specifically, when in all paths there is an MCU, which transcodes video streams, the capabilities of the MCU must trigger the behavior of the other devices in the call. Also, there could be multiple MCUs in a conference call that even complicate the setup.

Each terminal has a terminal-type value that should increase with its capabilities. H.323 proposes values for terminals, but this is a configuration choice. The system in a conference with the greatest terminal-type value will be the master. If there is no sole device with the highest number, the system randomly chooses between the endpoints with the greatest numbers. This process has the name Master – Slave Determination (MSD). After two endpoints have connected via H.323, each endpoint sends a H.245 *MSD-Request* to each other consisting of its terminal-type value and a random number. The endpoint with the higher number afterward sends an *MSD Ack-Master* message and the other endpoint an *MSD Ack-Slave* Message.

H.245 negotiates the capabilities between terminals and uses the Terminal Capabilities Set (TCS) messages for this purpose. A *TCS-Request* communicates the initial set of characteristics. The capabilities not only mention what is available but also which combinations are possible – for example, video and audio combinations. The possible capability values fill more than 20 pages in the H.245 standard. Possible capabilities are confirmed with a *TCS-Ack* and *TCS-Reject*.

After the agreement of the capabilities, the master sends Open Logical Channels (OLC) commands to the terminals. There exist unidirectional and bidirectional channels. Here, we show the OLC with a unidirectional channel and as a parameter the Multimedia System Control:

```
request : OpenLogicalChannel:
   {forwardLogicalChannelNumber   35,
    forwardLogicalChannelParameters
       {dataType audioData : g711Alaw64k : 20,
          multiplexParameters  h2250LogicalChannelParameters:
             {sessionID 1,
                mediaControlChannel unicastAddress : iPAddress :
                   {network 'AB136C3A' H, tsapIdentifier 4001}
                silenceSuppresion TRUE
                }
             }
          }
```

Most of the elements are self-explanatory apart from the IP Address. The message *OLC-Request* asks for a connection to the socket "171.19.108.58," Port 4001 – written in Hex "AB.13.6C.3A." The recipient answers with *OLC-Ack* message where most parts of the parameters are the same apart from its socket. Then, a final *OLC-Confirm* is sent from the originator of the *OLC-Request*. The Close Logical Channel (CLC) message closes the channels.

Channels can be opened and closed during an established connection. Call transfers use this functionality. For a call transfer, a call is put on hold but the original channel is closed. For Music-on-Hold (MOH), a new temporary connection to the gatekeeper gets established. The system builds the connection to the new partner, closes the MOH channel afterward, and opens the channels for the connection to the new partner.

When H.245 closes the last channel, an H.245 termination gets communicated by an *End Session*.

We can expand Figure 7.4 with the detailed OLC messages from H.245 – see Figure 7.5. However, the complete setup also requires the MSD and TCS messages. For speeding up this whole process, the option of a Fast-Connect exists. So, the H.323 messages *Setup* and *Connect* can already carry H.245 logical channel information. If the terminal does not react to this type of expanded message, the communication falls back to the regular procedure.

H.245 supports sending of DTMF tones by special System Control Messages.

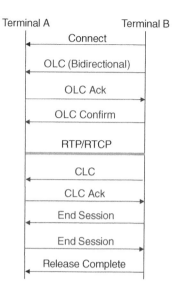

Figure 7.5 Details of the H.245 establishment and release illustrated in Figure 7.4

7.1.3 Comparing SIP and H.323

Here, we look at a number of different criteria that are summarized in Table 7.1.

Today, both standard sets are complete when looking at telephony. However, 15 years back, this statement was not true, and many parts of SIP were missing. So, any larger voice system that needed replacement between 2000 and 2005 was migrated to H.323, since this was the only comprehensive system. Nowadays, migration to SIP is also possible for major systems, but still this is not the case in corporate environments. Solely for videoconferences, SIP still misses detailed functionality for capability agreement and session setup.

From the examples of H.323, apparently, many resource management functions do not exist for SIP. However, this is not true, and we will see examples of additional functionality in Section 7.4.

SIP is an easily expandable system while changes to H.323 require touching many ISDN-related elements and cannot be done smoothly. As a consequence, newer functionalities such as Presence only exist with SIP, and this will probably intensify over the next few years to come.

H.323 was already a complex system when introduced in the 1990s due to the built-in ISDN protocols. In comparison, SIP was simple and straightforward. However, SIP has also become complicated by adding more and more RFCs. So, in Chapter 6, we mentioned over 70 RFCs, and many more will be mentioned in Section 7.4. Consequently, when comparing the amount of written documentation between the two systems, their quantity would be close.

H.323 complexity is partly related to a merging of layers due to reuse of ISDN and mixing of H.245 into H.323. In contrast, SIP has a clean structure and is easier to oversee.

The closeness of H.323 with ISDN always had the advantage that an integration into the PSTN world was easier than with SIP. Also, the migration from old PBXs with ISDN functionality to a new VoIP system was more straightforward with H.323.

Today, every major telephony switch product implements H.323 completely and SIP partly. Any big telephony system in companies uses H.323. End customers outside companies use SIP and the old ISDN. SIP will gain more and more usage, while H.323 will start to shrink soon but will coexist for a long time.

For further comparisons dating back over 10 years, see Refs [408, 409].

7.2   T.120 Data Applications

In the overview of H.323 – see Figure 7.1 – the data applications were already mentioned. T.120 [410] is the umbrella specification for a number of such protocols. Figure 7.6 provides a graphical overview of those seven substandards T.122–T.128 – see also Ref. [411].

The description of the functionality of those standards starts with the ones of the lower layer:

T.123:   This standard describes profiles with characteristics of the transport network to support the T.120 suite of protocols. This is mainly of interest when transported over ISDN.

T.122/T.125:   The Multipoint Communication Service (MCS) [412] is based on domains. A domain is the combination of all channels for a conference. The standard defines actions for associating transport connections with an MCS domain. The MCS channels or providers are created as a tree of channels. Figure 7.7 shows such a join process.

The additional specification T.125 [413] describes the protocol implementing the primitives. Appendix A of this standard contains as a curiosity. The Multicast Adaptation Protocol (MAP) distributes conferences media in a multicast manner. This later part was probably never implemented.

Table 7.1   Summarized comparison of H.323 and SIP

Criteria	SIP	H.323
Completeness	Requires additional RFCs in addition to basics to be used for conferencing and advanced topics	Complete for standard telephony and (video)conferencing
Expandable	Doable in increments without touching basics – reflects IETF approach: Start simple	Only by standard committee working on the overall standard
Complexity	Simple when not considering additions	Complex
Debugging	Straightforward when not considering all additions	Complex
Structuring	Clean and layers are separated	It is not a clean structure. The reuse of former standards created duplicity of functions. Some included functions will never be used
Integrability	Easier into new systems	Easier into existing ISDN telephony
Efficiency	Very similar – see Ref. [408]	

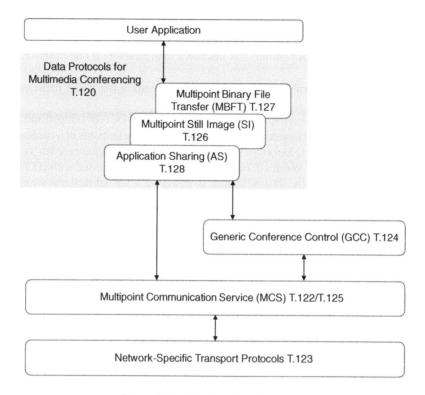

Figure 7.6 T.120 family of standards

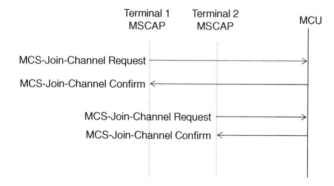

Figure 7.7 Multipoint communication services

T.124: The Generic Conference Control (GCC) is responsible on top of MCS to control the association with a conference. In addition, GCC manages conference resources, supports capability exchange, and maintains a conference-wide roster. GCC allows creation and resolution of conferences. Also, join and leave primitives are part of the specification.

T.126: The "Multipoint Still Image and Annotation" is based on either the JPEG or the JBIG standard to transport pictures for whiteboards. Whiteboards, today, are often implemented with touch screen on big computer displays and a screen sharing application – for example, T.128 – between participating computers. Solutions exist from major providers such as Cisco or small ones such as Oblong. It is unknown to the authors if one of those solutions uses T.126.

T.127: The Multipoint Binary File Transfer (MBFT) sends a file to all participants of a conference.

T.128: This multipoint Application Sharing (AS) protocol [411] was developed and proposed by Microsoft and describes how screens are shared between participants of a conference. It is part of the Microsoft Remote Desktop Protocol (RDP), which is implemented in the products Netmeeting (Microsoft) or WebEx (Cisco). The products were extended with HTTP support to easily pass over firewalls. The definition is based on a virtual desktop with included subwindows. A window is hosted if it is owned by an application executing on the local system. For each hosted window, there exists a shadow window within the virtual desktop on other systems. In addition, local windows are possibly available on a system that is not shared. Even if a local window is not shared, the application sharing conference entity needs to manage it since it may overlap hosted or shadowed windows.

Text and graphics are sent in a vector format and not as bitmaps. During application sharing, the contents of the window increments dynamically – that is, updates are sent only for those portions of the virtual desktop that has changed. The approach of sharing presentations by sharing the desktop is often questioned. One option would be to send the slides in some saved form to all clients and then send commands to move from slide to slide. Another option would be the distribution by a version of compressed video information [414, 415].

These products are used by millions of users to accompany telephone conferences with shared slides or documents. These tools are an essential element for effective global virtual groups [416]. They are operated as outsourced service or within companies.

7.3 Gateway Control

Media gateways were already mentioned in the previous sections. We have seen examples when crossing the border between service providers or between VoIP and PSTN as in Section 6.2. Media gateways are separated from proxies or gatekeepers because they have a dedicated function and an ongoing load during conversations. When the number of users in a VoIP installation increases typically, only gateways get added.

It is worthwhile for gateways to separate in a controller function and the true media gateway. This setup creates flexibility and scalability. The base configuration is shown in Figure 7.8. This figure mentions some of the protocols used. Digital Subscriber Signaling (DSS) is a part of the overall PSTN protocols, which faces the user system. The overall protocol for PSTN networks is SS7 (Signaling System 7). As shown in the figure, the media gateway controller interacts with the media gateway using the Media Gateway Control Protocol (MGCP) or MEGACO (MEdia GAteway COntrol protocol). The media gateway has to convert the internally compressed audio to the external standard.

The MGCP [417, 418] was a protocol developed solely by the IETF, but MEGACO (IETF name) or rather H.248 (ITU name) was an IETF-ITU codevelopment [419–421]. Later, ITU took over H.248, and the IETF protocol plus the name MEGACO was history. MGCP elements are similar to H.248, but H.248 has more features. Consequently, we will treat H.248, and at the end of Section 7.3.1, the differences to MGCP are listed.

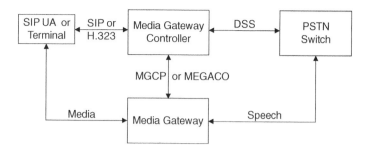

Figure 7.8   Media gateway with controller

7.3.1   H.248

The encoding of messages in this protocol is ASN.1 plus Augmented Backus – Naur Form (ABNF) [422]. ASN.1 was clearly expected by the ITU, but IETF apparently wanted a syntax out of their own space. For better readability, we use ABNF for later examples.

The overall task of this protocol is to control media gateways that interconnect media streams in a reasonable manner. The endpoints of media streams are called terminations. A number of media streams can belong to the same termination – for example, an audio and a voice part. Terminations are combined in a *Context* together by *Add*, *Subtract*, or *Move* commands. The specification of contexts and their terminations are sent as { }-bracketed trans-actions from the Media Gateway Controller (MGC) to the gateway and vice versa. The request message contains a *Transaction* command, which is answered by a *Reply*. Each message between the two systems begins with *MEGACO/3* with IP and port of the sender. The Media Gateway (MG) might inform the controller of an event by *Notify*. The controller adapts prop-erties of termination by *Modify* and retrieves properties by *AuditValue*.

Now we can understand the example illustrated in Figure 7.9. The controller requests to add two terminations to the *Context=111* and the creation of a new context with the termination *A777* initially in it. A newly required context is distinguishable by a "$." The gateway answers by repeating the transaction number in its *Reply* and only adds the number for the new context.

For the specification of the properties, the following main possibilities exist:

- **Media** is a structured field. The *TerminationState* specifies the status of a termination – for example *test, in service* – plus an event behavior. A *TerminationState* is valid for a number of *Streams*. Each *Stream* consists of the direction of sending – that is, *LocalControl* plus specifications for *Local* or *Remote*. *Local* or *Remote* are SDP descriptions.
- **Events** are detailed for which notifications are expected. However, the gateway typically reports all noticed events in any case.
- **Audit** descriptors specify the values that the MG should send in a response.
- **Digitmap** specifies a valid digit combination based on an existing dial plan – examples are *8xxxxxx* for a local number dialed in the United States. The MG captures the numbers based on the dial plan and sends complete numbers to the controller. So, the main function is to avoid sending digit per digit in separate messages.
- **Observed Events** describe events for terminations.
- **Signals** send specific signals for termination – for example, *off, dial tone, busy*.

```
From MGC: MEGACO/3 [212.3.3.10]:30001
        Transaction = 1234 {
                Context= 111 {
                        Add=A555
                        Add=A666
                        }
                Context=$ {
                        Add=A777
                        }
                }
From MG:   MEGACO/3 [212.3.8.15]: 5678
        Reply = 1234 {
                Context= 111 {
                        Add=A555
                        Add=A666
                        }
                Context=222 {
                        Add=A777
                        }
                }
```

Figure 7.9   Simple H.264 commands

- **Topology** specifies for a context how the flow – *isolate, oneway, bothway* – between the terminations shall work by triples *(Termination-1, Termination-2, flow)*. This is useful for transferring calls where one party has been put on hold.

Knowing those properties, we can now understand a more complicated communication as shown in Figure 7.10 – adopted from Ref. [420]. We do not show the confirm messages in this figure. Each statement in the example has the following meaning:

1. The MG informs the MGC that an off-hook event has occurred on an analog line with the termination identification *A4444* at a timestamp in the year 2015.
2. The controller sends the dial plan to the gateway. Note that, currently, there are no terminations in the context, and the context is empty as denoted by "-." The controller tells the gateway to play a dial tone *dt* using the tone generator package *cg* within the *Signals* part.
3. The gateway reports the captured number according to the dial plan with an unambiguous match *Meth=UM* and again an updated time.
4. The controller advises the gateway building up a new context where *A4444* is one part, and the other termination identification will be created by the gateway. Since new identifications are required, a "$" is mentioned for the *Context* and the *Add*. Since it is the only *Stream*, it receives the number 1. Because the other participant parameters are currently not specified, the *Mode* is only receiving. The specification refers to an *nt* network package, which is a summary of parameters and adds a specific jitter value. The SDP contains the information on possible acceptable audio codecs – PCMU or G723 as specified in Ref. [16]. The SDP also contains "$" where the gateway has to select a value. In the answer from the gateway – not shown here – the gateway will add its IP address and port for the connection and selects the preferred codec.

(1) *MG1 to MGC:MEGACO/3 [67.34.23.11]:55555*
> *Transaction = 3000 {*
>> *Context = - {*
>>> *Notify = A4444 {ObservedEvents =2222 {*
>>>> *20150418T17010000:al/of}}*
>>> *}*
>> *}*

(2) *MGC to MG1:MEGACO/3 [67.34.25.171]:55555*
> *Transaction = 3001 {*
>> *Context = - {*
>>> *Modify = A4444 {*
>>>> *Events = 2223 {*
>>>>> *al/on(strict=state), dd/ce {DigitMap=Dialplan0}*
>>>>> *},*
>>>> *Signals {cg/dt},*
>>>> *DigitMap= Dialplan0{*
>>>>> *(0| 00|[1-7]xxx|8xxxxxxx|Fxxxxxxx|Exx|91xxxxxxxxxx|9011x.)}*
>>> *}*
>>> *}*
>> *}*

(3) *MG1 to MGC:MEGACO/3 [67.34.23.11]:55555*
> *Transaction = 3002 {*
>> *Context = - {*
>>> *Notify = A4444 {ObservedEvents =2223 {*
>>>> *20150418T17010203:dd/ce{ds="916135551212",Meth=UM}}}*
>> *}*

(4) *MGC to MG1:MEGACO/3 [67.34.25.171]:55555*
> *Transaction = 3003 {*
>> *Context = $ {*
>>> *Add = A4444,*
>>> *Add = $ {*
>>>> *Media {*
>>>>> *Stream = 1 {*
>>>>>> *LocalControl {*
>>>>>>> *Mode = RecvOnly,*
>>>>>>> *nt/jit=40; in ms*
>>>>>>> *},*
>>>>>> *Local {*
>>>>> *v=0*
>>>>> *c=IN IP4 $*
>>>>> *m=audio $ RTP/AVP 4*
>>>>> *v=0*
>>>>> *c=IN IP4 $*
>>>>> *m=audio $ RTP/AVP 0*
>>>>> *}}}*
>>> *}}*
>> *}*

Figure 7.10   More complicated H.264 commands

Table 7.2 Comparison MGCP and H.248

	H.248	MGCP
Syntax	ASN.1 and ABNF including SDP	ABNF and SDP
Connection types	Call is termination within context	Call is endpoints within connections
	Straightforward	Complicated
	The subdivision of terminations in streams is only available here	
Standard control	ITU	IETF, many providers have specific variants
Supported transport protocols	TCP, UDP, SCTP – treated in Section 7.3.2	UDP
Functionality of commands	Comparable to MGCP but transactions exist	Comparable to H.248 but without transactions

Table 7.2 provides an overview of the main differences between MGCP and H.248. Implementations of both standards are available from every major provider.

7.3.2 Signal Control

Assume that we have a central VoIP solution, with a central VoIP switch and, at remote sites in other countries, media gateways to the PSTN network. Of course, we would at best like to have a single media controller at the central side. This setup works fine for controlling the media gateway via H.248 or MGCP. However, the signaling between SIP/H.323 and DSS – see Figure 7.8 – will not work. In this section, we will develop the solution for this, but need more background from the PSTN world.

We recall from H.225 that it uses Q.931 to create the overall functions. Here, we see that protocol Q.921 is the transport layer for Q.931. Q.921 contains a protocol named Link Access Procedure D-Channel (LAPD). LAPD is an extension of the more common HDLC used for serial line transmission. The D in LAPD stands for Data Channel. Note that unlike what one may assume, the Data Channel does the signaling but the Bearer Channel or B-Channel does the media transport. The media transport is neither treated here nor shown in Figure 7.8. Media transport in the PSTN world is done typically by SDH/SONET (Synchronous Digital Hierarchy/Synchronous Optical NETwork).

The left part of Figure 7.11 is not so interesting since these are the internals of the SS7 layering, which is used within the PSTN provider network. However, it is clear that the three Message Transfer Parts 1/2/3 (MTP1, MTP2, MTP3) correspond to Q.921. ISDN User Part (ISUP) initializes and tears down calls and maps one-to-one to Q.931. Since we know the functionality of H.225, we roughly know the functionality of ISUP as well. The Signal Connection Control Part (SCCP) is a protocol to inform about changes, and the Transaction Capabilities Application Part (TCAP) is a protocol that supports remote procedure calls.

SS7		DSS
Transaction Capabilities Application Part (TCAP)	ISDN User Part (ISUP)	Q.931
Signal Connection Control Part (SCCP)		
Message Transfer Part 3 (MTP3)		Q.921
Message Transfer Part 2 (MTP2)		
Message Transfer Part 1 (MTP1)		

Figure 7.11 Simplified SS7/DSS signaling protocol stack

We still want to run the signaling of a remote media gateway at a central location with up to 150 milliseconds delay. One option would be to have a signaling gateway that gets its commands from a central signaling gateway controller. This would be a solution similar to the media part with H.248. However, this setup option is too complex, and the solution is to have a central call agent. The central call agent sends and receives Q.931 messages to the remote PSTN switch via an intermediate signaling gateway as visible in Figure 7.12. Note that the shading in Figure 7.12 explains where the location of different components in this interaction typically is.

Q.931 needs to get transported to the central location. The regular underlying protocol Q.921 cannot be used within an IP network. Consequently, a replacement is required. The replacement consists of an ISDN Q.921–User Adaptation layer (IUA) [423] and a Stream Control Transmission Protocol (SCTP) [424–427]. With this information, we can draw the protocol stacks of the new components. The protocol stacks of the PSTN switch, the signaling gateway, and the call agent in Figure 7.12 are shown in Figure 7.13. Note that the arc in the signaling gateway copies the packets between the stacks and is sometimes denoted as Nodal Interface Function (NIF). A common name for the overall function is PRI Backhaul. PRI stands for Primary Rate Interface and delivers 23 B-channels with 64 kbit and one 64 kbit D-Channel. PRI is the standard interface for connections with the PSTN network. By the way, there also exists a similar shorthand BRI (Basic Rate Interface) with two B-channels and one 16 kbit D-channel for private households.

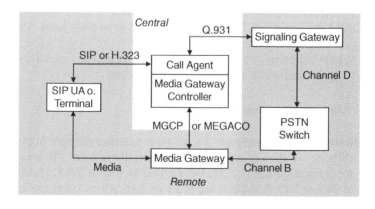

Figure 7.12 Central Q.931 call agent for remote signaling gateway

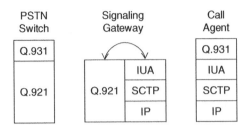

Figure 7.13 Protocol stacks for PSTN switch, signaling gateway, and call agent

SCTP, IUA, and IP have to mimic most of the functionality of Q.921 to work as a replacement. Also, it was claimed that it needs to be faster than TCP, but this seems not to be true [428]. It is interesting to see what is specific in SCTP and IUA because we will get an idea on the requirements of telephony systems.

SCTP has the following characteristics:

Handshake: TCP has the vulnerability of half-open connections, which means that it may wait infinitely for the last acknowledgment of its connection partner. This weakness was removed in SCTP by an additional acknowledgment and triggers when an answer is missing.

Multiple receivers per association: Endpoints in SCTP are a combination of one or more IP addresses and port numbers. An association establishes a relation between endpoints, which is another word for a connection. However, an association can be multihomed. Multihoming means that transported data arrives at multiple hosts at once. The reason for this functionality is to cope with failover scenarios. To achieve high availability, one must use failover systems. For a fast failover, both systems must be in the same state and receive all messages in parallel.

Multiplexed streams: Associations are initialized when the system comes up, and the initialization contains a number of unidirectional streams. The later use of those streams does not require a new setup. One chunk transports data of one individual stream. One or more chunks create a packet within an association. Figure 7.14 shows an SCTP packet, which consists of chunks. Only for *Chunk 2*, the *Chunk Value* is expanded. Chunks can be control data, but as shown in the figure, it is a data chunk identified by *00000000*. The *B* marks the first chunk, and *E* the last chunk. TSN counts all chunks in an association. Then, the chunk is assigned to a stream *S* with a sequence number *N*.

Reliability: Acknowledgments are transported in special chunks and mention which packets were missed – gaps – and which were received as duplicates.

IUA has the following characteristics:

Multiple application server processes: The protocol manages multiple Application Server Processes (ASP). As we already explained, this is done for failover. The messages to manage this are, for example, *M-ASP-UP, M-ASP-DOWN, M-ASP-ACTIVE, M-ASP-INACTIVE, M-NTFY-ASP*. Figure 7.15 shows the case of a communication to stop application server 1 and instead start application server 2. SCTP does, in addition, actively produce heartbeat to see if all endpoints are still active.

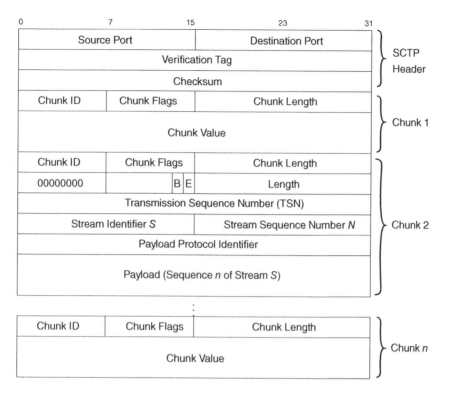

Figure 7.14   SCTP packet

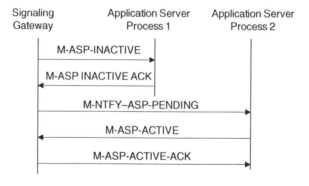

Figure 7.15   Example of IUA communication to application server processes

Channel to associations/streams: Within one SCTP association each existing D-channel is mapped to exactly one of the multiple streams and in addition one stream is used for IUA management functions.

Handing over Q.931 messages: All messages need to be delivered from the assigned interfaces to the streams and vice versa. Error messages also need proper resolutions.

In this section, we show signaling control or the PRI-Backhaul function with IETF standards, which would fit for SIP. The ITU handles this in Annex M1 of H.323, and it is defined as the tunneling of Q SIGnaling protocol (QSIG). However, the IETF version had to translate Q.921 characteristics in IP protocols. This allowed us to better understand the characteristics of telephony protocols, which emphasize stability and availability much more than the usual Internet protocols.

Any major voice system deployed in companies has support for IUA/SCTP and QSIG.

7.4   Mobile VoIP

In this section, we first treat the IP Multimedia System (IMS), which is a specification how mobile telephony networks using SIP. The second subsection handles VoLTE (Voice over Long-Term Evolution), which is the global standard implementing IMS. VoLTE is in deployment in over 40 countries worldwide.

7.4.1   IP Multimedia Subsystem

European Technical Standard Institute (ETSI) and ATIS (Alliance for Telecommunication Industry Solutions) created the 3rd Generation Partnership Project (3GPP) and 3GPP2 in 2004. North American and some Asian networks then, based on CDMA (Code Division Multiple Access), created the 3GPP2 project. The latest version of CDMA is EVDO (Evolution Data Optimized). The ETSI countries used UTMS (Universal Mobile Telecommunications System) with HSPA+ (evolved High-Speed Packet Access). 3GPP and 3GPP2 had the task to create SIP-based specifications for mobile networks. The Next-Generation Network (NGN) initiative of the ITU and the ETSI initiative named TISPAN (Telecommunication and Internet converged Services and Protocols Advanced Networks) are part of this standard process as well. The IETF documented this collaboration in two RFCs [429, 430]. The work led to an ITU standard ITU Y.2021 [431] in 2006. However, 3GPP has continued with newer versions, and ETSI has published the latest version in 2015 [432]. The latest versions also included the use of WIFI. The book by Camarillo and García-Martín [393] is an excellent source on IMS but is based on a 2008 version.

IMS does not use new SIP methods but has substantially increased additional proxies and message headers. As we will see later, a number of standardized IETF functionalities so far not treated are included. Figure 7.16 shows a rough overview of the architecture. Note that, for the communication with the mobile phone, additional networks are required, which we show as Radio Access Network (RAN) and the Core Network (CN) – see also Section 7.4.2.

The basis of the IMS architecture is the Call Session Control Function (CSCF), which is responsible for all signaling and coordination with other network entities. The IMS architecture as shown in Figure 7.16 consists of

P-CSCF: The Proxy-CSCF is the first contact point for any user agent or User Equipment (UE) in IMS. The UE can find the P-CSCF via DHCP and other methods. It passes the SIP registrations to the correct home network and other messages to the responsible S-CSCF.

I-CSCF: The Interrogating-CSCF is an SIP Proxy, which is located at the edge of each administrative domain and sends SIP messages to the right destination if they are allowed to get forwarded. It may need the Subscribe Location Function (SLF) for identifying the home domain of a UE.

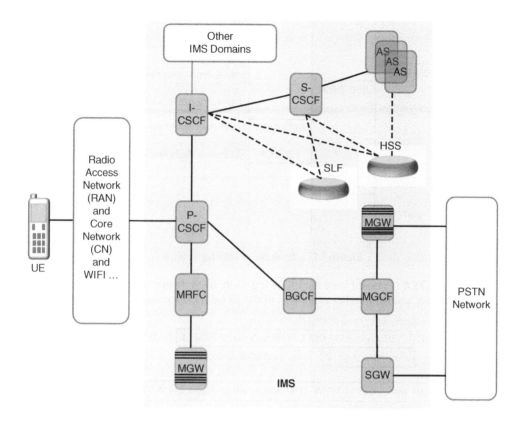

Figure 7.16   Overview of IMS and its connections

S-CSCF: The Serving-CSCF is the registrar of a domain. The S-CSCF registers users if they
are maintained in the Home Subscriber Server (HSS). The HSS and the SLF implement the
IETF Diameter protocol [433, 434] for a secure access. In addition, the S-CSCF manages
the application services that are associated with a UE.

MGCF/MGW/SGW: The Media Gateway Control Function (MGCF) controls the Media
GateWay (MGW) with H.248 and communicates with SCTP plus IUA to the Signaling
GateWay (SGW). This setup is the identical configuration we detailed in Section 7.3 and is
illustrated in Figure 7.12.

BGCF: The Breakout Gateway Control Function determines one of the multiple gateways
to the PSTN for routing a message. Figure 7.16 shows only one gateway to the PSTN. It is
the Default Outbound Proxy – see Section 6.2.

MRFC: Media Resource Function Control controls another media gateway with H.248 for
providing dial tones or any type of Announcement.

AS: In addition, Application Servers can provide different services such as instant messaging
or conferencing [435].

Let us now look for a registration as shown in Figure 7.17. In this registration, we have
left out DNS and any message exchanged on the RAN as well as the DHCP-based lookup for
P-CSCF.

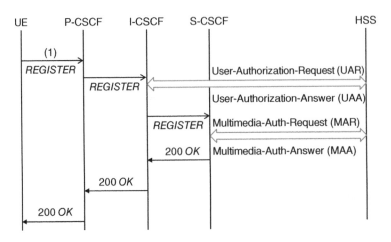

Figure 7.17 Example of IMS registration

The *REGISTER* (1) could have the following form where boxed items with a number are commented later and the number is not part of the *REGISTER* message:

REGISTER sip:registrar.ims234.net SIP/2.0

*Via: SIP/2.0/UDP [2001:0ab8:85a3:08b3:1319:8a2e:0a70:7324 **1**]:1357;*

             *comp=sigcomp **2** ; branch=z9hG4bKnasfs47*

Max-Forwards: 70

*P-Access-Network-Info: 3GPP-UTRAN-TDD; utran-cell-id-3gpp=234562D0FAE11 **3***

From: <sip:ue1_public@ims234.net>;tag=4fa3

To: <sip:ue1_public1@ims234.net>

*Contact: <sip:[2001:0ab8:85a3:08b3:1319:8a2e:0a70:7324]:1357; comp=sigcomp **2** >;*

            expires=600000

Call-ID: apb03a0s0956238kj49111

*Authorization: Digest username="ue1_private@ims234.net" **4***

            *realm="registrar.ims234.net" **5** ,*

            *nonce="ace56b53aacc5750d2ac53848ac32ccc245" **7** ,*

            *algorithm=AKAv1-MD5 **8** ,*

            uri="sip:registrar.ims234.net",

            *response="3769fae49393ab5697450937635c4abe" **6** ,*

            integrityprotected="yes"

*Security-Verify: ipsec-3gpp; q=0.1; alg=hmac-sha-1-96; **9***

            *spi-c=98765432; spi-s=3725321; **9***

            *port-c=8642;port-s=7531 **9***

*Require: sec-agree **10***

*Proxy-Require: sec-agree **10***

CSeq: 3 REGISTER

Supported: path

Content-Length: 0

The registration command has some unknown parts. IMS uses only IP6 addresses, and we see such an address in (**1**). In (**2**), a statement *comp=sigcom* is included, and this specifies that the signaling should get compressed [145]. The UE delivers in (**3**) the access type and access info of the access network. UTRAN-TDD stands for UMTS Terrestrial Radio Access Network with Time Division Duplexing. The private user identity (**4**) contains the *username* field of the digest Authentication and Key Agreement (AKA) protocol [436]. The realm field (**5**) is the network name for the authentication of the user. The USIM (Universal Subscriber Identity Module) contains the data for the user authentication that is an extension of a SIM card. The field *response* (**6**) in the *Authorization* message shows that we have not seen a former registration trial that was challenged by a server for the arbitrarily chosen *nonce* (**7**). This communication is a standard HTTP authentication [379] with a specified *algorithm* (**8**). The *Security-Verify* is a SIP header [437], which describes the possible security methods of the UE (**9**). *spi* stands for the Security Parameter Index of IPsec [438]. The field *sec-agree* in *Require* and *Proxy-Require* states that a security agreement needs to be set up (**10**) [437].

On the hops of the *REGISTER* message, the I-CSCF asks for authorization by the HSS. The authorization requests *UAR* and *UAA* – *MAR*, *MAA* – are Diameter commands as specified for SIP [434]. The I-CSCF expands the *REGISTER* not only by an additional *VIA* command but also by

```
P-Visited-Network-ID: "Visited Network Number 1"
P-Charging-Vector: icid-value="Bfgd36yU0dm+602"
```

The header *P-Visited-Network-ID* lists the IMS domains where this message passed through. The IMS Charging IDentifier (ICID) is created during registration at the P-CSCF. The Event Charging Function (ECF) will collect the records that also cover interoperator information. Call duration is delivered into the overall charging system by either the supporting access networks or by the S-CSCF since the *BYE* is forced through this proxy. The S-CSCF is a back-to-back user agent.

Figure 7.18 shows the first parts of an *INVITE* message. In this case, both UEs are not in their home networks but in visited networks – that is, both UEs are roaming. The *INVITE* of the calling UE_0 passes through the S-CSCF of its home network and then steps into the home network of its roaming partner UE_T. Each session initiation for a UE has to pass through its S-CSCF since even a roaming user should have all its services functioning. The I-CSCF in the home network of the UE_T has to check with HSS the address of the responsible S-CSCF. The I-CSCF sends a Location-Info-Request (LIR) to HSS and receives a Location-Info-Answer (LIA). Passing through the P-CSCF of the visited network, the *INVITE* reaches the UE_T.

The *INVITE* (1) could look as follows and the boxed parts are commented later. Again the numbers in the boxes do not belong to the *INVITE* but help to associate with the comments.

INVITE tel:+41-44-632-1111 SIP/2.0
Via: SIP/2.0/UDP [2001:0ab8:85a3:08b3:1319:8a2e:0a70:7324]:1357;
                comp=sigcomp; branch=z9hG4bKnasfs47
Max-Forwards: 70
P-Access-Network-Info: 3GPP-UTRAN-TDD; utran-cell-id-3gpp=234562D0FAE11
Privacy: none **1**
Contact: <sip:[2001:0ab8:85a3:08b3:1319:8a2e:0a70:7324]:1357;comp=sigcomp>;
           expires=600000

From: <sip:ue1_public@ims234.net>;tag=4ba3
To: <tel:+41-44-632-1111>
Call-ID: apb03a0s03456238kj49111
Cseq: 127 INVITE
Require: sec-agree

Supported: precondition, 100rel, gruu, 199 **2**

Accept: application/sdp,application/3gpp-ims+xml
Proxy-Require: sec-agree
Security-Verify: ipsec-3gpp; q=0.1; alg=hmac-sha-1-96;
            spi-c=98765432; spi-s=3725321;
            port-c=8642;port-s=7531
Contact: <sip:[2001:0ab8:85a3:08b3:1319:8a2e:0a70:7324]:1357;

            gr=urn:uuid:f81d4bae-7dec-11d0-a765-00a0c91e6bf6; **3** *comp=sigcomp>;*
            expires=600000

Allow: INVITE, ACK, CANCEL, BYE, PRACK, UPDATE, REFER, MESSAGE
Content-Type: application/sdp
Content-Length: 669
v=0
o=- 2987933615 2987933615 IN IP6 2001:0ab8:85a3:08b3:1319:8a2e:0a70:7324
s=-
c=IN IP6 2001:0ab8:85a3:08b3:1319:8a2e:0a70:7324
t=0 0

m=video 3400 RTP/AVP 98 99 **4**

a=tcap:1 RTP/AVPF **9**

a=pcfg:1 t=1 **10**

b=AS:75 **11**

a=curr:qos local none **12**

a=curr:qos remote none **12**

a=des:qos mandatory local sendrecv **13**

a=des:qos none remote sendrecv **13**

a=rtpmap:98 H263

a=fmtp:98 profile-level-id=0 **8**

a=rtpmap:99 MP4V-ES

m=audio 3456 RTP/AVP 97 96 **5**

a=tcap:1 RTP/AVPF **9**

a=pcfg:1 t=1 **10**

b=AS:25.4 **11**

a=curr:qos local none **12**

a=curr:qos remote none **12**

a=des:qos mandatory local sendrecv **13**

a=des:qos none remote sendrecv **13**

a=rtpmap:97 AMR **7**

a=fmtp:97 mode-set=0,2,5,7; maxframes=2 **8**

a=rtpmap:96 telephone-event **6**

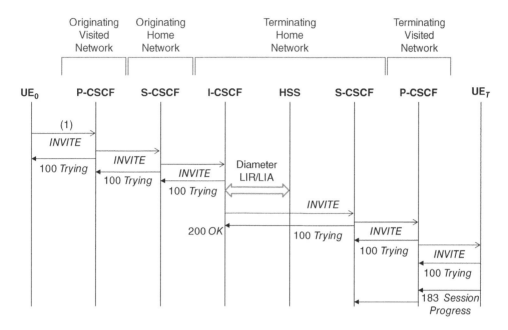

Figure 7.18   Example of IMS invite with two roaming UEs

Due to the former *REGISTER*, we know most parts of the message header as shown in
Figure 7.18 already. However, there are a few more elements that require explanation. The
Privacy=none (**1**) says that the user does not wish to hide the identity. The *Supported* (**2**)
has a number of attributes. The *100rel* states that the UE$_0$ will acknowledge 1xx messages
and supports the 199 "Early Dialog Terminated." The attribute *gruu* specifies that the client
supports Globally Routable User Agent URIs (GRUU) [439], which help to identify multiple
instances of registered users with the same name. This GRUU identification is visible in the
Contact header as *gr=urn* … (**3**).

The attribute *precondition* in *Supported* (**2**) states that the call setup must get delayed until
the agreement of the resources [440]. All UEs in IMS must support this feature.

We now can step through the SDP but will not explain well-known elements – see also
Section 4.2. SDP for a video has two optional codecs (**4**) and an audio part with one codec
plus a *telephone event* (**5**). The *telephone event* in the = rtpmap (**6**) describes the transport of
DTMF tones in RTP and *AMR* stands for Adaptive Multi-Rate (**7**) or G.722.2 [441] – see also
Section 3.1.3. The *a=fmtp* (**8**) specifies parameters for a media specification – one time for *98*
and the other time for *97*.

The element *a=tcap:1 RTP/AVPF* (**9**) expresses that the parties need to use the SDP capability negotiation [139] – see Section 4.2.2 with the feedback profile AVPF for RTCP [146]. The
next element *a=pcfg:1 t=1* (**10**) is also part of the capability negotiation and provides a list of
possible configurations. Here, it lists only one capability given by the above line. The statement
B=AS (**11**) is a bandwidth modifier [142] stating that for each stream, the required bandwidth of
75 kbit/second or rather 25.4 kbit/second. The last missing elements of RTP concern resources
based upon [440]. The resource reservation states the current resources *a=curr* (**12**) and the
desired resources *a=des* (**13**). At the time of the *INVITE*, no capacity is reserved locally and

remotely. The UE only states the mandatory capacities for the sending part but not for the receiving remote part. When the called UE returns the *183 Session Progress*, it changes the SDP to state that the receiving parts also need a mandatory reservation.

For the whole resource reservation, an additional Policy and Charging Control Function (PCRF) is responsible. The reservation is for the media stream, and the critical resources are with the physical and data link layer. So, the reservation is part of the core network, and we will see this again in Section 7.4.2.

Complete examples can be found in Ref. [442].

Most of the additional fields are explained before and all mentioned protocols are IETF standards. So, IMS has used the wealth of IETF standards to complement the required functionality for the mobile usage.

7.4.2 VoLTE

Until 2008, it was unclear whether the whole IMS architecture as specified by the 3GPP would ever come to life. The question was: why should the mobile provider change to an SIP-based infrastructure? At the end of 2009, the One Voice initiative with AT&T, Vodafone, Verizon, Nokia, and others declared that the IMS-based solution is their strategy for the future and named it VoLTE. A few months later, the Global System for Mobile Association (GSMA) announced that they adopted the proposal of One Voice. Since 2010, the GSMA has been maintaining a profile for IMS, which specifies in detail what needs to be supported by any deployment in provider networks [443].

Figure 7.19 shows the reason for implementing VoIP by the providers. First, note that this figure is the left side of Figure 7.16. Until 2000, only the circuit-switched GSM network existed. With General Packet Radio Service (GPRS), a packet-switched domain was introduced at the same time that runs over UTRAN. With LTE (Long-Term Evolution), from 2008 onward, another Radio Access Network (RAN) became operational. For LTE, the bolded nodes within the packet-switched domain were added. So, substantial new investments were made to expand the packet-switched capabilities. However, the majority of the revenues for the operators are still voice calls and SMS (Short Message Service) but with a decreasing tendency. The current combination of networks and components is quite large and requires high ongoing costs. The goal is to simplify the infrastructure over time. VoIP runs over a packet-switched network, and with VoIP, the circuit-switched domain could be eliminated over time. Terminating the circuit-switched network would also free valuable voice frequencies for reuse with higher capacity by the new packet-switched networks.

Why are mobile telcos moving to SIP?

Very simple: There must be a business case. Only with SIP, they can simplify their infrastructure and save cost.

Figure 7.20 shows the interworking between IMS and EPC (Enhanced Packet Core) using an *INVITE* as we have discussed this in detail for IMS in Section 7.4.1. All abbreviations in Figure 7.19 are explained in Table 7.3. The IMS part is simplified in this figure since we have seen all the details in Figure 7.18 and in the associated *INVITE* call. The *(1) INVITE* sends the

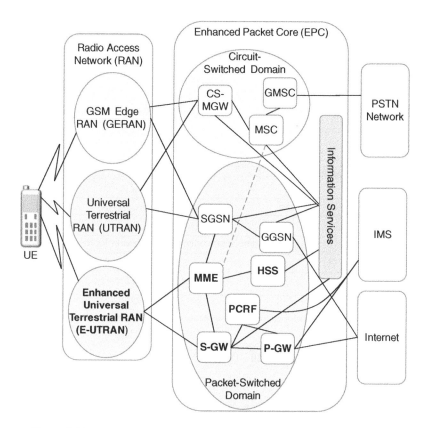

Figure 7.19   3GPP enhanced core and radio access network – very simplified

Table 7.3   3GPP enhanced core and radio access network – abbreviations

Technology	Shorthand	Meaning
GSM	CS-MGW	Circuit-Switched Mobile GateWay to the PSTN
	MSC	Mobile Switching Center controls distribution and bearer data.
	GMSC	Gateway Mobile Switching Center
UMTS/ GPRS	(U)SGSN	(UMTS) Serving GPRS Support Node controls all the UTMS/GPRS network.
	GGSN	Gateway GPRS Support Node is the interface to the Internet.
LTE	MME	Mobility Management Entity is central call control.
	HSS	Home Subscriber Server is the database server.
	S-GW	Serving GateWay does the routing and the bearer control.
	P-GW	Packet data network GateWay is the default router and the interface to IMS. It assigns IP addresses.
	PCRF	Policy and Charging Control Function handles QoS regulation and charging.

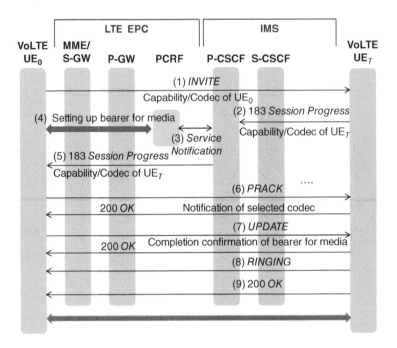

Figure 7.20   IMS and VoLTE INVITE

requested capabilities and the possible codecs as we have seen in the former SDP. In the *(2) Session Progress*, the P-CSCF receives the capabilities of the UE_T and exchanges the required capabilities with the PCRF. The PCRF communicates with the P-/S-GW and MME to get approval for the media reservation. If this all works out, the PCRF will confirm the request and the P-CSCF forwards the *(5) Session Progress* to the UE_0. Then, the User Equipment informs each other on the acknowledgments and now the terminating UE can ring. The original *INVITE* required that the session setup first has to terminate the resource setup.

If the connection to LTE is good, VoLTE immediately offers the following advantages for the users:

- HD Quality for voice communication
- Phones require up to 40% less power
- Much faster connection setup.

The migration to VoLTE is challenging since coverage of LTE by far is not as excellent as the GSM coverage. So, a number of coexistence and migration solutions exist:

CSFB: The Circuit-Switched Fallback option in case of phone conversations of an LTE UE using the GSM infrastructure. When the UE registers in the LTE/IMS network, the MME does a parallel registration at the MSC in the GSM network – see the dotted connection of those two units in Figure 7.19. When the UE initiates or receives a call, the device is switched over in the GERAN network. At the end of a call, the device is switched back. Existing connections to data networks at calling times may, however, drop. SMS exchanges do not require a change of the infrastructure since the MME and the MSC handle this. This solution is standardized by the 3GPP [444].

SRVCC: The Single Radio Voice Call Continuity is a standardized 3GPP solution that is based on IMS. If SIP cannot maintain a call, the IMS system hands this call over to GSM [445].

SVLTE: Simultaneous Voice and LTE uses a handset that currently supports two connections and uses GSM for voice and LTE for data transport. No specification exists for this variant since it can be created by each handset company at their discretion.

OTT VoIP: Of course, installation of any VoIP client on the phone is possible, which runs over-the-top. However, this software cannot provide handover when changing cells and the quality is probably not sufficient.

VoLGA: Voice over LTE via Generic Access uses its own protocol to access a new server in the Enhanced Packet Core (EPC) – see Figure 7.19 – which tunnels the request to a call server in the circuit-switched domain for further handling. 3GPP did not accept the VoLGA proposal.

For details and comparisons of these variants for coexistence, see Ref. [446].

The rollout for VoLTE is announced or ongoing in the United States, nearly all major European countries, and all major Asian countries. So, SIP might get utilized soon by millions of users if the rollout is successful, and coexistence solutions do not continue to be used.

7.5 Skype

Skype – originally "Sky peer-to-peer" – is not specified by any standard, but over 700 million users downloaded the software, and this forms a standard of its own. The software was launched in 2003 by a small Swedish/Danish company. In 2011, Microsoft bought Skype.

With a lot of effort, the Skype developers hinder the code analysis by binary packing, code obfuscation, encryption of packets, and antidebugging techniques [447]. As a consequence, only fragments of the overall system are known. Apparently, the different Skype versions exhibit different behaviors, and consequently, older information might be outdated. Here, we list a number of the most recently described characteristics [448]:

Peer-to-peer: After the transfer to Microsoft, the system is no longer a peer-to-peer application since the supernodes of Skype are now run by Microsoft. The Skype architect of Microsoft also stated that the peer-to-peer behavior created an increasing number of problems. The huge number of mobile devices – more than 2 billion in 2016 – cannot any longer be used as peers since they are not reliably reachable.

Audio: Skype in 2012 standardized their audio codec SILK under the new name OPUS with the IETF [449]. Good but not excellent results were analyzed for this codec [450]. The bandwidth for audio is variable between 6 and 150 kbit/second and typically around 7.7 kbit/second.

Video: The video codec is VP8 [81]. The measured maximum video rate was 150 kbit/second in 2012. For videoconferences, Skype does not use transcoding MCUs but the sender has to send multiple resolutions – maximum 3. The central servers per receiver distribute one of the resolutions. When the sender of the video sits on a slow link or has weak computer resources, the client only sends one codec. Mean video roundtrip times for global calls were measured to be around 1 second. However, video calls in the United States or between Europe and the United States involve much smaller delays since the global servers are situated on the US East Coast. Skype extensively uses FEC to stabilize the received video, and no retransmission is performed [451].

Firewall/NAT traversal: Skype has always been known to work reliably over firewalls. This function results from a high flexibility to work over multiple ports. Each client tries various ports to build connections. The last resort is the use of the HTTP Port 80. If both clients sit behind NATs, Skype applies a media relay on the Internet if nothing else works [452]. The use of a media relay is similar to the IETF TURN solution [370] as mentioned in Section 6.5. Skype also utilizes a relay for the signaling part.

Signaling: Skype uses TCP for the signaling and UDP/TCP for the media connection. The signaling connection stays during the connection and heartbeats are regularly exchanged. No fixed ports are associated with each service. The listening port for the responder is chosen during the Skype installation. The client sends status information about calls via HTTP to Skype central servers. Due to the high variety of used ports, the Skype traffic is hard to detect on the network, but signatures for traffic analyzes were published [453, 454].

8

WebRTC

For many years, all multimedia use in browsers was achieved by plug-ins such as Adobe's Flash or Microsoft Silverlight. Recently, the HTML5 standard has added support for playing video and audio without a third-party addition [455]. Web Real-Time Communication (WebRTC) adds real-time communication to the browser with a focus on peer-to-peer communication. In 2011, a first implementation was released by Google [456], which was based on the code of the acquired company Global IP Solutions. W3C is responsible to create the APIs with additional programming methods integrated into HTML5 [457, 458]. The IETF is adding multiple new standards to WebRTC. By the time of writing of this book, very few RFCs have been approved and the W3C definitions have not been finalized. The reasons why the IETF standards take so long are evident in Section 8.1. However, the first implementations are available based on some preliminary versions. Apart from the diverse referenced papers and RFC drafts, the book by Johnston [292] gives more insight into the subject.

Upon considering the focus of this book, we will only briefly touch on the programming but concentrate on the protocol side. Figure 8.1 shows an overview of the standard. The transport-specific details are left out in this initial figure, but Figure 8.2 shows this later together with more information. The audio and video standards listed in the figure were taken from IETF documents [459, 460], and these are the mandatory ones. Google's WebRTC website, in May 2015, showed additional audio codecs and left out H.264. Google tries to incorporate its own, license-free codecs into WebRTC. To cope with the license-free VP8 (VP9) codec, Cisco has provided its H.264 implementation code license-free[1] – see Section 3.5 for a comparison of H.264, VP8, and VP9. As mentioned in Chapter 3, the license prices even for H.265 are rather moderate [88]: US $0.20 per unit after the first 100,000 units each year and no fees beyond these number of units.

Some elements shown in Figure 8.1 need additional comments:

Signaling: The WebRTC standard does not include the signaling for building a session. Typically, the browser provider will include signaling code with an API in the browser. An additional standard SIP over WebSockets plus an SDP extension was approved recently [461, 462]. With this extension, plug-ins can be created to connect to any signaling SIP server on the Internet.

[1] http://www.openh264.org/, Last retrieved on May 22, 2015.

Multimedia Networks: Protocols, Design, and Applications, First Edition. Hans W. Barz and Gregory A. Bassett.
© 2016 John Wiley & Sons, Ltd. Published 2016 by John Wiley & Sons, Ltd.

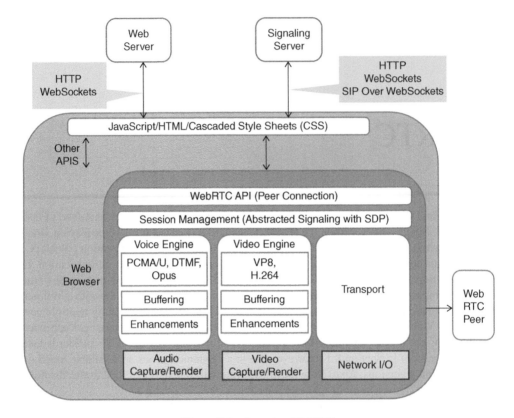

Figure 8.1 Overview WebRTC

Since the signaling is not standardized, browser providers could only allow implementation of their signaling client, which interconnects solely to the browser company infrastructure. This use could generate separate clouds of browser providers that, in the worst case, would require gateways to interwork even between browsers [463].

WebRTC, however, includes one element of the signaling: the SDP descriptions in a standard offer/answer model. The SDP descriptions are hidden in JavaScript calls that are, from a functional point of view, described in an IETF standard [147] and are exposed to the programmer by diverse APIs. This JavaScript description has the name JSEP (JavaScript Session Establishment Protocol) and contains more than the SDP creation as we see in Section 8.1.1.

Within WebRTC documentation, the later replacement of SDP by Java Script Object Notation (JSON) [464] is mentioned.

Capture/Render/Network: These three blocks have a gray background as shown in Figure 8.1, which means that it is the decision of the browser provider on which devices get supported. This situation could initially be a problem with older devices.

Audio standards: The chosen mandatory audio standard G.711, which covers PCMA/U, needs 64 kbit/second – see Table 3.3. The Opus codec is much more efficient and requires between 6 and 150 kbit/second – see Section 7.5. For mobile communication, Opus would be the only possible option for WebRTC. For standard mobile voice, the providers use G.722.2, which is probably the more efficient vocoder – see Section 3.1.3. In the current worldwide rollout of VoLTE, G.722.2 will deliver HD quality to mobile calls. However, G.722.2 is not royalty-free, and this is not acceptable for browser providers. Royalty-freeness is the paradigm that rules everything even for some of the richest IT companies. Transcoding is possible, but the results will be quite notable [465].

The application developer knows the following main elements from WebRTC:

GetUserMedia: This call returns a local media stream – for example, a camera attached to the computer.

RTCPeerConnection: This call builds a communication channel with a peer. An ICE agent starts locally and at the peer's site. The setup will be clarified in Section 8.1.1.

Adding a media stream to a communication channel triggers a negotiation with the peer on diverse parameters such as format or resolution and, finally, an RTP stream with a dedicated SSRC – see also Section 4.1.2.

RTCDataChannel: This interface represents a bidirectional data channel that is created by an RTCPeerConnection variant. In the next section, we discuss the transport for those data channels inclusive of the establishment message for the protocol stack.

In addition to the APIs of WebRTC, object-oriented APIs are also under development by the W3C named ORTC (Object Real-Time Communication) [466]. ORTC seems not to have an impact on network functionalities. So, we will not treat this further. Microsoft and Apple are not participating in the WebRTC standardization. Microsoft has proposed the "Customizable, Ubiquitous Real-Time Communication over the Web" (CU-RTC-Web) [467]. In addition, Microsoft announced that it will support ORTC with Internet Explorer 11. CU-RTC-Web only uses RTP and RTCP from the WebRTC components. There are no statements available as to when Apple is going to support WebRTC in its browsers.

8.1   WebRTC Transport

Figure 8.2 shows the network stack for WebRTC, which fits in the transport box shown in Figure 8.1. This network stack is a surprise when first looking into it, and it will have more surprises when looking into details. We already treated all of the protocols mentioned before – apart from one. The security elements SRTP, DTLS-SRTP for key exchanges, and DTLS have been covered in Section 6.7. DTLS is the Datagram Transport Layer Security on top of UDP similar to HTTPS. SRTP is the Secure RTP, and the figure shows possible different streams by SSRC1$\cdots$ SSRCn. We handled STUN, ICE, and TURN in Section 6.5 on NAT, but this is expanded in Section 8.1.1. SCTP is the Stream Control Transmission Protocol as treated in detail in Section 7.3.2 on H.248. WebRTC newly developed the *Data* protocol, and we treat it next.

The development of WebRTC is based on the requirement that firewalls and NATs should no longer pose any problem. The ease of passing through firewalls and NATs has been a major reason for the success of Skype. The possibility to use PMP, PCP, or the UPnP port control (Sections 6.5 and 10.2) to open ports at the NATs was not considered because not every possible

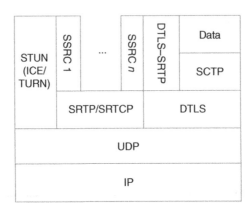

Figure 8.2 Overview WebRTC protocols

model supports it. Also, port opening protocols are normally not allowed for firewalls and NATs within universities and companies. So, the WebRTC groups decided to use ICE, which we explained shortly in Section 6.7 and revisit in Section 8.1.1. For the moment, only recall that ICE includes STUN.

The fewer the connections one needs to communicate over NATs, the greater the chance for successful traversal. The WebRTC group decided to use only one connection and multiplex all traffic over this connection. This approach was already employed in another RFC [468] when combing SRTP with DTLS. The first byte of the payload of a UDP packet tells which protocol is used on top. With this information, packets are forward to (S)RTP (128–191), DTLS (20–63), TURN Channel (64–79) or STUN (0–3) based on an updated IETF proposal [469]. DTLS codepoints are further subdivided based on [380] and the IANA registry for "TLS ContentType" – for example, application data (23).

Let SIP work from everywhere

SIP has always been hampered by NAT/firewall traversal problems. Skype is such a successful VoIP implementation because it works from "everywhere." WebRTC uses a number of methods to let SIP work the same way. These methods, however, made the WebRTC protocols complex and less elegant.

Apart from RTP communication, WebRTC also foresees data communication over the chosen connection. For multiple data connections multiplexing, guaranteed delivery and a secure transmission are required. DTLS delivers the secure transmission. The SCTP is built for multiplexing and guaranteed transmission as shown in Section 7.3.2. As the reader may recall, SCTP was created to replace the ISDN Q.921 protocol to transport Q.931 over IP. By the inclusion of SCTP in WebRTC, previous ISDN communication methods have found their way in the new world. However, on top of SCTP, another adaptation protocol IUA builds – among other functionality – associations between the SCTP streams and ISDN channels. The WebRTC DATA

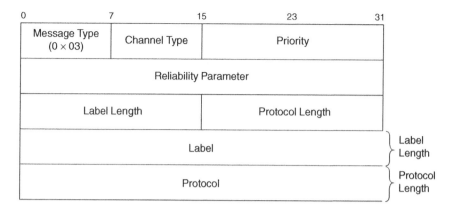

Figure 8.3   DCEP: data channel open message

protocol handles exactly these associations – see next. Originally, SCTP was foreseen for shorter command messages in Q.931, but now larger data blocks need to be sent. Consequently, for SCTP, a new chunk type was defined to transport large data blocks in fragments [470].

The Data Channel Establishment Protocol (DCEP) [471, 472] creates a new channel that consists of a pair of unidirectional data streams. The DCEP Data Channel Open message is shown in Figure 8.3. The *Message Type* has the value *0x03* for the open message. There exists only one other value for the acknowledgment. The *Channel Type* specifies the reliability type (reliable, partially reliable) and the ordering (strict, unordered). *Priority* is the priority of this new channel. The *Reliability Parameter* can specify the maximum number of retransmissions or the lifetime of partially reliable messages. The *Label* is a Unicode name. The *Protocol* is a name specified as a subprotocol in the "WebSocket Protocol Registries" with the IANA based on [473]. There is neither a channel termination message nor a possibility to change channel attributes. An Open message needs to be acknowledged. If no acknowledgment is received, the channel is not established.

We have seen that SCTP has its own built-in priority scheduling. It was pointed out that the internal SCTP scheduler needs to be reworked to cope with real-time traffic [474]. This scheduling is, however, solely within the SCTP protocol function. There are no proposals on how to specify the importance of a UDP packet with specific content. On the network, we can differentiate using the previously mentioned codepoints for a UDP packet. This way we can give priority to RTP, but it is unknown which media traffic is transported. Since the media could also be video, we cannot assign it to a voice QoS class over the WAN.

8.1.1   ICE Revisited

We shortly treated Interactive Connectivity Establishment (ICE) in Section 6.5 as the IETF universal solution to the NAT problem [367–369]. As we recall, ICE uses hole punching, STUN (Session Traversal Utilities for NAT), and TURN (Traversal Using Relays around NAT). ICE only works if both clients support it, and this was one of the reasons why ICE never gained much ground so far. However, with WebRTC, it is mandatory.

The first element is the hole punching. Here, each client creates a list of possible reflexive addresses to connect to its peer. This list is created by:

- Using home addresses of the client by opening a few sockets and storing the addresses as candidates.
- Contacting STUN and TURN servers, using the home addresses, and storing the reflexive addresses as further candidates. The protocol for TURN [361, 370] is also the STUN protocol. Each STUN binding request has to wait several milliseconds before the next one can be sent to avoid overload.
- The candidates get priorities: home addresses very high, reflexive addresses at the NAT middle, and relay addresses from TURN the lowest. The priority is evident since the best variant is a direct connection without an NAT and relaying the stream over a media TURN server is the worst variant.
- After the removal of duplicates, the candidate list is included in SDP using a format *a=candidate*, which accompanies the *INVITE*. An SDP excerpt for ICE is seen in Figure 8.4. Message integrity of STUN requests is kept by a password **(1)**. **(2)** is the STUN username since situations such as two different senders send STUN messages to the same IP address can occur. The candidate description **(3)** mentions first foundation (1), component (1), transport protocol (*UDP*), priority (21 …), IP address plus port, and type (*host*). Foundation indicates all candidates of the same type, the same interface, and the same contacted STUN server.

The signaling uses the signaling server as shown in Figure 8.1. As stated before, other signaling protocols might be applied as well. Jingle [475] is often mentioned as another candidate that is used in Google's Talk application. Jingle is based on Jabber [343].

During the whole process until the setup of the session, all open reflexive and home addresses must stay intact, and the client has to maintain the existence of the reflexive addresses.

The peer receives the candidate list and starts its corresponding candidate search and sends its SDP response back to the inviter with *200 OK*. This *200 OK* shall be sent over a reliable

```
v=0
o=- 2987933615 2987933615 IN IP4 196.0.2.17
s=
c=IN IP4 53.0.2.1
t=0 0
a=ice-pwd:ghd4092pdd7acekzjYfgiqg (1)
a=ice-ufrag:8hhY (2)
m=audio 45664 RTP/AVP 0
b=RS:0
b=RR:0
a=rtpmap:0 PCMU/8000
a=candidate:1 1 UDP 2130706431 196.0.2.17 8998 typ host (3)
a=candidate:2 1 UDP 1694498815 53.0.2.1 45664 typ srflx raddr
```

Figure 8.4 Example of an SDP description for ICE

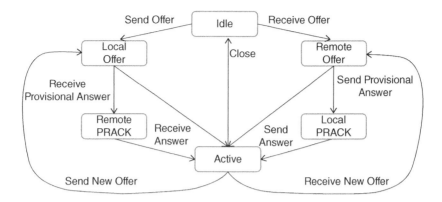

Figure 8.5 JSEP state diagram for the SDP answer/offering

PRACK – see Section 6.1.3. Thereafter, both peers try to reach each other using the sent list. STUN does the check for reachability. If such a STUN binding message reaches one peer, the answer includes the received IP address and port pair. Each successful STUN check creates a valid address pair. At the end of the trials, both endpoints have the same list of valid pairs. Typically, the inviter will then decide on the "use candidate." This information is again sent by STUN.

The whole ICE establishment process can take quite some time, even if the maximum number of candidates is restricted to 100. The Trickle ICE proposal [476] is already part of the WebRTC drafts, and the idea is to send incremental information on candidates during the establishment phase.

Note that STUN messages do not use the signaling path but the future media path. The aforementioned ICE procedure runs into a problem when media proxies exist in this path. Such a situation, for example, occurs with SPEERMINT – see Section 6.2. Consequently, media gateways must be enhanced also to proxy STUN messages and may need to be informed when ICE negotiations are taking place.

All the previously mentioned activities apart from sending STUN messages need software in the client. The JavaScript methods JSEP [147] cover this in addition to creating the SDP information. Due to the described APIs, WebRTC streams are added to the RTP connection over time. Consequently, the SDP negotiations can occur during the whole connection time, and JSEP maintains an answer/offering system as shown in Figure 8.5. An IETF document describes use cases for SDP within WebRTC [477].

Finally, it should be noted that there is an ICE Lite option in Ref. [369]. With ICE Lite, a client only uses its host address as a candidate and does not execute connectivity checks. It is mandatory in WebRTC to support the communication with a peer that uses ICE Lite [147].

8.2 RTP/SDP Adaptations

The WebRTC standard contains several adaptations and profiling to RTP, RTCP, and SDP.

In the previous section, we already mentioned that different media streams are transported by a single RTP session. The original RTP specifications [16] mandated that an RTP connection can only transport the same type of media in different SSRCs. The reason for this restriction was that RTP sends data in a strict timing sequence defined by the transport media stream characteristics. The draft proposal [478] removes this limitation. The draft states that, today, it is more important to send multiple flows on a single connection. Since media streams will not maintain their synchronicity on the network, more buffering is the consequence. More buffering results in a longer end-to-end delay.

In the previous chapters, we have seen that there are many additional RFCs for RTP and RTCP. An additional RFC [479] defines the mandatory elements out of those RFCs for implementing a WebRTC client. This profile also contains some conference extensions in previous RFCs that we have not introduced so far but are worthwhile mentioning here:

Full intra request: An RTP mixer may request (FIR) a participant to send an I-frame – originally from Ref. [480].

Picture loss indication: A receiver sends a PLI to ask a sender to resolve a picture loss – originally from Ref. [146].

Slice loss indication: A receiver sends an SLI to ask a sender to resolve a slice loss – originally from Ref. [146].

Reference picture selection indication: A receiver sends an RPSI to ask a sender to base future increments on an older reference picture – originally from Ref. [146]. This functionality is used later in Section 12.4.2.

Temporal–spatial trade-off request: The receiver sends a TSTR to indicate to the video encoder to change the relation between quality and frame rate – originally from Ref. [480].

Temporary maximum media stream bit rate request: A media receiver informs the sender that it has a temporary limitation to receive the transmitted bandwidth (TMMBR) – originally from Ref. [480].

All these aforementioned functionalities are optional or recommended for senders within WebRTC.

The work of the WebRTC groups has also triggered some summarizing RFCs on the usage of different RTP technologies:

Topologies: An overview of the many different used topologies for RTP [481].

B2BUA: A taxonomy of Back-to-Back User Agents was published in Ref. [482]. This RFC triggered an additional one on the usage of RTCP in B2BUA [483].

8.3   Interworking

As we have seen in the previous sections, there are many specifics of SIP and RTP within WebRTC, which do not exist outside of WebRTC. Consequently, the interworking with existing SIP communities might initially need gateways. To summarize another common form of the WebRTC architecture, we show in Figure 8.6 the so-called WebRTC Trapezoid.

When interworking with a standard SIP client, a number of issues arise. The chosen communication over SRTP/SRTCP and the session setup with the mandatory inclusion of ICE respective ICE Lite will not work with a standard client. The missing communication for the data channel with DTLS/SCTP is not necessary for SIP, but all communication arrives

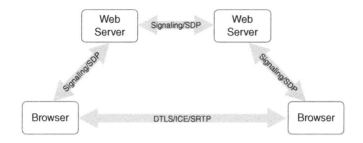

Figure 8.6 The often shown WebRTC trapezoid

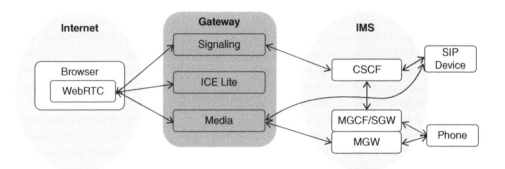

Figure 8.7 A typical gateway between WebRTC and IMS

on the same UDP port. The usage of a single port will cause incompatibilities. The chosen audio codecs (Opus, PCMA/U) are also not supported by typical SIP clients. It is obvious that a gateway with transcoding is required, or specialized new WebRTC adapted clients must become available.

Some gateways have been developed by smaller companies or within the research community [484, 485]. In 2012, Ericsson reported on such a gateway [486] that, in addition, works with IMS.

When checking with the current global rollout of IMS/VoLTE, one sees some stumbling stones. VoLTE will only implement Opus if the patent issues are settled. Patent claims exist from Qualcomm and Huawei, but browser companies are confident to ship Opus. Similarly to VoLTE, WebRTC uses SRTP, but the key exchanges are not compatible. WebRTC exchanges the master keys over a DTLS secured channel. VoLTE exchanges the master key in a clear-text message during signaling – see Section 7.4.1. A typical gateway setup between IMS and WebRTC is shown in Figure 8.7 [463]. A short study of the 3GPP has developed a description with a similar setup [487]. Section 7.4.1 explains the mentioned control elements in IMS shown in this figure.

When deploying WebRTC in companies, security is a major concern. Companies may reluctantly allow SIP-based phone calls to pass into their network, but they will restrict the use of uncontrolled data applications on browsers. As mentioned before, application data can be identified by a codepoint in DTLS packets, and this may work to control the use of applications. The literature also proposes to install TURN gateways at the borderline between the internal and the external network [488] to control SIP telephony. The article also mentions that a man-in-the-middle attack might also be used for the SRTP key generation. This setup might be a challenge for security groups in companies.

9

Streaming and Over-the-Top TV

In this chapter, we discuss the current and most widely deployed streaming technologies and how they are implemented in the marketplace. The technology review will start with proprietary implementations. Section 9.1 reviews Apple's HTTP Live Streaming protocol. Microsoft's Smooth Streaming is then covered in Section 9.2. The last proprietary technology is Adobe's HTTP Dynamic Streaming (HDS) treated in Section 9.3. The standards-based Dynamic Adaptive Streaming over HTTP (DASH) is treated in Section 9.4 with the network interaction of streaming discussed in Section 9.5. We then shift, in Section 9.6, to technologies used to deliver this content cost-effectively over the Internet. Finally, Section 9.7 looks at the current top content providers and describes their implementations.

Over-the-Top TV

The term *Over-the-Top TV* refers to video content that is provided over the Internet. Stated another way, the content provider has no direct control over the network (Internet). A well-known example is YouTube.

An important aspect of each of these implementations is the use of HTTP as a transport. Although Real-Time Protocol (RTP) was widely accepted, see Section 4.3, several factors have led to the shift to HTTP.

- The significant increase of bandwidth available to the average consumer has minimized the impact of using a connection-based protocol (TCP). If packets are lost, retransmission need not significantly impact the stream if there is sufficient bandwidth.
- Virtually all home and many corporate networks use Network Address Translation (NAT) to access the Internet. Using HTTP over TCP eliminates almost all problems of NAT traversal. See Section 6.5.
- Using HTTP as a transport minimizes or eliminates the need for special server software. Standard web servers (sometimes with enhancements), proxies, and caching servers can be used to deliver content.

Multimedia Networks: Protocols, Design, and Applications, First Edition. Hans W. Barz and Gregory A. Bassett.
© 2016 John Wiley & Sons, Ltd. Published 2016 by John Wiley & Sons, Ltd.

You should note that HTTP streaming technologies we cover are quite similar. All are based on HTTP, and some use virtually the same encoding and file structure schemes. It is also important to keep in mind that this particular technology is changing rapidly with almost constant challenges to the incumbent providers.

9.1 HTTP Live Streaming – Apple

HTTP Live Streaming (HLS) is an adaptive streaming protocol developed by Apple to support their family of devices from iOS and Apple TV to Macintosh running OSX. HLS supports both on-demand and live streams. It is the only streaming protocol natively supported on Apple's platforms. Due to the popularity of Apple devices, there has been wide acceptance of HLS by both streaming server vendors and other client platforms. Apple requires any iPhone or iPad application that sends large amounts of audio or video data over the cellular network to use HLS. Apple has submitted HTTP Live Streaming as an Internet Draft [489].

HLS is comprised of three major components, as seen in Figure 9.1:

Media preparation: This system is responsible for accepting audio and video input, encoding it using H.264 and AAC or MP3, placing the encoded content into an MPEG Transport Stream container (see Section 3.4.1), and segmenting that stream into individual files (or chunks) of normally 5–10 seconds of content. The input can be encoded at different quality

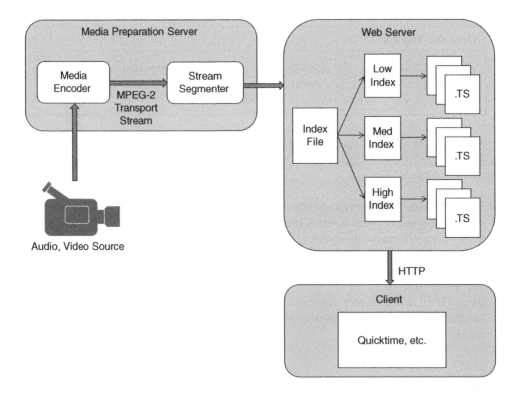

Figure 9.1　HLS structure

levels or bitrates. This system also creates a hierarchical set of index files that describe the available .TS files at the various quality levels, as shown in Figure 9.1. This process can also occur in real time with .TS and index files created as live content is encoded and segmented.

Web server: After the .TS and index files are created, they are placed on a normal HTTP server and published for distribution. The web server is responsible only for hosting and delivering requested files.

Client player: The playback of content is totally controlled by the client player. It requests the index and specific .TS files to download the content with the desired quality. The client is also responsible for changing the quality of the stream, if appropriate. Under normal conditions, this decision is made by estimating the available bandwidth to the server and choosing the best quality that can be downloaded without interrupting playback. It does this transparently to the user by downloading the appropriate index file and requesting different .TS files.

In Figure 9.2, we can see an example of a resulting Index File. The extension of this file is .M3U8. This simple example shows a single stream with a total of 29.1 seconds of content, separated into three consecutive segments. In this case, there is only one quality level (bit rate) offered.

Although HLS does not natively support the protection of content using Digital Rights Management (DRM), Apple recommends a method to encrypt the content with Advanced Encryption Standard (AES-128) encryption and including a cipher key in the index file.

In summary, HLS has the following key characteristics:

- Defines a highly standardized method to prepare video and audio content, resulting in a set of index and content files.
- Has native support in all appropriate Apple products. Many other platforms (Windows, Android, etc.) can also play HLS-streamed content.
- Uses standard HTTP web servers, proxies, caching devices, and other equipment to distribute content.
- Uses an intelligent client player to dynamically select content stream to maximize quality of playback.

HLS Index file for 3 segments of 10 seconds each

```
#EXT-X-VERSION:3
#EXTM3U
#EXT-X-TARGETDURATION:10
#EXT-X-MEDIA-SEQUENCE:1

#EXTINF:10.0,
http://media-server.example.com/segment0.ts
#EXTINF:10.0,
http://media-server.example.com/segment1.ts
#EXTINF:9.1,
http://media-server.example.com/segment2.ts
#EXT-X-ENDLIST
```

Figure 9.2 Simple sample of HLS index file

9.2   Smooth Streaming – Microsoft

Smooth Streaming is Microsoft's offering in the HTTP adaptive streaming space. It was intro-
duced in 2008 as part of the Silverlight Architecture [490]. Smooth Streaming is a Media
Services extension to Microsoft's Internet Information Services (IIS) extensible web server.
Microsoft now brands Silverlight, IIS, and Smooth Streaming under the name Microsoft
Media Platform.

Smooth Streaming is encoded with H.264/AAC and uses the Microsoft's Protected Interoperable File Format (PIFF) based on the ISO Base Media File Format (ISO/IEC 14496-12) as
its disk (storage) and wire (transport) format – see Section 5.9. This file format is also known
as *fragmented MP4* format. The reason for this specific file format is based on Microsoft's
special approach to file storage on the web server. You will recall that Apple's HLS segments
the media into (normally) 5-second chunks. This results in a large number of individual files
stored on the web server. For example, a 2 hour movie encoded with three different bit rates
results in more than 4000 files. Since the HLS client requests a chunk in the form of a single
file, the web server needs no special knowledge of the content of the file and simply delivers
it to the client.

Microsoft took the approach of creating a single content file per bit rate encoding. Using
the ISO/IEC 14496-12 file format, the content is internally segmented and a manifest describing the content is created. During delivery, the IIS server extracts a virtual segment out of
the appropriate content file and streams it to the client. This requires some knowledge of the
content and processing power on the IIS server, but simplifies file storage.

Another significant difference with Smooth Streaming is the use of a one-time manifest and
timecodes to request specific segments in the stream. This is as opposed to the continuously
updated index file of HLS that must be frequently downloaded. The timecode is used to specify
a specific location within the presentation. If the stream is live, the timecode is incremented to
match the live stream. The manifest does not need to be updated or downloaded.

In an effort to capture more of the Apple device market, Microsoft introduced a feature
in IIS Media Services 4.0, which enables Live Smooth Streaming H.264/AAC videos to be
dynamically repackaged into the Apple HTTP Adaptive Streaming format and delivered to
iOS devices without the need for re-encoding.

The three types of files stored on the IIS server are:

- The server manifest file (.ism) describing the relationships between the media tracks, bit
  rates, and files on disk. This file is used by the server to provide segments when requested
  by the clients.
- The client manifest file (.ismc) provides several key types of metadata necessary for the
  client to download, decode, and render the media.
- MP4 files (.ismv) containing audio and/or video. There is one file per encoded video bit rate.

The server and client manifest files are relatively large and complex. Therefore, samples are
not included in this book. Detailed information and samples can be obtained on Microsoft's
Developer Network website.[1]

Figure 9.3 illustrates the overall structure and data flow of Smooth Streaming. As with
Apple's HLS (see Figure 9.1), Smooth Streaming has three components:

Media preparation: The encoder accepts either a fixed media file or a live stream and produces one fragmented MP4 file for each offered bit rate. Each fragment with the MP4 file

---

[1] https://msdn.microsoft.com/en-us/library/ff436055(v=vs.90).aspx [accessed 18 April 2015].

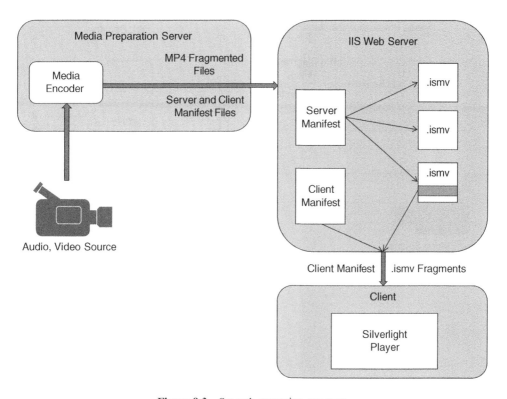

Figure 9.3 Smooth streaming structure

is normally 2 seconds long. This system also creates manifest files for both the server and the client.

IIS web server: The MP4 and manifest files are placed on an IIS HTTP server and published for distribution. The web server is responsible for a number of functions:

- Download the Client Manifest file when requested. This happens only before the beginning of the stream.
- Accept a URL from the client specifying which bit rate and fragment to download.
- Extract the appropriate fragment from the MP4 file using the information contained in the server manifest.
- Download the fragment to the client.

Client player: The playback of content is totally controlled by the Silverlight client player. It requests the client manifest file to determine the available bit rates, fragment sizes, and appropriate URLs. As with HLS, the client is also responsible for changing the quality of the stream, if appropriate.

9.3 HTTP Dynamic Streaming – Adobe

Similar to Apple and Microsoft, Adobe also introduced multimedia streaming based on HTTP. As with Smooth Streaming, HDS uses H.264 encoding with an MPEG-4 Part 14 (ISO/IEC 14496-12) file format. The content is stored in .f4f files with one file per segment per encoded bit rate. HDS also uses a manifest file (.f4m) to describe the content to the client and an index

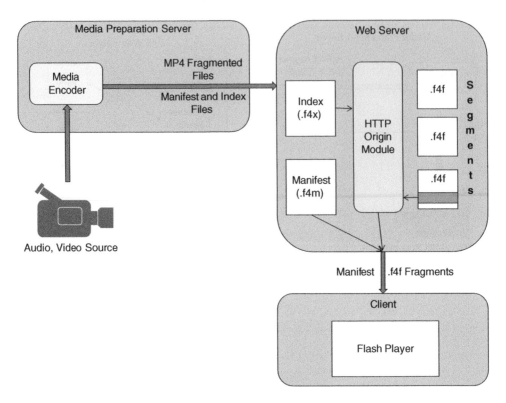

Figure 9.4   HDS structure

file (.f4x) to describe how the server should extract fragments from the .f4f files. On the server side, Adobe provides a plug-in called the "HTTP Origin Module" to transfer and interpret the manifest and index files.

Figure 9.4 shows the relationship between the various modules of HDS. Note that the different bit rate encodings are called segments. The following list demonstrates the flow of events to play a video using HDS:

- The HDS Player sends an HTTP request to the Web server;
    for example, http://www.server.com/media/sample-video.f4m.
- The web server sends the request to the HTTP Origin Module.
- The HTTP Origin Module returns the manifest file (.f4m) to the client.
- The client receives the manifest file and uses the contained data to translate the encoding time to the segment # / fragment # / format.
- The client sends a second HTTP request to the web server and requests a specific content fragment; for example, http://www.server.com/media/sample-video-seg1-frag1.
- The web server receives the request and transmits it to the HTTP Origin Module.
- The HTTP Origin Module uses the index file (.f4x) to define the offset (in bytes) of the media file (.f4f) and sends the appropriate fragment to the client.

The previous three sections described the major proprietary HTTP Adaptive Streaming technologies. Note that although some implementation differences exist, the overall structure and

Table 9.1 Proprietary streaming protocols

Features – see Ref. [149]	Windows Smooth Streaming	Apple's HTTP Live Streaming	Adobe's HTTP Dynamic Streaming
Specific HTTP server required?	Yes: chunks are embedded in specially structured files	No: since chunks are individual files	Yes: fragments are embedded in .f4f segment files
Index file format	Proprietary XML file	M3U8 format for playlists	Proprietary (.f4m)
Formats supported	Fragmented MP4	No restriction	Fragmented MP4

delivery methods are quite similar. Table 9.1 describes these three technologies and highlights the key differences. For a more complete comparison, see Ref. [491].

9.4 Dynamic Adaptive Streaming over HTTP – DASH

Dynamic Adaptive Streaming over HTTP, or *DASH*, is a general term to describe any HTTP streaming method that allows the client to select the quality of a media stream from those offered by the server. There is also a standard that has been defined for a specific instance of DASH. The ISO/IEC FCD 23009-1 standard is better known as *MPEG-DASH* [492]. In this section, we discuss *MPEG-DASH*.

9.4.1 History of MPEG-DASH

The first adaptive standards were published using RTP as a transport and are described in the profile for SVC [114, 115]. Adaptive Streaming over RTP was also standardized by ETSI 3GPP for use over wireless mobile networks [493]. In order to promote standardization, the Open IPTV Forum (OIPF) extended 3GPP's Release 9 Adaptive HTTP Streaming (AHS) format to include additional features and support for MPEG-2 Transport Streams [494].

MPEG began work on an HTTP-based standard (MPEG-DASH) in 2010, and the first version was published as an ISO standard in 2012 [492]. It was based on the 3GPP, Open IPTV, and proprietary standards mentioned. The second edition of this standard was published in 2014. The latest version of the Open IPTV standard aligned its specifications with MPEG-DASH and 3GPP Release 9 standards [494].

As mentioned earlier in this chapter, the first major commercial implementations of adaptive streaming were proprietary implementations such as Microsoft's Smooth Streaming, Apple's HTTP Live Streaming, and Adobe's HTTP Dynamic Streaming [495].

9.4.2 Description of MPEG-DASH

MPEG-DASH shares a number of key similarities with the three proprietary technologies already discussed.

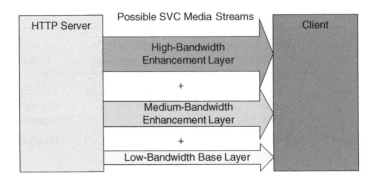

Figure 9.5   SVC layers

- HTTP is used for transport, permitting the use of standard web components and distribution technologies.
- Although not strictly specified, the most commonly used encoding scheme is H.264.
- The client chooses and changes the stream quality/bit rate.
- Segments are stored as individual files as in Apple's HLS.
- All have some form of index and stream description file(s).

MPEG-DASH differs from those streaming methods in one key area; the encoded streams. MPEG-DASH enables the adjustment of the quality of the stream by offering the client a number of SVC layers, see Sections 3.4.5 and 4.1.3. As shown in Figure 9.5, these layers are additive. It means that the client will always receive the *base layer*. To increase the quality of the broadcast, additional layers can be added. In this example, the highest possible quality consists of the base layer and two additional enhancement layers.

To see how this works over time, refer to Figure 9.6. Here, the same three layers as in Figure 9.5 are available. To offer the user a rapid initial display of the stream, only the base layer is transmitted at the start. As desired and with appropriate network conditions, additional layers can be transmitted to enhance the quality of the stream. If network congestion is detected, the transmission of the enhanced layers can be stopped to bring the bandwidth usage down. When the congestion no longer exists, the additional enhancement layers can again be added to the stream.

To reiterate, MPEG-DASH is not a protocol, system, or client specification. It is a set of formats that can be used with an HTTP transport to deliver media streams. There are two major components in MPEG-DASH: the *Media Presentation Description* (*MPD*) and the format of the media segments. The MPD defines three major items in XML format: *Periods, Representations, and Segments*, including time, format, and size. To see how these elements relate to each other, refer to Figure 9.7. You can see a sample of an MPD in Figure 9.8. Note how the representations and segments are structured and the fact that, in this example, there are two audio streams available.

Period: This is the largest segment of a media stream. All of the Periods in a contiguous stream constitute the complete media presentation.

Representation: Each Period can have multiple encodings, resolutions, or frame rates. These different versions of the content are called Representations. These are also called layers.

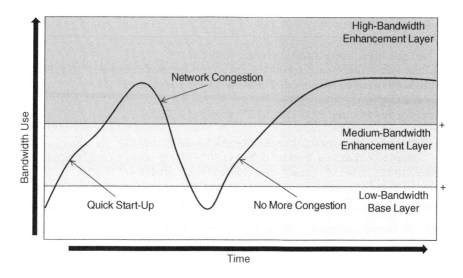

Figure 9.6 Dynamically SVC adapting streams

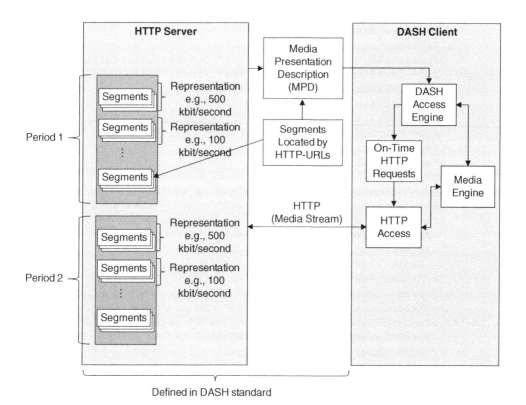

Figure 9.7 MPEG-DASH format and message exchange

MPD Sample

```
<MPD xmlns="urn:mpeg:DASH:schema:MPD:2011"
mediaPresentationDuration="PT0H3M1.63S" minBufferTime="PT1.5S"
profiles="urn:mpeg:dash:profile:isoff-on-demand:2011"
type="static">
  <Period duration="PT0H3M1.63S" start="PT0S">
    <AdaptationSet>
      <ContentComponent contentType="video" id="1" />
      <Representation bandwidth="4190760" codecs="avc1.640028"
height="1080" id="1" mimeType="video/mp4" width="1920">
        <BaseURL>sample1-89.mp4</BaseURL>
        <SegmentBase indexRange="674-1149">
          <Initialization range="0-673" />
        </SegmentBase>
      </Representation>
      <Representation bandwidth="869460" codecs="avc1.4d401e"
height="480" id="3" mimeType="video/mp4" width="854">
        <BaseURL>sample1-87.mp4</BaseURL>
        <SegmentBase indexRange="708-1183">
          <Initialization range="0-707" />
        </SegmentBase>
      </Representation>
      <Representation bandwidth="264835" codecs="avc1.4d4015"
height="240" id="5" mimeType="video/mp4" width="426">
        <BaseURL>sample1-85.mp4</BaseURL>
        <SegmentBase indexRange="672-1147">
          <Initialization range="0-671" />
        </SegmentBase>
      </Representation>
    </AdaptationSet>
    <AdaptationSet>
      <ContentComponent contentType="audio" id="2" />
      <Representation bandwidth="127236" codecs="mp4a.40.2"
id="6" mimeType="audio/mp4" numChannels="2" sampleRate="44100">
        <BaseURL>sample1-8c.mp4</BaseURL>
        <SegmentBase indexRange="592-851">
          <Initialization range="0-591" />
        </SegmentBase>
      </Representation>
      <Representation bandwidth="31749" codecs="mp4a.40.5"
 id="8" mimeType="audio/mp4" numChannels="1" sampleRate="22050">
        <BaseURL>sample1-8b.mp4</BaseURL>
        <SegmentBase indexRange="592-851">
          <Initialization range="0-591" />
        </SegmentBase>
      </Representation>
    </AdaptationSet>
  </Period>
</MPD>
```

Figure 9.8   MPD sample

Segment: Each Representation consists of a collection of Segments, each of which can be uniquely identified by a URL. This URL either can be explicitly listed or can be derived using a template.

Normally, the client will send a request to the server (using HTTP) and will receive an MPD with details about content and URLs used to access the media streams. The client will use the provided URLs to download the media stream. The Initialization range and index range within the Segment Base section will tell the client where to find the header information and content within the stream. The client can use a different set of URLs at any point to change the request to a different representation of the media stream (e.g., higher or lower bitrate/resolution) [496]. If the client requests a change to a different representation, the change will only occur at the end of the current segment. It is important to note that MPEG-DASH only states the capabilities the client must have – as described in Figure 9.7. It does not state how well the client implements these capabilities. For example, a good client might be expected to verify that local CPU power and network bandwidth are sufficient *before* requesting an increase in stream quality. This may not always be the case.

There are special considerations when MPEG-DASH is used for live streams. From an end-user viewpoint, the most important consideration is the end-to-end delay. For example, in the broadcast of a sports event, it is annoying to hear a cheer from nearby people several seconds before the user sees a goal from the media stream. Another consideration is the generation and refresh of the MPD. Since the segments are generated in real time, the MPD defines the access URLs either at the end of the broadcast or until the next MPD refresh. The client is expected to retrieve the MPD as recommended in the MPD. Since the MPD defines the broadcast start time in terms of the wall-clock time, the client's clock must be synchronized to properly calculate the latest media segment to request [497].

It is interesting to note that, as mentioned in Section 2.2, Telecoms and television broadcasters in some countries often restrict streaming to either pay-per-view or certain geographies. This is most prominent with the streaming of live sporting events.

9.5   DASH and Network Interaction

As Internet capabilities, consumer broadband connections, and streaming technologies have improved, the amount of Internet traffic generated by streaming video has increased substantially. In 2011, 29.7% of all peak downstream traffic in the United States and Canada was generated by Netflix [498]. In 2012, this increased to 32.25% in the United States. Internet video is now the single biggest traffic generator [499]. To offer the end user a consistent, high-quality experience under changing network conditions and various client systems, content providers have started implementing DASH.

Much research has been performed to understand the impact of different DASH implementations on the playback and on the network. Much of this research has concentrated on three major themes:

- Reaction of Player to Changing Network Conditions
- Fairness, Efficiency, and Stability
- "Bufferbloat" in network routers.

As we introduce this topic, it is important to reiterate that the client player makes all decisions and takes actions regarding rate adaption. Each player implementation has its own bandwidth measurement and rate adaption algorithms. Each of these implementations has a different impact on user experience and the network. A key challenge is the interaction of the client's rate adaptation algorithm with TCP congestion control. This creates a *nested double feedback loop* that is extremely difficult to model or predict.

9.5.1 Player Reaction to Network Conditions

Some research, such as Ref. [500], measured how a variety of players reacted to varying network conditions. What they observed is that the algorithms implemented in the different players radically impacted how the players reacted. For example, the Smooth Streaming player quickly converges to the highest sustainable bit rate and fills its large playback buffer as quickly as bandwidth permits. Since it smoothes the bandwidth calculation over time, it has a tendency to react more slowly to increases and decreases of available bandwidth. This leads to lower effective use of bandwidth.

The Netflix player is similar to the Smooth Streaming Player, with both based on Silverlight. However, the algorithms used are quite different and result in a much more aggressive rate adaption behavior with the Netflix Player. It attempts to provide the highest quality possible and may trigger additional bit rate changes to accomplish this. This player also uses a very large playback buffer of up to several minutes. It will even switch to bit rates higher than the available bandwidth if the playback buffer is almost full.

The referenced research [500] provides substantial additional detail on the tests and conclusions. It is recommended reading for anyone interested in this topic. As Silverlight includes a software development kit, implementers can build their own control algorithms for playback.[2]

9.5.2 Fairness, Efficiency, and Stability

With a rapidly increasing number of video streams and players, it is important to examine the interaction across multiple, simultaneous streams that will be competing at potential chokepoints within the network. Research in this area has concentrated on three major (and sometimes conflicting) design goals [501]:

Fairness: Multiple competing streams should converge on an equitable allocation of networking resources at potential bottleneck links.
Efficiency: Players should maximize the user experience by converging on the highest sustainable bit rates.
Stability: Players should minimize unnecessary bit rate changes, which can adversely impact user experience.

Measurements have shown that the existing commercial players fail in one or more of these goals when competing with other streams over a network bottleneck [502].

Figure 9.9 shows the major algorithms used by an Adaptive Rate player.

Bandwidth estimation: One root cause of the issues observed is the fact that players measure the network state during the download of a chunk. When the player is in a steady state

---

[2] https://msdn.microsoft.com/en-us/Silverlight.

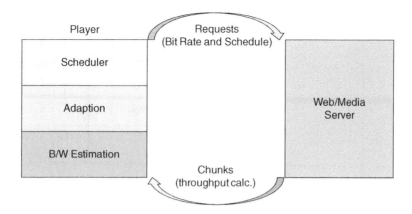

Figure 9.9   Adaptive player framework

and downloading chunks on a periodic basis, no measurement is possible during the idle period between requests. This means that the player does not have a true picture of the network state.

Adaption: In current player implementations, the calculation of desired bit rate is stateless. The calculation is based on available bandwidth and offered bit rates. If one of the players is currently streaming a higher bit rate, it will observe a higher bit rate. This can result in multiple players "locking in" to bit rates that are not evenly distributed. In addition, unless decisions to switch bit rates are smoothed and delayed, unnecessary switches may occur.

Scheduler: When players reach a steady state, they generally request and receive chunks on a periodic basis. With multiple players, the requests can become synchronized and result in inefficient use of bottleneck link resources.

As with much of the cited research in this area, Ref. [500] introduces possible algorithms and template players that could overcome some of the identified issues. For example, Ref. [501] proposes the following improvements:

- A harmonic bandwidth estimator based on a number of past estimates will provide a more accurate picture of the true bandwidth availability.
- Stateful and delayed bit rate decisions can be based on the bandwidth and the current bit rate selection. Aggressive changes in the beginning of a stream will slow down as the bit rate becomes higher. The conflicting goals here are efficiency (i.e., maximum bandwidth utilization) and stability (minimum changes).
- Introducing randomized scheduling for downloading chunks during steady state will significantly reduce multiplayer download conflicts.

9.5.3   Bufferbloat

The concept and impact of bufferbloat were treated in detail in Section 5.7. Since bufferbloat is caused in part by TCP and the streaming in this chapter is all based on HTTP over TCP, we will summarize and re-emphasize the important and relevant points here.

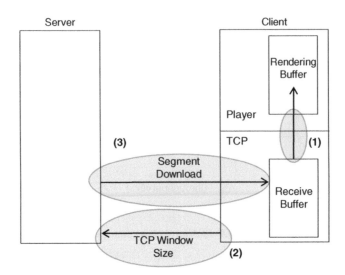

Figure 9.10   SABR bandwidth control technique

The basic cause of bufferbloat is the interaction of TCP and very large packet buffers in the network. When there is a network bottleneck, these large buffers can "hide" the congestion for a time and delay the dropping of packets. Since each TCP flow can have a number of packets in flight, a dropped packet can cause the retransmission of a significant number of packets in the stream. Although TCP can adapt to bandwidth restrictions, the delay in loss detection can cause TCP to miscalculate the available bandwidth. This impact is made worse by the fact that HTTP streaming tends to be very bursty in its steady state. For example, a 2–10 seconds of video segment is downloaded from the server, then the client pauses for several seconds before the next segment is downloaded. Since the client TCP buffers are normally empty just before the next segment is downloaded, TCP offers a large value for the bytes-in-flight or receive window and the transfer starts at a high data rate. All of this results in bufferbloat and significant delays for other ongoing Internet applications in a typical residential network.

As discussed in Section 5.7, a number of methods have been proposed to mitigate this problem. Each proposal to date has its limitations and implementation issues. A common issue with most proposals is that they address the problem by implementing measures in Internet routers. Due to the number of routers and vendors involved, adoption would likely take a very long time.

In the context of HTTP Streaming, there is one previously mentioned proposal worth detailing. In Reference [261], the authors propose a method of mitigating bufferbloat by controlling the download speed of video segments at the application layer. They call their technique *Smooth Adaptive Bit Rate (SABR)*. A very simple representation of this technique is shown in Figure 9.10. By carefully controlling the rate at which the receive buffer is emptied and leaving it partially filled (1), TCP calculates and sends a smaller receive window size (2), and downloads become less "bursty" (3). In the measurements, bufferbloat was significantly reduced and resulted in minimal impact on other networked applications. Special modes are used to permit the player to adapt bit rates and use the best profile for the stream.

This proposal is interesting because it does not require changes to the network infrastructure, only to the media player.

9.6   Content Delivery Networks

In Section 4.7, we discussed the general topic of caching. In the past, caching of video content required systems that understood standard and proprietary streaming protocols, such as RTP and the proprietary Real-Time Messaging Protocol (RTMP) from Adobe [503]. As the transport of video shifted more to HTTP, standardized web server implementations made distributed content delivery more economical. In this section, we focus on the technologies and implementations of Content Delivery Networks (CDN) that deliver video content primarily using HTTP.

9.6.1   CDN Technology

A CDN, sometimes called a Content Distribution Network, is a collection of geographically dispersed servers implemented to provide high-performance and reliable delivery of content to end users. This content can include any downloadable object such as web pages, images, applications, social networking content, and both on-demand and live streaming media. Table 9.2 summarizes the driving force behind the deployment of CDNs [504]. This table shows the impact of increasing distance on the download of a large video (or any other) file.

CDNs can be implemented on internal networks by Enterprises, in networks of a specific Internet Service Provider (ISP), or on the Internet by CDN Service Providers. Although Enterprises can and do implement internal CDNs, the buildup requires significant expertise and resources [505]. Enterprise CDNs are frequently used in companies that stream a large number of internal Webcasts.

ISPs implement CDNs to reduce the load on their networks and, if they offer additional services such as IPTV, host content close to their end users. The Digital Video Broadcast (DVB) standard [506] details how a content provider, such as an ISP, could implement a CDN.

Table 9.2   Effect of distance on throughput and download times

Distance server to user	Network latency	Typ. packet loss	Throughput (quality)	4 GB DVD download time
Local: <100 miles	1.6 milliseconds	0.6%	44 Mbit/second (HDTV)	12 minutes
Regional: 500–1000 miles	16 milliseconds	0.7%	4 Mbit/second (not quite DVD)	2.2 hours
Cross-continent: 3000 miles	48 milliseconds	1.0%	1 Mbit/second (not quite SD)	8.2 hours
Multi-continent: 6000 miles	96 milliseconds	1.4%	0.4 Mbit/second (poor)	20 hours

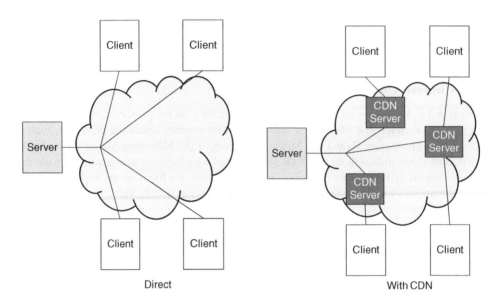

Figure 9.11   Comparison of direct versus CDN delivery of content to client

Mobile operators also implement CDNs for the same reasons as ISPs. This is particularly true for Mobile operators that provide premium content to their users [507].

In its most common form, a CDN is implemented on the Internet by a CDN Service Provider. Content Providers then pay the CDN Provider to deliver their content to their end users. The content is loaded onto multiple CDN servers and delivered to end users closest to each server. The overall structure of a CDN implementation is relatively straightforward and is illustrated in Figure 9.11. This structure serves multiple purposes:

- It reduces the load on the Content Provider's servers and network links.
- With the delivery server closer (in networking terms), the user experiences less packet loss, lower latency, and less jitter.
- With multiple sources of data, both reliability and resistance to Denial-of-Service attacks are higher.
- It reduces the overall bandwidth usage within the network.
- Scaling to very large numbers of users is much easier.

Although the overall structure of CDNs is fairly simple, there are a number of very complex technologies involved [508].

Content loading: Content to be loaded must be identified from the Provider's available media on the *origin server*. Normally, not all content is served by the CDNs. There are many algorithms available, but popularity-based caching algorithms are frequently used by the larger Content Providers. Content from the Content Provider can either be preloaded or cached in the traditional sense after the first user access. The choice of which content to load (and when) is more complex. For example, a provider may choose to host a lower, more commonly accessed bit rate stream on one CDN server, but host a higher quality stream on a different, possibly less expensive, server. Alternatively, some providers host popular content on the CDN network but stream lesser accessed content centrally. The retention period of content on the CDNs must also be decided.

Content distribution/synchronization within the CDN: If the CDN servers are acting as pure caching servers, redistribution of content to other servers does not happen. That is, if a CDN server receives a request for an object and does not have that object, it requests it from the Content Provider origin server, caches it, and delivers it to the client. However, if content needs to be replicated on more than a single server, there exist a number of methods to accomplish this. The content can be pushed or pulled from the original server to each CDN server. Alternatively, the CDN servers can exchange content using peer-to-peer connections. This is illustrated in Figure 9.12.

CDN server selection: Perhaps the most important step in this process is to direct the client to the appropriate CDN server. The client begins by sending a request to the Content Provider's server to download or stream specific content. The Content Provider can use a number of different criteria to select a CDN server.

- The *Geolocation* of the client's IP address can been looked up and a CDN server that is geographically the closest can be assigned.
- The number of network hops and latency can be used to select a CDN server that is closest in terms of network proximity.
- A *static IP address* lookup table can be used, although this is not common.
- CDN servers can also be selected based on availability, load, or other criteria of other CDN servers in the network.
- The aforementioned criteria can be combined in almost any algorithm to select a CDN server. This means that the CDN server assigned may not necessarily be the closest in either physical or network terms.

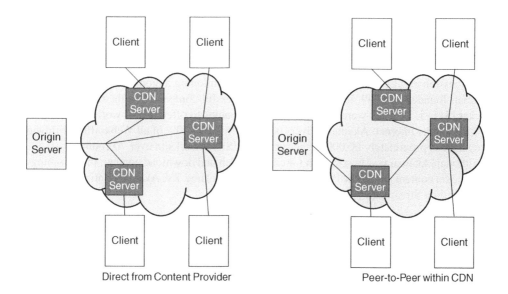

Figure 9.12 Distribution of content: direct versus peer-to-peer

Once the appropriate CDN server is identified, the Content Provider must direct the client to this server. There are two methods used to accomplish this:

- The URL sent by the client to the Content Provider to request the content can be answered with an HTTP 30× redirect, sending the client to the selected CDN server.
- Before the client sends a URL to a server, it must look up the IP address of that server using Domain Name Service (DNS). For example, for the URL http://content.provider.com/video/segment1.mp4, the client must resolve the name *content.provider.com*. The DNS servers that are authoritative for that domain reply with the IP address of the selected CDN server.

Inter-CDN connections: As mentioned earlier, CDNs are sometimes provided by ISPs and other entities. In an effort to maximize the quality of experience for their end users, many Content Providers use multiple CDNs to deliver content. At present, since there are no standards for the direct interconnection of CDNs, the Content Providers must treat each CDN as a separate, independent entity. In an effort to define standards in this space and allow the exchange of content, there is now an IETF Working Group for *Content Delivery Networks Interconnection* (CDNI) [509]. RFCs have been published to define the problem statement, use cases, requirements, and a framework. Draft RFCs exist for many CDNI technical areas.

Live streaming: Earlier in this list we discussed the loading of content to the edge servers streaming content to the end users. This applies to Video on Demand (VoD) where content can be preloaded (if desired) with optimization concentrating on the last hop to the user. When CDN providers stream live content, the delivery paradigm changes significantly. Now the paths from the ingest server (that receives content from the provider) to all of the edge servers must be optimized for the real-time delivery of what is often a 3 Mbit/second stream [510]. All major CD providers, including Akamai, Level 3, and Limelight, offer live streaming already but must prepare for it in a significantly different way than VoD content.

9.6.2   Akamai

No discussion of CDN would be complete without mentioning a pioneer of CDN technology. Akamai, founded in 1999, is one of the most successful companies in the Content Caching market [511–513]. They were one of the first companies to offer almost worldwide "caching for hire" on the Internet. Akamai delivers between 15% and 20% of all Web traffic worldwide. They have approximately 60,000 cache servers in ISP networks in over 70 countries.

Although Akamai was best known for caching Web and download content, their caching of multimedia content started in 1999 with Apple's QuickTime TV. Akamai continues to be one of the leading Streaming Video CDN providers.

9.6.3   The Future of CDNs

The CDN marketplace is changing at an extremely rapid pace. What started as a caching platform to help web pages load faster has now become the primary distribution method for video over the Internet. Here, we concentrate on two key topics.

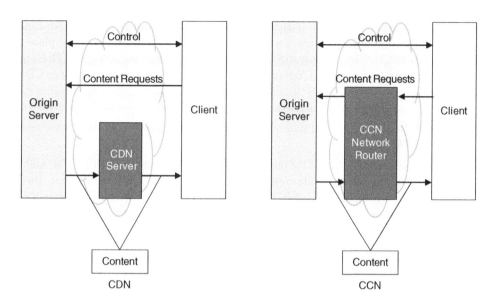

Figure 9.13   Content delivery networks versus content centric networks

9.6.3.1   Content Centric Networks

At present, DASH and similar Adaptive HTTP Streaming methods retrieve segments of video using a URL that points to the server containing the content. This means the client is asking for a resource based on *where* that resource is stored. For example, the URL http://content .provider.com/video/segment1.mp4 defines the IP name of the server (content.provider.com) as well as the location of the content on the server (/video/segment1.mp4). Over the past several years, there has been much research into the concept of Content Centric Networks (CCN), a type of Information-Centric Network [514]. The basic premise of CCN is that instead of a client specifying a location (*where*), the client specifies instead *what* it is interested in, called a *named data* object. The network then places this named data request into a *Pending Internet Table (PIT)*, determines the location of that content, retrieves it, and streams it to the client. See Figure 9.13 to compare the high-level differences in how the client obtains the content. Although this technology is in its infancy, the overall paradigm of routing packets based on content request and delivery fits the future of the Internet significantly better than existing point-to-point location-based connections. In some ways, this technology can be considered an extension of CDNs. The CCN Network Routers shown in Figure 9.13 provide, in addition to other functions, exactly the same service as CDN servers. The key difference is how that data is requested by the client. Delivery of the content remains much the same.

9.6.3.2   Managing Multiple CDNs

As discussed earlier, large Content Providers frequently contract with multiple CDN service providers to offer their clients high-availability and extended geographical reach. At the present, each of the CDN provided must be managed separately. We also discussed research to develop standard methods to interconnect CDNs [509]. Another research direction examines the possibility of creating a global measurement-driven video control plane to select appropriate CDN and bit rates [515]. While control planes exist within CDNs, this research examines a global, multi-CDN implementation with the possibility to change CDNs mid-stream.

9.7 Providers

In this section, we discuss some of the current top video content providers. These are not the IPTV vendors, but those that provide content "Over The Top" (OTT) via the Internet. This is not intended to be a comprehensive list, but a discussion on how these vendors prepare and distribute content to their end users. We focus on the technical aspects of their implementation. In some cases, limited information exists about the technical implementation. It is important to note that the following information is current as of the publication of this book. Rapid advances in the marketplace, such as consolidations, HTML5, and other technical advances, will render some of this content out-of-date over time.

9.7.1 Amazon Instant Video

Amazon first introduced streaming video in 2006 as Amazon Unbox. In 2008, it was renamed Amazon Video on Demand and again renamed Amazon Instant Video in 2011. Amazon has continually added content to better compete in the market and offers this service either as pay-per-view or as part of their Amazon Prime service.

Support for Amazon Instant Video is either built-in or downloadable for a number of televisions, game consoles, set-top boxes, Blu-ray players, and others. Support for PCs and Macintosh is through Silverlight or Flash. Tablets and smartphones are supported with dedicated applications. Amazon uses only Akamai to provide CDN services.

9.7.2 YouTube

YouTube was founded in 2005 as a video repository for users with the ability to share content [516]. Its success was immediate and spectacular. YouTube was purchased by Google in 2006 and continued its rapid growth. By 2012, the number of videos viewed per day had reached 4 billion. During this time, the delivery infrastructure evolved almost constantly. The third-party distribution infrastructure is now wholly operated and maintained by Google. Google has provided limited information about the architecture and technologies behind YouTube. However, as with most major providers, there has been extensive investigation and research.

As of 2015, YouTube supports Flash and HTML5 containers for its streaming video [517]. For Flash streamed videos, a dedicated Shockwave Flash player is downloaded to control the browser's Flash plug-in. During start-up of a Flash format video stream, YouTube aggressively downloads the video file until the play-out buffer is full (40 seconds for Flash and 10–15 MB for HTML5) [518]. In January 2015, YouTube changed the default container to HTML5 with MPEG-DASH for those browsers that support it.

The YouTube cache servers are organized in a three-tier hierarchy [519]. Some of the cache servers are colocated inside ISPs. Based on measurements in the cited research, YouTube assigns cache servers based on network proximity (in terms of round-trip delay) and internal load balancing. In addition, it was observed that the cache servers assigned depended on the time of day and the client's ISP. YouTube uses a combination of HTTP redirect messages and DNS entries to direct the client to the appropriate cache server.

9.7.3 Netflix

Netflix is an on-demand streaming video provider distributing content with over 60 million subscribers in over 50 countries (as of 2015) [520]. Netflix expanded from a DVD rental company to on-line streaming in 2007 as the sales of DVDs began to drop. In 2014, Netflix accounted for than 30% of prime-time (evenings) Internet traffic [521]. Due to Netflix's size and dominance in the video streaming market, significant interest has been shown and research has been done on its strategies, architectures, and technologies.

Netflix maintains a minimal data center internally, which hosts their primary website for user registration and payment [498]. The servers used for content storage and distribution are hosted in the Amazon cloud [522]. For distribution, Netflix contracts with three CDN providers: Akamai, Level 3, and Limelight.

Netflix started by streaming their content to the client using Adobe Flash. They later migrated to Microsoft's Silverlight plug-in for playing their content in web browsers. Netflix is currently shifting to HTML5 in place of Silverlight, which is supported on Apple's Safari Browser on late model Macs, Internet Explorer 11 on Windows 8.1, and modern versions of the Chrome browser on PCs and Macs. Netflix requires support for certain HTML5 options, such as Media Source Extensions to support DASH adaptive streaming, and Encrypted Media Extensions to support Digital Rights Management encryption of media [264]. For special devices, such as gaming consoles and set-top boxes, special applications are implemented. All streaming videos use the DASH protocol.

As mentioned earlier, there has been a significant amount of research on Netflix. Adhikari et al. [498] is a good example of this research. They examined the behavior of the Netflix ecosystem on how the clients dealt with multiple CDNs. This research observed that Netflix appears to rank the three available CDN providers on a user account basis. That is, under normal conditions, a user will always be assigned to a specific CDN provider regardless of the content, time, or location. Netflix offers multiple formats and bit rates for its content. When a client requests a manifest of a specific movie, the Netflix server will return a manifest based on the client's capabilities, such as formats that can be played and computer performance.

The most significant plans for Netflix at this time (2015) are the continued migration to HTML5 and the continued introduction of higher resolution content, such as 4K/UltraHD.

9.7.4 Hulu

Hulu is a company providing streaming video that was founded in 2007. It provides services only in the United States and its territories. Its content concentrates on TV shows and movies. It offers (at present) ad-supported free and paid subscription services.

Hulu streams its content in Flash format, HLS for Apple devices [523], and on supported devices, MPEG-DASH. Hulu has announced that all new deployments will use MPEG-DASH [524]. It is supported on most computing and mobile devices as well as gaming consoles, set-top boxes, and directly in many new Smart TVs.

Hulu uses the same three CDN providers as Netflix to deliver content: Akamai, Limelight, and Level 3. Measurement in Ref. [525] showed that a CDN server assigned to a client at the beginning of a stream would normally not change unless performance dropped below the lowest available bit rate. If the client subsequently started a new stream, even if the stream is the same video, a new CDN server would be assigned. It also appeared that the chosen CDN vendor was not dependent on the location of the client.

As with the other major content providers, one of the largest technical advancements is the implementation of MPEG-DASH for as many client devices as possible.

9.7.5   Common Issues for all Providers

Until now, we have concentrated mostly on streaming technologies and how they are deployed by various content providers. Although the future appears very promising for OTT Streaming Video, there are substantial issues facing the content providers. In Ref. [526], a wide range of issues were researched and analyzed. Here are several key findings from this report (unless otherwise noted, statistics are from 2014):

- 75% of consumers will abandon a stream if they are dissatisfied for more than 4 minutes.
- From 2013 to 2014, the average bit rate streamed to clients increased 30%.
- Viewers experienced a Full Start Failure (failure to fully start a video stream) 2.6% of the time.
- Viewers were streamed lower resolution than desired 58.4% of the time.
- The % of views that experienced buffering (pause in the play-out) was 28.8%.
- No one is really ready for 4K mass deployment.
- There are substantial variations in overall streaming performance for different platforms and geography. Here are some examples of buffering interruptions:
  - Streaming Flash in Australia interrupts 1.79% of the time.
  - Sliverlight in Denmark 0.26%.
  - Mobile streaming to Android in Argentina buffers 7.63% of the time.
- Platform capabilities vary significantly, even within a platform type. For example, Ref. [527] reports that 18,796 (HW/SW) distinct variants of Android were seen in 2014.

Some of the conclusions drawn from the data are:

- The tolerance of users for poor quality has decreased, and they will abandon a stream and even a provider more quickly than before. Therefore, content providers must focus heavily on user satisfaction.
- Content providers must consider multiple dimensions when optimizing global user experience: Platform (i.e. Android/iOS), device size (desktop, laptop, smartphone, tablet, TV), ISPs, and geography (North/South America, Europe, Asia Pacific).

10

Home Networks

So far, we have discussed protocols, infrastructure, and methods to distribute content through a Service Provider or over the Internet. In this chapter, we examine the protocols and infrastructure required to support the playing of content within the home.

Because of the lack of trained home network administrators, the overall setup requires simplicity and high interoperability between devices. Since interoperability can hardly get addressed by the home user, someone else has to address this. All "normal" standard committees – such as IETF, IEEE, and others – have shown limited interest to standardize all elements of the home network. We treat the available IETF home network standards in Section 10.1 and already handled the Port Mapping Protocol (PMP) for NAT in Section 6.5 [364]. Otherwise, home networks got "standardized" by the industry organizations driven by Microsoft and Apple as main end-user software manufacturers.

The special standard bodies in the home network area are:

UPnP: The Universal Plug and Play organization has specified one basic protocol plus many methods and device descriptions. This is treated in Section 10.2.
DLNA: The Digital Living Network Alliance uses UPnP and other standards to create profiles for devices – see Section 10.3. Devices get certified and the DLNA standard label is visible to the end customer on home electronics equipment.
HGI: The Home Gateway Initiative has publicized recommendations for residential gateways. We treat a few of these proposals in Section 10.4.
I3A: The International Imaging Industry Association (I3A) had defined standards for exchanging pictures. The organization does not seem to exist any longer.

The I3A-developed Picture Transfer Protocol (PTP) handles the exchange of pictures – typically JPEG – between cameras and computers and other devices. Later versions of the initial standard also support the transport of videos as well as streaming. The standard is now maintained by the ISO [528]. The protocol was expanded to support transfer over IP [529], and this standard is now maintained by CIPA (Camera and Imaging Products Association) [530]. Most implementations use USB, Bluetooth, and WiFi. Microsoft and Apple support

Multimedia Networks: Protocols, Design, and Applications, First Edition. Hans W. Barz and Gregory A. Bassett.
© 2016 John Wiley & Sons, Ltd. Published 2016 by John Wiley & Sons, Ltd.

the standard. I3A also once developed the Internet Imaging Protocol (IIP) [531] to fetch by application image tiles from a server.

Apple does support the Zeroconf standards as described in Section 10.1 under the name Bonjour. Other multimedia home network standards are proprietary, such as Digital Photo Access Protocol (DPAP), Digital Audio Access Protocol (DAAP) on top of Digital Media Access Protocol (DMAP). The DMAP protocols have the same functionality as the UPnP AV Architecture – see Section 10.2.2.

Microsoft uses, for example, the Media Server (MS) and Media Transfer Protocol (MTP). MTP is an extension of the Picture Transfer Protocol (PTP) [532] for the home network. The specifications of these Microsoft protocols are available from Microsoft [533].

10.1 IETF Home Standards

IETF has defined a number of standards to support Home Networking. The two major standards we treat in this section are Zeroconf and the Service Location Protocol (SLP). Zeroconf is a term used to describe a collection of services that can be used to provide a usable network of devices without the need for manual configuration or outside services [534]. In order for devices in a home network to communicate and either provide or consume services, three core technologies are required [535]:

IP address assignment: Each networked device requires a unique numeric IP address.

Name resolution: Since numeric IP addresses are difficult for users to understand, a user-friendly name is assigned to each device. These names must be resolved to numeric IP addresses before they can be used.

Service discovery: There must be a way to discover what services are available within the network and how they can be accessed.

In the early development and deployment of IP networks, virtually all systems were manually configured and based in universities and corporations. Two key configuration activities were the assignment of the unique IP addresses and creation of a table that converted a user-friendly node names to their respective numeric IP address. The need to automate these activities led to the development of Dynamic Host Configuration Protocol (DHCP) for the assignment of IP addresses and Domain Name System (DNS) for the resolution of host names. Both of these services were developed to provide centralized administration of these resources.

During this same time period, the major vendors developed proprietary methods to automate network configuration over Local Area Networks (LAN). Examples of these suites are Apple's AppleTalk, Novell's IPX, and Microsoft's NETBIOS/SMB.

To address the need for IP network automatic configuration, the IETF created the Zeroconf working group. This working group existed from 1999 to 2004 and developed a draft RFC *Dynamic Configuration of IPv4 Link-Local Addresses* [536]. Since that time, work has continued and other standards in this area have been proposed and accepted. Apple is the largest proponent of Zeroconf. UPnP is the most significant alternative technology but with an enlarged scope, see Section 10.2.

10.1.1   IP Address Assignment

When a networked host is not manually configured or does not have access to a DHCP server, it must choose its own IP address. Zeroconf calls this *address autoconfiguration* or AutoIP and created a proposed standard for its implementation [536]. Both IPv4 and IPv6 have address blocks defined for these link-local addresses, 169.254.0.0/16 and FE80::0 [537], respectively. In general, the IP address is derived using the network interface MAC address, which is normally unique.

An important characteristic of these link-local address blocks is that they are never forwarded outside of the local LAN segment. That is, each of the hosts with a link-local address is assumed to be directly connected to the same LAN segment. After an IP address is selected, the host must then use the Address Resolution Protocol (ARP) to query the network and detect if an address conflict exists. Only after this probe is completed and conflicts (if any) are resolved is the host permitted to communicate on the LAN.

10.1.2   Name Resolution

To resolve host names to numeric IP address on the Internet, a hierarchical domain name structure exists and is supported by domain name servers (DNS). For hosts connected to the Internet, names can be registered and resolved by sending queries to these DNS servers.

If hosts are not connected to the Internet or do not have the ability to register their hostname with a DNS server, a different, cooperative method is needed. Multicast DNS [538] (mDNS) is based primarily on unicast DNS using the existing message structure, name syntax, and resource record types. Instead of querying a centralized DNS server with a database of host names and IP address, the client uses a multicast UDP packet to send a DNS query to all network devices on the LAN. The target system with the requested name replies with its IP address. Since the reply is also sent using multicast, all systems on the LAN can populate their mDNS cache to avoid the need for future queries. The multicast addresses used for IPv4 and IPv6 are 224.0.0.251 and FF02::FB, respectively. Both operate on port 5353.

Host names on the Internet have the form *myhost.example.com*. Host names used in local networks using mDNS must use the form *myhost.local*, with *.local* as the Top Level Domain. These local names are valid only on the local network. Cheshire and Krochmal [538] recommends the use of a flat namespace. That means subdomains, such as *myhost.net1.local*, should not be used. Figure 10.1 shows the exchange of messages for an mDNS query.

Now that we have introduced the general mDNS query and response sequence, we can examine how a host creates and claims its own hostname. Similar to address autoconfiguration, the host chooses a potential hostname and issues an mDNS query for that name. After three queries without a response, the name is considered available. The host can then issue a gratuitous mDNS response announcing its IP address and DNS entries. Using the network described in Figure 10.1, the messages can be seen in Figure 10.2.

Although we have only discussed hostname queries, DNS A records, most DNS record types are supported by mDNS. The most important are the following:

- CNAME records for name aliases
- AAAA records for IPv6 hostnames

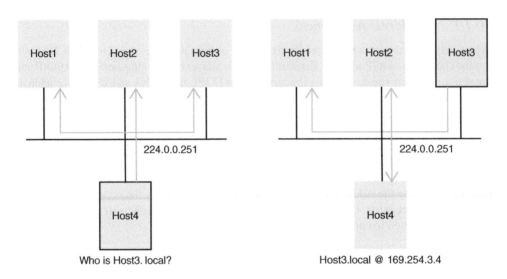

Figure 10.1   mDNS query and reply

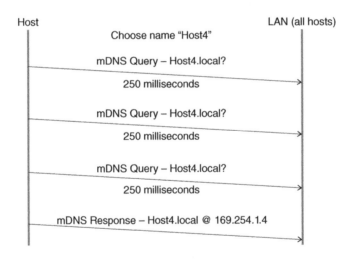

Figure 10.2   Claiming a hostname with mDNS

- PTR records for reverse address mapping (i.e., requesting a hostname assigned to a specific IP address)
- SRV and TXT records for describing available services, see Section 10.1.3.

It should be noted that Microsoft has its own implementation for name resolution called Link-Local Multicast Name Resolution (LLMNR) [539]. Although it provides similar functions to mDNS, it is not compatible with either DNS or mDNS. It uses multicast for both IPv4 and IPv6 (224.0.0.252 and FF02::1:3, respectively) on port 5355. Responses to queries are sent via unicast packets. It is supported on Windows systems starting with the Vista release.

10.1.3 Service Discovery – Zeroconf and Others

The previous two services discussed, IP address assignment and name resolution, have a long history of development on the Internet. Service discovery has somewhat different roots. Until recently, the discovery of services happened mostly on a LAN, not over the Internet. The previously mentioned protocols from Apple, Novell, and Microsoft all included service discovery in their suites.

There are three significant implementations for the announcement and discovery of services:

Service Location Protocol (SLP): This method is an IETF standard and is treated later in this section.

Simple Service Discovery Protocol (SSDP): This protocol is the discovery protocol of the UPnP standard. It uses HTTP over UDP (HTTPU) with a specific multicast address and port number. This is Microsoft's implementation of a Service Discovery protocol. This is treated in Section 10.2.

DNS-based Service Discovery (DNS-SD): This mechanism uses DNS PTR, SRV, and TXT Resource Records [300] and permits clients to query available services using mDNS, see Section 10.1.1. It is standardized in Ref. [540]. It is the third service included in the Zeroconf set of technologies and is treated next.

10.1.3.1 DNS-SD

DNS-SD can be used with both traditional unicast DNS and mDNS. Here, we only discuss the use of mDNS, as used in Zeroconf. Each service offered has three DNS records:

PTR record: Each service has a DNS PTR record of the form:
<instance>.<service>.<transport>.local. For example, a printer offering an Internet Printing Protocol (IPP) service might have a PTR record that looks like this: *printer1._ipp._tcp.local.* Note that by convention, services and transports always begin with a "_".

SRV record: This record connects a specific services offering to a hostname and port. For example, *printer1._ipp._tcp SRV 0 0 631 printer1.local.* shows the hostname and port (printer1 and 631, respectively) of the service being offered by the printer.

TXT record: This record is used to provide additional information about the service. It consists of key–value attribute pairs. Each attribute pair consists of a byte count of the key pair substring, followed by the text of the pair. Figure 10.3 shows an example of additional information about a printer service. In this case, the service offers A4 paper size and color printing.

The actual query can be seen in Figure 10.4. Here, a client is requesting services for Internet Printing (IPP). The query is sent out using mDNS to the LAN. The service provider (or any host with the cached information) replies with the list of PTR records for any service matching the query. The client then selects the desired instance (in this case, *printer1*) and requests the SRV and TXT records for that service. The SRV response contains the hostname and port information, while the TXT record provides any additional information about the service.

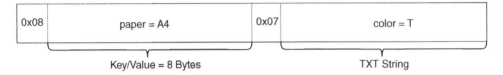

Figure 10.3   Example of DNS-SD TXT resource record

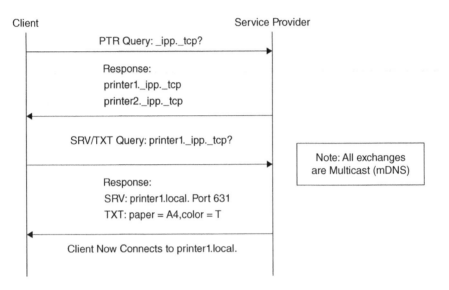

Figure 10.4   Example of DNS-SD query

10.1.3.2   Service Location Protocol

The next method of service discovery is called the SLP. This protocol was designed to scale from small and unmanaged networks to large enterprise networks. SLP is defined by IETF RFCs [541, 542]. The standard is well developed and deployed on most LAN-attached printers.

SLP defines three possible roles for devices:

User Agents (UA):   These are devices that search for and use services.

Service Agents (SA):   These devices offer one or more services.

Directory Agents (DA):   These optional devices cache service information and can reduce the traffic in large implementations. If DAs are implemented, then UAs and SAs both must use the DA for the exchange of service information.

Services are identified by a URL and can have an unlimited number of name/value pairs that describe its attributes. The URL format and fields are defined in service templates [543]. For example, here is a URL for a printer:

service:printer:lpr://color-printer/printqueue

The first three fields of this URL *service:printer:lpr* define the abstract service type. Although SLP includes the possibility to digitally sign messages using public key cryptography, it is rarely used.

SLP has the following message types:

Directory agent advertisement: The DA advertises its availability via multicast.

Service agent advertisement: If no DA is configured, the UA can directly solicit SA Advertisements. The request is multicast and the query is based on a service template [543].

Service request: The UA sends a unicast and a multicast message to the DA and SA requesting services that match the search criteria. The criteria are specified in Lightweight Directory Access Protocol or LDAP-style queries [544] (Boolean operators on name–value pairs).

Service reply: The DA or SA replies to a request with all addresses where services matching the criteria are available.

Service registration/Acknowledgment/Deregister/Update: Sent from the SA to DA, which acknowledges receipt.

Optional others: Service Type Request/Reply to retrieve all services, Attribute Request/ Reply to retrieve all attributes of a service.

All messages are scoped. Scope strings are registered by administrators for UA, SA, and DA. Integration into DNS for remote queries has been proposed [545]. SLP is implemented in Netware.

Other Service Discovery protocols exist for other environments like SUN's Jini, Bluetooth SDP, see Refs [546, 547].

10.1.4   Zeroconf Implementations

There are two major implementations of the IETF Zeroconf technologies available.

Bonjour: Apple's implementation of Zeroconf was originally named Rendezvous and renamed in 2005 to Bonjour. It includes the IP address assignment, host name resolution, and service discovery. It is included with Apple's OS X and iOS operating systems and can also be installed on Windows systems. It is included within some Apple products such as iTunes and Safari.

Avahi: A public, free Zeroconf implementation project was started in 2004 and resulted in the Avahi implementation. The package includes mDNS and DNS-SD. It is highly compatible with Apple's Bonjour. It runs on most Linux systems and is included in many mainstream distributions.

10.2   UPnP

UPnP is a set of networking protocols designed to enable the easy interconnection and exchange of services for a variety of devices in the home. It allows devices seeking services (Control Points) to discover and use services offered by Controlled Devices. The standards are promoted by the UPnP Forum, which was formed in 1999. The forum consists of more than 1000 companies in a wide variety of industries.

The architecture is based on established standards such as IP, HTTP, XML, and SOAP [548] and consists of six major areas:

Addressing: UPnP uses either DHCP, if a server is available or AutoIP, which selects a link-local address – see Section 10.1.1.

Discovery: SSDP is used to announce and find services. This is treated in Section 10.2.1.

Description: Once the existence of a service is known, the URL from the discovery message is used by the Control Point to obtain detailed information in XML form from the device. This information covers the Device Control Protocols (DSPs), which include all exposed actions of a device. Each such action is a SOAP-transported remote procedure call. The standard documents are mostly those device templates that a provider may augment.

Control: With the detailed information from the device, the Control Point can now invoke actions on the device using SOAP.

Event notification: This service provides the ability to communicate specific events from a controlled device to a Control Point – the General Event Notification Architecture (GENA). The device standards typically contain few events apart from sensor management systems.

Presentation: Devices may offer a URL for Web interface for the control of the device. This is, however, not specified in the standard documents.

The UPnP Device Architecture standard Version 1.1 [549] was recently updated to Version 2 [550], but the actual underlying device standard still uses, at the time of writing of this book, Version 1.1. The UPnP Version 1.1. and over 20 DSPs were also adopted as ISO/IEC standards 29341-x. The architecture is illustrated in Figure 10.5.

The abbreviations HTTPU and HTTPMU are explained in Section 10.2.1. GENA is defined in the overall architecture standard [549]. The UPnP Forum approves additional DCPs, which contain services. Vendors can augment the DCPs. The standards do not contain any security layer since it is assumed that this is not required in the home network. However, in the meantime, UPnP is discussing to include cloud services, and this would require an adaptation of the base standards.

An overview of existing multimedia DCPs is contained in Table 10.1 [552]. In addition, there exist numerous nonmultimedia services such as remote access, sensor management, and system management.

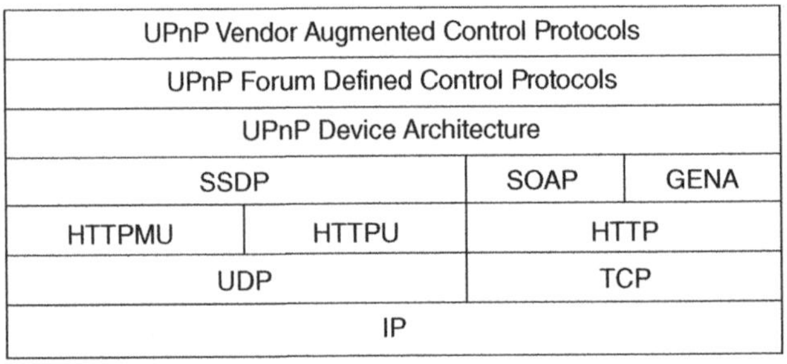

UPnP Vendor Augmented Control Protocols			
UPnP Forum Defined Control Protocols			
UPnP Device Architecture			
SSDP		SOAP	GENA
HTTPMU	HTTPU	HTTP	
UDP		TCP	
IP			

Figure 10.5 UPnP device architecture

Table 10.1   UPnP media related service areas and functionality

Area	Service elements	Main functionality
Audio/video	Control Point, Connection Manager, Content Directory, Rendering Control	Play media from an MS on a Media Renderer (MR) – see Section 10.7
Home automation	Digital security service, motion image, still image	Controls camera for the delivery of security services
Gateway	WAN connection, WAN Firewall Control	Control residential gateways and support opening ports for traffic (pinhole)
Printing	Printer enhanced	Printing by pushing document or retrieve by sent URL; precise printing using XHTML and CSS [553]
Telephony [554]	Telephony client, Telephony server, address book, calendar, call management, messaging, presence	External calls arrive at telephony server – method not specified. Integration of address book, calendar, messaging, and presence. If RFC with XML exist, the descriptions is aligned – e.g., presence [555].
Companion screen	Screen device	Sharing media between screens [556] – see Section 11.7

10.2.1   Service Discovery – UPnP

As was discussed in Section 10.1, once IP network connectivity is established, the next step is to either offer or search for services of interest. DNS-SD, SLP, and others were reviewed previously.

The UPnP protocol suite uses the SSDP to provide service announcement and discovery. SSDP was originally developed by Microsoft and Hewlett-Packard and submitted as an unapproved RFC draft in 1999 [557]. Although the draft expired without adoption in 2000, SSDP was included in the UPnP protocol stack.

The UPnP suite uses SSDP for discovering available services in a LAN environment, such as home or small office. It uses HTTPU, which is the HTTP protocol with UDP as the transport based on an unapproved RFC draft [558]. HTTPMU is the multicast version of HTTPU. Services are announced to an assigned multicast address using UDP port number 1900 [549].

The basic SSDP mechanism starts with the devices offering a service (commonly just called "devices") advertising service availability using an HTTPU *NOTIFY* message multicast to the LAN. When a Control Point (devices seeking a service) joins the network, it requests interesting services by multicasting a query. Devices offering a service that matches the criteria of the query then send a unicast HTTPU *NOTIFY* message back to the Control Point. Announcements are made for the device itself (root device), any embedded devices, and for the services offered by the device(s).

Services are individually identified with a Unique Service Name (USN). The USN is composed of a Unique Device Name (UDN), (usually the UUID of the device hosting the service), together with a *service* and a *service-type*.

```
NOTIFY * HTTP/1.1
HOST: 239.255.255.250:1900
CACHE-CONTROL: max-age=60
LOCATION: http://192.168.0.50:5200/Printer.xml
NT: urn:schemas-upnp-org:service:PrinterBasic:1
NTS: ssdp:alive
SERVER: Network Printer Service UPNP/1.2
USN: <uuid>:<vendor>-Printer-1.1::urn:schemas-upnp-org:service:PrinterBasic:1
```

Figure 10.6   Example of SSDP NOTIFY message

An example of the *NOTIFY* advertisement message is shown in Figure 10.6. In this case, a printer is advertising a basic print service to the LAN using multicast. The key lines in this message are:

NOTIFY: This identifies the message as an announcement.
HOST: The multicast address and port used for the HTTPU message.
LOCATION: This is the URL an interested Control Point can use to obtain detailed information about the service.
NT: Notification Type. This describes what is being advertised (device or service). In this case, a printer service.
NTS: Notification SubType. The "alive" indicates the device or service is available, "byebye" would indicate the termination of the service, and "update" would mean a change to the device or service.
USN: The Unique Service Name as mentioned earlier.

The *M-SEARCH* message has a similar structure and is sent by a Control Point to find interesting services. It uses the same USN structure as earlier to identify desired services.

10.2.2   AV Architecture and its Elements

The audio video (AV) architecture [559] and its components are the basic system for UPnP. UPnP specified first the AV architecture and all other service descriptions use it and build on top of it. The AV DCPs have also undergone the most revisions. Figure 10.7 shows the AV architecture with its main elements.

The functionality of each UPnP service is described by state variables and actions (or remote procedure calls) that work on those states.

The Control Point (CP) is the instance that controls all activities. In essence, the MS has all the content, and the MR will play it. The decoder, the creation of content at the MS, and the media transfer between the MR and MS are not in the scope of UPnP. For the media transfer, any IP transport is supported. The CP does not show any service because the CP does not expose any. Of course, the CP receives messages but those are only the result of a formerly called SOAP call. However, the CP has built-in functionality to use the exposed services of the MR and the MS correctly.

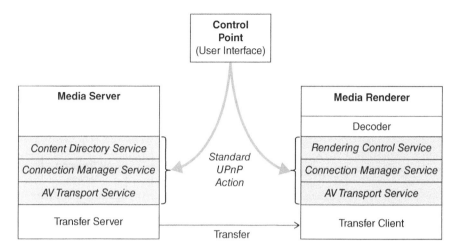

Figure 10.7   UPnP AV architecture

As we see in Figure 10.7, the AV devices support four service areas:

Connection manager service   matches capabilities, exchange information, setup, and tear-down connections – see Section 10.2.2.1.

Rendering control service   has the task to render one or multiple instances of a device – see Section 10.2.2.2.

Content directory service   provides content lookup and storage. We do not expand on this standard [551] because it is mainly around information retrieval and storage. The representation of the content structure is done with a variant of the standard XML Digital Item Declaration Language (DIDL) [560].

AV transport service   controls the transport and the playback of media using different transfer methods – see Section 10.2.2.3.

We can now look at the functional elements of MR and MS in Figure 10.8 – see Refs [561, 562]. The boxes with a dot are not part of the standard – for example, the *Content Creation*. We see which services influence the overall outcome and the dependencies. The optional *Scheduled Recording* service is not shown here. This service delivers input to the *Content Creation* and *Content Storage* but otherwise has no new concepts, and the functionality is straightforward. Consequently, this service is not covered.

Note that all control element names of UPnP are written in the standards as one contiguous word that we will mostly avoid to ease readability.

10.2.2.1   Connection Manager Service

We show now the most important Connection Manager Service (CMS) [563] state variables and actions that work on those states:

SourceProtocolInfo: This variable is a colon separated list of protocols and formats, which the source supports. The source is always the MS. The detailed format of

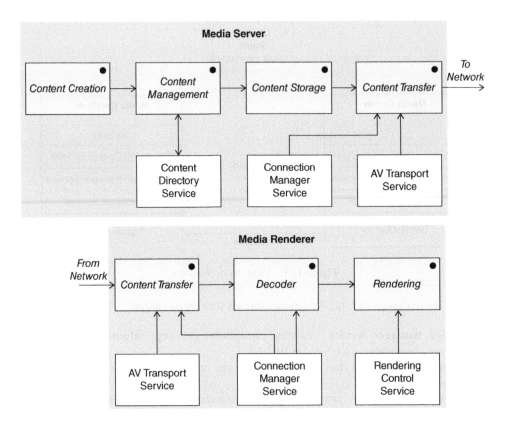

Figure 10.8   Media Renderer and Server functional diagrams

this list is *<protocol>:<network>:<contentFormat>:<addInfo>*. An example is *rtsp-rtp-udp:*:MPV:**. The *<contentFormat> MPV* stands for an MPEG-2 transPorted Video. The use of the fourth term *<addInfo>* is seen in Section 10.3.

SinkProtocolInfo: The same information as given for the sink – that is, the MR.

CurrentConnectionID: This is a list of currently active connections.

ConnectionStatus: The status of a connection is either *OK* or a marking for an error status such as *UnreliableChannel*.

ConnectionManager: This is a description for the peer connection manager device when building a connection. It has the form *<Uniform Device Name (UDN)>:<vendor>*. SSDP can deliver this information – see Section 10.2.1.

ConnectionID: This is a unique local identifier for a connection.

AVTransportID: This is a unique local identifier for an AV transport connection.

With those state variables, we can understand the following main shortened actions of the CMS. We need to remember that UPnP is implemented by SOAP, which is a remote procedure call. A SOAP call cares for the remote execution for a locally generated call. The transfer over the network works by sending a SOAP XML envelope over HTTP [564]. This means that the listed actions or calls are the "packets," which are transmitted in both directions. Note that only the CP can issue an action.

GetProtocolInfo (Out: *SourceProtocolInfo, SinkProtocolInfo*): This action retrieves the values of the *ProtocolInfo* for the source and the sink.

PrepareForConnection (In: *RemoteProtocolInfo, PeerConnectionManager,* Out: *ConnectID, AVTransportID, RcsID*): This action asks a device to prepare for a connection. For this purpose, the device receives the *ProtocolInfo* from its peer device. The former action *Get-ProtocolInfo* retrieved this information. The *PeerConnectionManager* is stored in the state variable *ConnectionManager* and SSDP has delivered it. The return values *ConnectionID* and *AVTransportID* describe the prepared connection and are stored in the corresponding state variables. The return variable *RcsID* is explained in Section 10.2.2.2. Note that the *ConnectionID* of peers of a connection is not the same at both ends. If the calls to both peers are successful, the connection is usable. A call is not successful if the *ConnectionID* is −1.

ConnectionCompleted (In: *ConnectionID*): This call closes the connection with the *ConnectionID*.

GetCurrentConnectionInfo (In: *ConnectionID*, Out: *AV TransportID, Protocolinfo, Status*): This call delivers the status information of the *CurrentConnectionID*.

10.2.2.2   Rendering Control Service

Only the MR operates the rendering control services [565]. The Rendering Control Service (RCS) can work with virtual streams as shown in Figure 10.9 sharing the same device. The streams are numbered by the variable *Rcs_ID*, and this variable is also an additional argument when connections are built as shown in Section 10.2.2.1. A typical example of such functionality is the Picture-in-Picture (PiP) for TV screens.

The rendering control has many states variables since there are a lot of characteristics in rendering. The main state variables with their ranges are:

- Mute is a Boolean state variable.
- Many state variables have the value range [*0:<vendor_defined>*] such as *Brightness, Contrast, Sharpness, Volume, Loudness* for the audio part and *Keystone, Vertical Keystone* for the video part.
- The state *Channel* has the value range {Master, LF,RF, ... ,R}.
- Some states are strings such as *Instance ID* for the identification of streams and *Device UDN*, which is a unique name for a device. The UDN begins with *uuid* and is assigned by the vendor uniquely for the lifetime of each device.

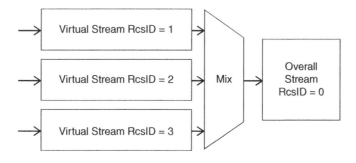

Figure 10.9   Rendering control service mixing three virtual streams

For each state variable, two actions exist. *Get·· ·* retrieves the value of the state variable and *Set·· ·* assigns a value.

10.2.2.3 AV Transport Service

The Audio Video Transport (AVT) service [559] does the actual transport by RTSP, HTTP, or other IP-based transfer protocols. Figure 10.10 shows the possible actions for the main state variable *Transport State*.

The values *Paused_Playback, Recording,* and *Paused Recording* only exist as a state if the *Pause* and *Record* actions are implemented on a specific device. If a concrete device supports recording, AV Transport stores the content of the stream on a device-dependent storage. The arguments for the actions *Seek* and *Previous* are *(In: InstanceID, Unit, Target)*. If the *unit="TRACK_NR"*, then *Target* is a track number. In this way, one can jump to tracks on devices that have tracks as identification (TV set, Audio-CD player). For a tuner, a track is an index into a given channel list. As one can see, each usual device play-out can be controlled by an AVT service. Of course, timing behavior might be different – that is, transitioning for a tape drive takes much longer than for a tuner.

The action *SetAVTransportURI* has as one argument *CurrentURIMetaData*, which describes a specific DIDL XML fragment in the content store. With this argument, the action can set the transport to any storage item in the content store for playback or recording.

Vendors can add extensions to actions and state variables. Extensions always begin with an *X_* followed by a registered domain name of the supplier.

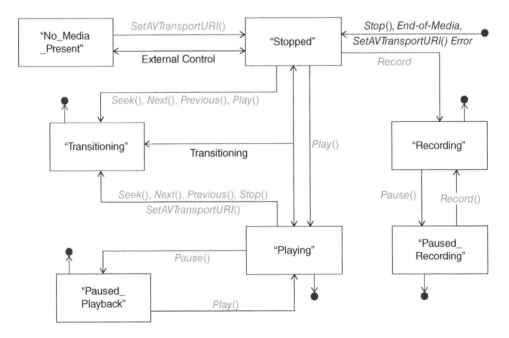

Figure 10.10 Status diagram for AV transport service

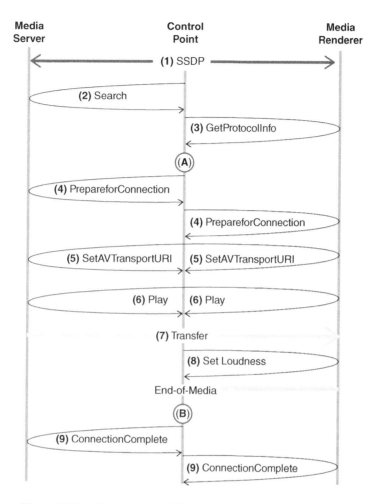

Figure 10.11   Combining the different services to an overall example

10.2.2.4   Summarizing Example

Figure 10.11 shows an overall session communication between Control Point, Media Server, and Media Renderer. The remote procedure calls are shown as curved arcs since a call cannot get separated into directions – the call always returns values for provided arguments.

In the previous sections, we have seen how the different UPnP AV services function. Figure 10.11 combines the information in all these sections together – see also Ref. [559]:

(1): Initially, the CP discovers with SSDP the MS and MR – see also Section 10.2.1. Note that a CP itself is never discovered.

(2): The Control Point executes a *Search()* with the Content Directory Service. A *Search()* as a result delivers back the UPnP DIDL XML Document as a reference for the content. The *Search()* also delivers the information of a *GetProtocolInfo()* action – see Section 10.2.2.1.

(3): The CP needs, in addition, the characteristics of the MR and the explicit CMS *GetProto-colInfo()* action delivers this information.

(A): The CP now has the supported protocols and formats from the MR and the MS. The CP has decided on the protocol and format. If the information is not sufficient, the CP can ask the MR for more detailed information by CMS *GetRendererItemInfo()*.

(4): When the CP has decided on the protocol and format, it executes a CMS *Preparefor-Connection()* for the MR and MS – see Section 10.2.2.1. The return values *AVTransportID, RcSID* are needed for the next step.

(5): If both connection setups were successful, the CP can execute the AVT action *SetAVTrans-portURI()* for the sink and source – see Section 10.2.2.3. This action uses the arguments for the DIDL XML fragment in the CS from step **(2)** and also *AVTransportID, RcSID* for the connection and the stream.

(6): The RCS action *Play()* starts the play-out and the arguments are again *AVTransportID, RcSID* for the identification – see Section 10.2.2.2.

(7): Thereafter, the out-of-band transfer of the media starts.

(8): The CP might send rendering service actions to the MR – in this case, an adjustment of the loudness – see Section 10.2.2.2.

(B): The media has ended and the *Transport State* is *"STOPPED"* – see Figure 10.10. The CP could now restart the content by issuing another AVT *Play()*.

(9): With sending the CMS actions *ConnectionComplete()* for the MS and MR, the connection ends.

10.3 DLNA

The DLNA has more than 200 members in the software and hardware industry, which covers all prominent names except Apple and browser companies. DLNA does not create new protocols but combines and restricts existing protocol suites and technologies in profiles. DLNA certifies devices against those profiles. A certified device gets a DLNA label that we often see on home networked components – such as NAS, networked TV sets, or remote controls.

The building blocks of DLNA are:

UPnP devices: Used UPnP services are CMS, CDS, AVT, RCS, and printer services from the UPnP Device Architecture.

Media formats: Included media formats are JPEG, AVC (H.264) encoded in MPEG-PES/TS, MP3, Windows Media Audio (WMA) similar to MP3, AAC, and LPCM (see Section 3.1.3.1).

Transport: The media and control communication is based on HTTP, RTP, RTSP over TCP/UDP/IP. The discovery uses SSDP.

Data link: DLNA includes WiFi (a,b,g), Ethernet, and Bluetooth.

Digital rights: Either Digital Transmission Content Protection over IP (DTCP/IP) by Intel et al. or WMDRM-ND (Windows Media Digital Rights Media for Network Devices) can be used [566].

By using the building blocks, the following DLNA devices define an overall functioning system, which is created out of all building block parts:

Network devices:

- Digital Media Server (DMS) contains UPnP CDS, AVT, and CMS.
- Digital Media Renderer (DMR) uses UPnP CMS, AVT, and RCS.
- Digital Media Controller (DMC) builds upon a UPnP CMS.
- Digital Media Player (DMP) combines a DMR and DMC. DMP and DMS together form a complete end-to-end system.
- Digital Media Printer (DMPr) is a device that is able to print images described in XHTML/CSS, which are transferred from other devices. For the transfer, the partner needs an optional Print Controller service capability.
- Devices may have further capabilities. If a Push Controller is installed in a DMR, it pushes content by initiating and controlling a playback session. The Upload and Download service capability allows upload from and download to a DMS.

Mobile handheld devices: Mobile devices are identical to network devices but carry an "M-" in front – for example, M-DMS, M-DMP. Within the detailed profile, the mobile devices have different connectivity specifications.

Infrastructure devices: The Mobile Network Connectivity Function (M-NCF) is a bridge between the mobile and fixed network – mainly developed for Bluetooth [567]. It contains no UPnP elements. The Media Interoperability Unit (MIU) is an MCU. The MIU is a virtual DMS offering the content of a DMS in other formats and implements CDS and CMS. When a CP looks for content, the MIU has to contact the CMS and then answers with the transcoded formats. Optimizations for this procedure were proposed in Ref. [568].

The DLNA combined different media, device categories, and regions for the creation of hundreds of profiles. When communication partners in a DLNA/UPnP network build a session, they will check for supported profiles to assure the interworking. A DLNA connection manager sends the profile identification in the *<addInfo>* field of *ProtocolInfo* (Section 10.2.2.1) when building the connection.

More details to DLNA can be found in Refs [569, 570].

10.4 Residential Gateway

The demarcation point between the service provider network (Telco, ISP) and the home network is the Residential Gateway (RG). This device has different names in the standards: Delivery Network Gateway (DNG) by DVB, Home Gateway by HGI (Home Gateway Initiative), Customer Network Gateway (CNG) by ETSI, and Customer-Premises Equipment (CPE) by ITU. Originally, it was only a simple NAT-capable router with DSL-modem and Ethernet interfaces. A residential gateway today can have many more functionalities such as:

- ISDN interface
- SIP-Proxy and SIP-B2BUA
- Answering machine and Fax gateway
- WLAN and DECT (Digital Enhanced Cordless Telecommunications) access points
- Network attached disk and UPnP media server
- VPN functionality
- Service-hosting gateway [571] which also covers caching and replication.

DSL providers are often selling the residential gateways as well. However, in many countries, the customers can decide about a device on their own.

The delivery of multimedia services to the home always passes through the residental gateway. If a Telco is able to deliver media till the gateway with QoS, the service quality is not longer guaranteed when entering the RG and the home network. If the home gateway was delivered by the DSL provider and remotely managed by the provider, the service guarantee might last till the RG's network interfaces.

The individual functions of an RG are governed by the corresponding standards (ITU, IETF, UPnP, IEEE).

The Home Gateway Initiative was founded in 2004 and has released recommendations for RG specifics. The HGI had initially over 70 members (Telcos, RG manufacturers) and was mainly active till 2010. We treat the HGI recommendations on IMS-enabled RG and RG Network Termination (NT).

10.4.1   IMS Integration

The first recommendation addresses the integration of IMS mobile phones in a home network. This recommendation is based on an ETSI TISPAN architecture for CNG [572, 573] and was further developed by HGI [574]. When an IMS-capable device – see Section 7.4.1 – is used inside a WiFi-equipped home, it can access the Internet. The SIP client on the device works over the mobile network, but it does not operate over the WLAN and the RG. It does not work because the service points cannot be reached – for example, the P-CSCF (Proxy Call Session Control Function).

Figure 10.12 shows the simplified proposal that avoids using new abbreviations as in the ETSI standard. The Plug-and-Play part (**1**) is a UPnP function for service discovery. The SIP user agent can arrange the signaling either directly with the P-CSCF or via B2BUA (**2**). The path through the B2BUA may be required for local services that could be a call transfer to another LAN SIP client. The media function in the UE sends media to the Access in the IMS infrastructure. The interworking client and gateway Firewall Function allows the signaling and media communication passing through the RG. In contrast to WebRTC (Chapter 8), this needs a particular implementation in the residential gateway. The overall system needs a location and configuration function for the device and the residential gateway (**4**). We have not shown additionally required security functions, and the authentication function (**5**) is only a placeholder for the missing parts.

The additional functionality can be integrated in residential gateways installed at the homes of the vendor's clients.

Also, 3GPP works on the topic of integrating VoLTE in Telcos-owned public WLANs [575], but the standard at best will be available in 2016. Note that 3GPP has more service elements in their solution, but we show in Figure 10.12 only the IMS version.

10.4.2   Network Separation

The Home Gateway Initiative has published recommendations on QoS in the home network. Without a trained administrator, these ideas are only usable with provider-managed residential gateways. A provider will primarily prioritize his services and probably try not to set priorities of other services. The NT proposal of HGI [576] is used in deployments of IPTV (Chapter 11)

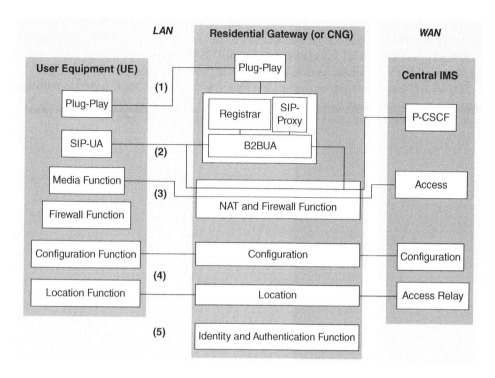

Figure 10.12   Simplified schema of the IMS home network integration

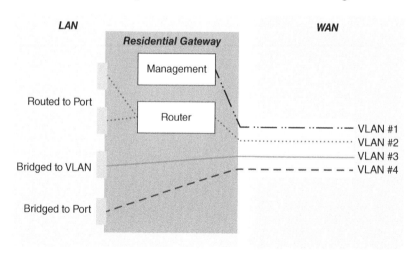

Figure 10.13   HGI network termination recommendation

and can achieve QoS improvements. HGI defines NT as the entity that terminates the OSI layers 1 and 2. HGI recommends certain port and VLAN (Section 4.5.4) settings as seen in Figure 10.13.

In Figure 10.13, certain VLANs are mapped from the WAN side of the RG to certain ports and VLANs at the LAN side. When the management of the VLANs in the RG stays with the provider, an improved in-house priority is achievable. The router with its additional functions – for example, VPN – is open for the management by the user. In some countries, IPTV deployments use a bridged port VLAN configuration. This VLAN port is preconfigured with higher priority in all IPTV-approved residential gateways in those countries.

This VLAN-based network separation is a first step to slice the home network in distinct parts and assign each slice to a certain service provider [577]. This configuration may help over time to achieve more quality for different services in the home network.

11

High-End IPTV

This chapter presents the standards and deployment aspects of high-end IPTV. High-end IPTV or, in the future, only IPTV delivers typical TV functions plus VoD with a guaranteed quality – for the requirements, see also Section 2.3. In Chapter 9, we have seen how video and audio can be delivered with streaming. Recent measurements from the United States, however, show [515] that 14% of the sessions have over 10 seconds of startup delay, and more than 40% of the streaming sessions have short pauses for rebuffering. Also, viewers quickly got upset when they had to wait for a start more than 2 seconds and did not return when they experienced rebuffering situations [578]. The average TV viewer might get upset even earlier since he/she is not so Internet savvy when accepting glitches of a new technology.

The first approach now might be to wait and hope for QoS plus better bandwidth to each TV viewer's home. However, this is not sufficient for a replacement of television because of the following two critical elements:

- Some TV programs have many millions of parallel users, and those numbers can even reach 110 millions of users in one country – for example, Super Bowl 2014.
- TV viewers are used to zapping between channels quickly, and less than few 100 milliseconds is the maximum for switching over.

To cope with the aforementioned requirements, it is evident that more need to be done than just adding QoS and not waiting for the days of real big pipes in each household.

This chapter describes the IPTV standard of the Digital Video Broadcast (DVB) association. Standards of the DVB are used in nearly all countries of the world apart from the United States and a few others. A number of bodies also work on IPTV standards such as ETSI TISPAN, ATIS (Alliance for Telecommunications Industry Standard), ITU IPTV Focus Groups, HbbTV, and Object Mobile Alliance (OMA). For a detailed overview, see Ref. [579]. We explain the noteworthy organizations in more detail next.

The ATIS IPTV Interoperability Forum (IIF) has publicized a whole set of IPTV standards numbered from ATIS-0800002 to ATIS-0800039. The paper [579] provides an overview of the ATIS standards, which we will not treat further here. HbbTV combines additional input sources in the user interface.

Multimedia Networks: Protocols, Design, and Applications, First Edition. Hans W. Barz and Gregory A. Bassett.
© 2016 John Wiley & Sons, Ltd. Published 2016 by John Wiley & Sons, Ltd.

The Digital Video Broadcast (DVB) organization has more than 200 members (Cisco, Microsoft, Telecoms), and the documents they created are standardized by ETSI. The following standard areas exist [580]:

- DVB-S for satellite broadcasting – S2, SMATV (Satellite Master Antenna TeleVision)
- DVB-T for terrestrial broadcasting (T2)
- DVB-C for coaxial broadcasting via Hybrid Fiber Coaxial (HFC)
- DVB-H/HSPA (High-Speed Packet Access) for terrestrial handheld broadcasting
- DVB-IP for Internet TV

As mentioned in Section 2.3, at the end of 2014, IPTV had more than 115 million subscribers worldwide in over 30 countries – see for a list of countries in the previous referenced section. The first IPTV deployments started in 2005, and the first DVB IPTV standard dates back to 2005 as well. Consequently, some of the deployments will not adhere to the current standards. However, legal requirements oblige European Telcos to adhere to ETSI standards over time.

We will mention the contents of the different sections of this chapter after the introduction of the overall architecture of DVB IPTV in Section 11.1. However, a few sections are not only related to IPTV. We will treat second screen applications in Section 11.7 because the richest and best specified one is related to DVB. Set-top-box functions are discussed in Section 11.8 plus the concept of virtual set-top boxes. Section 11.9 discusses not only IPTV but also integration of IMS, WebRTC, and DLNA.

There is no book that covers DVB IPTV technologies in a comprehensive way, and there are only very few articles on parts of the overall norm. The resulting ignorance of these technologies has created the belief that the future of Internet TV lies solely in HTTP streaming concepts, which probably is not true. In the future, both solutions will coexist for quite some time.

Learn more on multicast

DVB IPTV is the mostly widely used existing multicast application. Such a broad use of multicast has created new ideas such as the Fast Channel Switch – see Section 11.2.2 – or the multicast "Acknowledgment" – Section 11.5. This chapter helps to learn about multicast networking.

11.1 Overview of DVB IPTV

In the following sections, we treat the DVB IPTV standard with the latest version from 2014 [581] plus associated documents. The functional model of DVB is shown in Figure 11.1.

The functional model specifies the services that are available at the network interface of the Home Network End Device (HNED). HNED is typically a set-top box. In actual deployments, most providers require that the HNED has a cable to the Delivery Network Gateway (DNG). The DNG is usually the (A)DSL router to the network provider. Often, the IPTV functionality only works with routers provided by the local Telco. Other router brands may have to create adaptations to their product to deliver the required IPTV functionality. When providing a high-quality service, it is obvious that the uncontrolled path on the home network until the entry into the provider delivery network is critical.

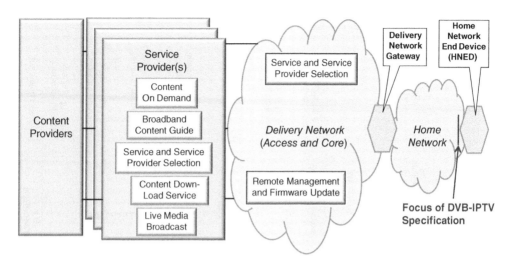

Figure 11.1 Overview of DVB IPTV functional architecture – adapted from Ref. [581]

The different services in the delivery network or from the different service providers are all used by the HNED to get IPTV working on a display device attached to it. The possibly different service providers behind the delivery network may work from the functional model but are not at all in the interest of the Telco that primarily offers the service. Perhaps, the original authors of this model included it to show federal network agencies how open the approach is for other providers. The model also shows the content providers who deliver the content to the service providers for inclusion in their overall offer.

The standards were written to be used by telecom providers and are deployed solely in their networks. Since Telcos only own the cabling to households in their home country, IPTV is deployed per country by the incumbent Telcos. The DVB IPTV standards are quite detailed, but they are not detailed enough that country implementations are interoperable.

Figure 11.2 shows an overview of the used protocols within the DVB IPTV system. For all the protocols, the chapters or sections where the protocols are explained are also mentioned in the figure. In this chapter, we treat the protocols with a shaded background in the figure. Some of the protocols are shown twice since they exist in a multicast and a unicast version.

We treat the Live Media Broadcast (LMB), which also includes technologies for retransmission and channel switch, in Section 11.2. Apart from protocol adaptations for live media, the DVB IPTV standard has introduced a new protocol DVB STP and integrated the IETF multicast protocol FLUTE. Both protocols are explained in Section 11.3. Also, in Section 11.3, we treat DSM-CC, which is a protocol implemented in all set-top boxes for years without being an IETF protocol. A number of supporting functions are required to keep such a system working, and we combined those in Section 11.4. This section consists of the Service Discovery and Selection (SD&S), the Remote Management System (RMS), Broadcast Content Guide (BCG), and Firmware Update System (FUS). The Content-on-Demand (CoD) service is handled in Section 11.5. After that, we show the main elements of actual deployments in Section 11.6. Already in the chapter on requirements (Section 2.3), we mentioned Companion Screen Applications (CSA), and DVB recently publicized a standard for a solution that we discuss in Section 11.7.

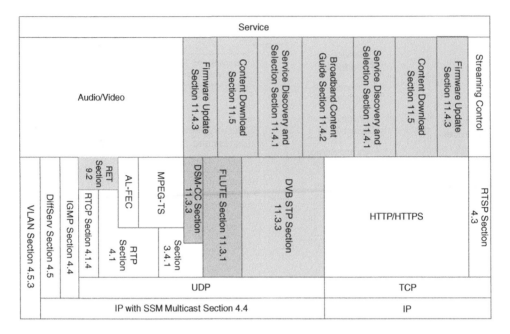

Figure 11.2   Protocol stack DVB IPTV

11.2   Live Media Broadcast

The live media is sent by a Source-Specific Multicast (SSM) in the provider's delivery network. At least for the last mile, the distribution network is shared with regular Internet access. The prioritization in the last mile is done by VLAN (IEEE 802.1Q), and the standard contains the priority settings in DSCP notation – see Sections 4.5.4 and 11.6. The specifications do not state which form of multicast routing protocol in the delivery network is used but state that IGMP version 3 needs to be implemented in the home network.

The transport of the media stream is performed via RTP/RTCP or just UDP with an MPEG transport stream encoding. A particular video encoding is not enforced. The standard mentions the possible use of RTSP, but for a live media stream, many commands – pause, forward, rewind – are not usable. So, RTSP is the method for controlling the output for CoD – see Section 11.5.

For error reductions, an Application Layer FEC (AL-FEC) can optionally be used. The standard of the Society of Motion Picture and Television Engineers (SMPTE) [582] is applicable for a single RTP flow with IETF extensions [583, 584]. Otherwise, a Raptor code is recommended [585, 586].

In addition, the standard also proposes a method for retransmission.

11.2.1   Retransmission

We discussed before that retransmission does not work for pure live video applications in the WAN – see Section 5.3. Despite the heading "Live Media Broadcast," television is not as "live" as pure video conferences. The broadcaster includes a so-called profanity delay – also named the 7-second delay – to hinder violence or bloopers from distribution. For IPTV, this delay is often expanded to around 10 seconds. Of course, in 10 or 7 seconds, retransmissions are possible. Another complication is that the information uses a multicast where error reporting and retransmission are not usually foreseen. However, we will see how this difficulty can be solved. Since retransmissions shall now be possible in some way, a significant part of the overall delay needs to occur on the client side. If a sufficient delay occurs on the client side, this allows retransmissions.

When looking at a typical provider infrastructure, the most errors occur as singular events in the last mile over "old" telephone cables. The last-mile problem zones led to the use of decentralized RETransmission (RET) servers. The RET server sits in the path from the headend of each MultiCast (MC) stream close to the clients. The RET server stores each multicast stream sent for the last minute. In the HNED, an RET client exists, which recognizes any incorrect or missing packets in the media stream.

Figure 11.3 shows the situation when the RET client has detected a missing packet. Here, the HNED sends **(1)** an RTCP FeedBack message (FB) to the LMB RET-server based on an extended feedback profile as specified by the IETF in Ref. [146] – see Figure 11.4. The first

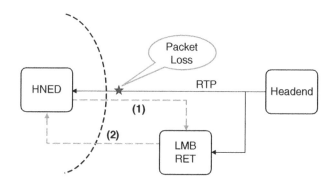

Figure 11.3   RET server in action

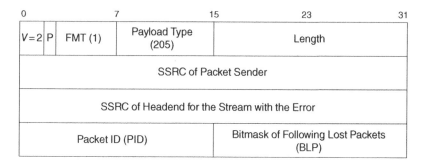

Figure 11.4   RTCP feedback message

three lines of this message are well known from the original RTCP messages – see Section 4.1.4.1. Only the second field in the first line is new: it is the "Feedback Message Type" (FMT). FMT describes how the Feedback Control Information (FCI), which follows after the third byte, looks like. Here, FMT is equal to 1, and we have a generic feedback as listed. PID refers to the RTP sequence number of the first lost packet, and BLP can mark up to 16 following lost packets.

When the RET server has received the RTCP packet (**1**), it sends via RTP a retransmission. For retransmissions, an additional RFC [587] exists, which specifies a slightly adapted RTP. The adaptation of RTP concerns the field Original Sequence Number (OSN). With this OSN, the HNED can place the corrected packet in the media stream. The ETSI implementation guideline [588] has more detailed examples on this topic inclusive of the SDP descriptions with some expansions.

If an error in a stream occurs on the path before the RET server, all the clients behind the RET server will receive this damaged information. Figure 11.5 depicts this situation. The error would trigger a mass of feedback messages from all HNEDs to the RET server. For this situation, the RET server can issue RTCP Forward Feedback (FF) messages in a multicast mode (**3**). The format of this communication is identical to RTCP Feedback message as shown in Figure 11.4 but are now sent from the RET server in a multicast mode. If the HNED receives this information, it should not send any error concerning this reported packet to the RET server. Since an HNED never knows if a detected error is a singular or a general error, the HNED has to wait for 200 milliseconds before sending a feedback message to the RET. The standard mentions that the RET server should contact the headend in such a situation (**4**) to receive a corrected packet. Since the norm concentrates on the interface to the HNED, this function is not detailed. If the communication between the headend and the RET server has retrieved the missing information, the RET can send it to all HNED as a multicast in the same way as before.

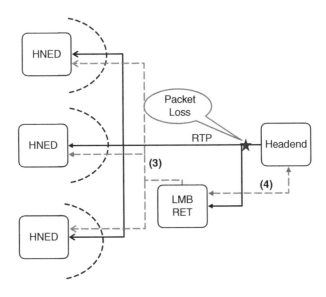

Figure 11.5   Error before RET server

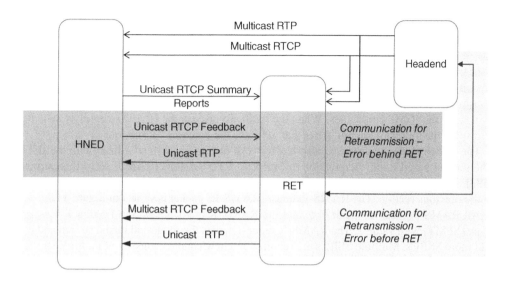

Figure 11.6   Summary of the communication with the RET server

RTCP in multicast environments with potentially millions of receivers needs to be adapted to avoid overloading the headend with messages from each HNED. In the standard, the HNED can only generate the so-called RTCP Receiver Summary Information (RSI) reports as specified in an RTCP extension for SSM [589, 590]. Normal RTCP Receiver Reports (RR) are not allowed from HNEDs. These RSI reports are sent as a unicast to the RET server and not to the headend, to avoid any unplanned heavy load on those units. Figure 11.6 gives a summary of the different communication forms for retransmission.

11.2.2   Channel Switch

The literature contains diverse recent articles on the channel switch problem – for an overview, see Refs [591, 592], but it has not been recognized that the DVB standards already provide a solution that has been deployed millions of times.

When using multicasts and IGMP, the obvious solution for a channel switch at the HNED is to issue IGMP *Leave* and *Join*. As discussed in Section 4.4.4, version 3 of IGMP was introduced to speed up the channel switch by combining the *Leave* and *Join* in one message. After the IGMP *Join*, the router – that is, Delivery Network Gateway (DNG) – still has to join the new multicast group as well. The router cannot receive all available multicasts at once since the transport speed on the last mile does not allow this.

The provider's goal was that this change has to occur within 30 milliseconds. In this context, we need to remember that the HNED, in addition, needs to wait for the end of next Group of Pictures (GOP) or rather a new I-slice. For the viewer, the two delays can quickly add up to 100 milliseconds. This challenging goal was not possible even with the changed IGMP Version 3. However, it is a critical requirement for the viewer.

As a consequence, a dedicated solution that uses the RET server again was created. This solution is named Fast Channel Change (FCC) in DVB terminology. The RET/FCC server receives all multicasts and stores the streams in such a way that it can deliver a new channel anytime starting immediately with a GOP. When viewers launch a channel switch, a new unicast RTP session is built from the RET server. Note that viewers with DVB IPTV can also issue channel switches in a Picture-in-Picture (PiP) mode, which means that the new channel is visible in a small picture within the former channel. If the viewer stays in the new unicast transmitted channel for typically 10 seconds, then HNED starts to change to a multicast mode by issuing the IGMP commands. When the multicast arrives, the HNED seamlessly switches to display the multicast stream. After that, it closes the unicast RTP session.

The switch to the new unicast also has to work fast and was standardized with the IETF. The unicast-based Rapid Acquisition of Multicast RTP Sessions (RAMS) serve this purpose [593]. The RAMS messages are expansions of the RTCP feedback messages as we have seen in the previous section. The RAMS Request (RAMS-R) packet is shown in Figure 11.7.

All RAMS packets have *FMT=6*, and the RAMS-R is visible as Sub-Feedback Message Type (SFMT) with value *1*. The RAMS-R request names the requested media stream and can add more SSRCs. Multiple SSRCs are required if a new channel consists of multiple streams or in different forms – for example, resolutions. In the optional fields, additional information can be communicated as:

- The minimum and maximum sizes of the buffer for the requested media stream in the HNED.
- Maximum bit rate for sending the requested media stream.

The buffer size and the transfer rate are critical because the RET server sends the unicast as a burst – that is, faster than the display speed. The data has to be transmitted in a burst since the actual transmission to other clients has already advanced, and the burst has to fill the missing gap in the local HNED buffer.

Each RAMS-R packet is answered by an RAMS-I information packet. The RAMS-I packet contains identification information in a TLV form: the sequence number of the first RTP packet to be sent, burst duration, and maximum transmit rate. Also, the earliest multicast

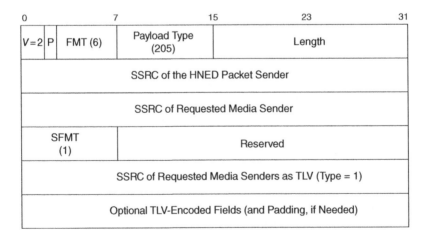

Figure 11.7   Structure of the RAMS request packet

join time is sent in the RAMS-I packet and is measured in milliseconds after the arrival of the first RTP packet. The first arrived packet at the HNED is foreseen to be played at the time of potential playing at other HNEDs. However, the parallel transmitting multicast stream has already advanced by the regular HNED buffering time. So, the HNED first needs all missing data before it can switch over. The earliest join time is close to the sum of the buffering time and the burst transmission time. By then, the RAMS-I packet arrives and the RTP burst traffic from the RET server starts. After the accelerated transmission, the RET server transmits with the normal transmission rate until told to stop by an RAMS-T termination packet.

Figure 11.8 now shows the overall behavior with one channel switch. After the port mapping setup, the RAMS-R message triggers the sending of the Unicast RTP stream and the RAMS-I message. After a short time, the HNED decides to switch back to multicast mode and sends the corresponding IGMP message that triggers the joining of the router to this multicast tree. Then, the RTP stream from the multicast source arrives. Soon afterward, the RTP stream from the RET server gets stopped by an RAMS-T message. The HNED may have to adjust the play-out of the two streams but with enough buffer and adaptive media play-out techniques – see Section 5.4.3 – this should work.

When an HNED user starts to zap through the channels, we would see, as shown in Figure 11.8, many sequences of RTCP RAMS-R, RTCP RAMS-I, and unicast RTP one after another.

Figure 11.8 shows an initial port mapping part. The reason for this setup part is that we mixed the usage of ports for unicast and multicast in the implementation of the retransmission and the channel switch. The RTCP RAMS refers to a multicast SSRC, and the following RTP is a unicast. Of course, a new unicast RTP session could get built, but this would take additional time. An additional Port Mapping Protocol (PMP) [594] was developed for this purpose. Note that there is also another PMP [364] (Section 6.5). With PMP, the RET server initially creates a token in the port mapping setup. The HNED includes this token into the RTCP packet with the RAMS part and a payload type 210. With this token, the RET server is permitted to send directly per unicast back to the sender port. With the initially prepared port mapping, the RTP flow can start after receiving a single RTCP RAMS packet.

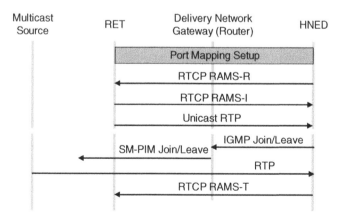

Figure 11.8   One channel switch with RAMS

11.3   Datacast Protocols

The reason for this heading is that DVB for multicast data distribution protocols uses the name IP DataCasting (IPDC) [595]. In this section, we treat three additional protocols, which are used in Section 11.4. Apart from the routing protocols, there exist very few specific multicast protocols. Out of the few IETF standardized ones, we now need the File deLivery over Uni-direcTional sErvice (FLUTE) [596], which is based upon the Asynchronous Layered Coding (ALC) [597]. Note that for Flute and ALC, newer documents exist [598, 599], which are, how-ever, not used by the DVB. The Internet multicast standards were created under the auspices of the Internet Research Task Force (IRTF), and the standards are atypically hard to interpret for RFCs. Another IETF multicast protocol is Pragmatic General Multicast (PGM) [176]. DVB has specified an additional new multicast protocol named DVBSTP for DVB SD&S Transport Protocol whereby SD&S stands for Service Discovery and Selection as seen in Figure 11.1.

11.3.1   Flute

Flute is a unidirectional file transfer protocol, which means that a single sender sends informa-tion to recipients without any confirmations and feedbacks. A typical use is for sending files or firmware updates, which are commonly needed for all multicast users. Flute is sent via UDP only. Since it is a multicast protocol, the number of recipients is unknown to the sender. If files get transferred without any feedback, the file content may be secured by FEC. Still data can arrive at the recipients in an unrecoverable manner. However, Flute resends files repeat-edly since it is a carousel protocol. When data is resent over and over, one needs to avoid any congestion in the network and the sender via the router need to regulate the amount of information flow.

In Figure 11.9, we see the overall structure of the Flute packet. The two bytes after the UDP header are a remnant from ALC. Since ALC describes building blocks for possible protocols, those two bytes represent the chosen structure for Flute in the usage by DVB [238], which we will take as given. The *Congestion Control* is not used.

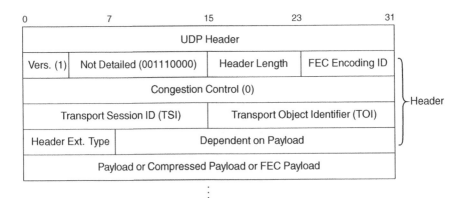

Figure 11.9   Structure of the FLUTE/ALC packet

The other fields shown in Figure 11.9 are defined as follows:

FEC encoding: The FEC encoding can be empty. If the encoding exists, two options can be used to create the FEC – Raptor [585, 586] or a source code block as specified in the Flute RFC [596]. The *FEC Payload* contains the *Payload* in the corresponding FEC format.

TSI: The *Transport Session ID* together with the sender's IP address defines one transport channel. The receiver must know by other methods which TSIs should be used. Also, the sender's IP address, multicast address, and other configuration parameters must be communicated in advance by other methods. When a receiver joins the multicast, it can receive packets of all TSIs, but only the relevant ones are selected for further use.

TOI: If the *Transport Object Identifier* is 0, the payload of the packet contains an XML File Delivery Table (FDT). An FDT describes a file that is identified by a file number. If the TOI is greater than 0, the payload belongs to a file specified by the TOI file number.

Header extensions: There exist header extensions for the regular payload, the compressed payload, and the FEC data for the payload. We only look at the regular payload that has the fields *Header Extension Type = 192, Version = 1, FDT Instance ID*. The *FDT Instance ID* provides the order of the received file payload parts.

With this information, we can look at how the receiver or HNED uses the logic as shown in Figure 11.10 to receive a file. We need to be aware that the receiver may join a multicast any time and some preceding information may not have been seen. However, the transmission is a carousel and missing information will be seen later. Now, let the HNED receive packets of the connected multicast stream. Packets are only retained if the *TSI* is of interest. When the receiver sees a *TOI* greater than 0, it stores the packet in the $Buffer_{TOI}$. The receiver can only decide whether the $File_{TOI}$ is complete if it has once received the FDT. A typical FDT may look as follows:

```
<?xml version="1.0" encoding="UTF-8"?>
<FDT-Instance xmlns:xsi=
"http://www.w3.org/2001/XMLSchema-instance"
xmlns:fl="http://www.example.com/flute"
xsi:schemaLocation="http://www.example.com/flute-fdt.xsd">
<File
    Content-Location="/m/trackl.html"
    Content-Length="2674"
    Expires="20150418T17010000"
    TOI="1"
    Content-Type="text/html"/>
    <File
    Content-Location="/t/track1.txt"
    TOI="2"
    Content-Length="6100"
    Expires="20150418T17010000"
    Content-Type="text"/>
</FDT-Instance>
```

Once this FDT has arrived, the receiver can decide for each $File_{TOI}$ by looking at the *Content Length* if it is complete. The receiver can recover each completed file based upon the FDT and

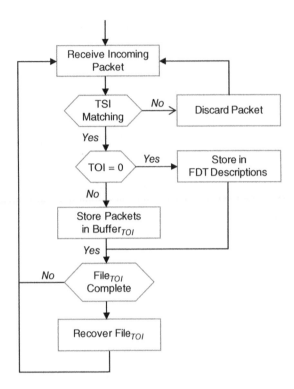

Figure 11.10   FLUTE receiver logic

the *FDT Instance ID*s. Note that the XML for each of the transferred files contains a mandatory expiration field. After the expiration time, the data needs to be deleted.

The Multimedia Broadcast Multicast Service (MBMS) within UTMS and eMBMS (evolved MBMS) [600] within LTE also use Flute. It is foreseen for the transmission of television in the mobile telephony networks.

11.3.2   DVB SD&S Transport Protocol

This protocol was specifically developed by the DVB for maintaining hierarchically structured information in a multicast mode. In Section 11.4, we will see where this protocol is used to maintain data elements. The structure that needs maintenance is shown in Figure 11.11. All parts apart from the root of this tree can change. Changes are visible by new *Segment Version*s or new *Last Section Number*s. The packet structure of DVBSTP is seen in Figure 11.12.

DVBSTP packets are sent with UDP. The payload of one packet contains exactly the value of one section. All sections of one segment always change together, and these changes are sent in as many UDP packets as there are sections in this segment. The less obvious fields shown in Figure 11.12 are defined as follows:

Enc: This field signals the usage of encryption for the payload.
Total segment size: Size of the payload of all sections of the segment that needs to be changed. Since the sections are single UDP packets, this size covers multiple UDP packets.

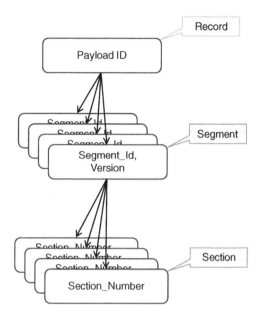

Figure 11.11   Data structure maintained by DVBSTP

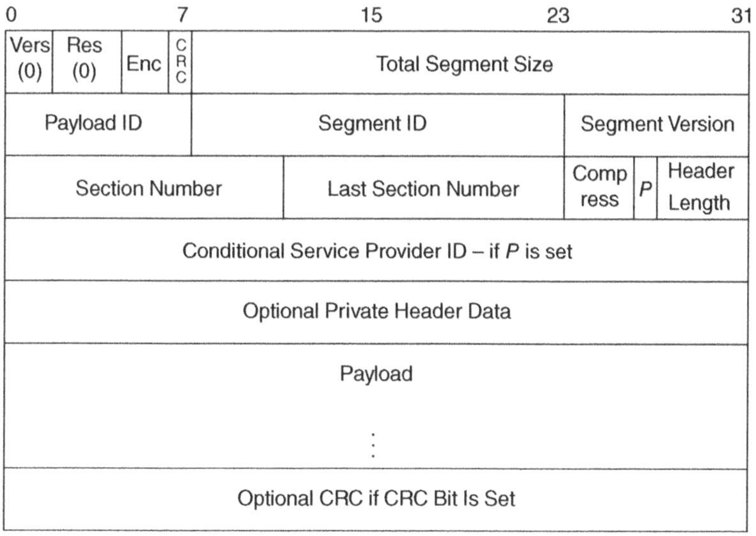

0		7		15	23		31
Vers (0)	Res (0)	Enc	C R C	Total Segment Size			
Payload ID				Segment ID		Segment Version	
Section Number				Last Section Number		Comp ress	P	Header Length
Conditional Service Provider ID – if *P* is set							
Optional Private Header Data							
Payload ⋮							
Optional CRC if CRC Bit Is Set							

Figure 11.12   Structure of the DVBSTP packet

The protocol does not show the size of a single DVBSTP packet since it is just the size of the UDP packet.

Payload ID: This number specifies the record that has to be updated.

Segment ID/Version: A number for the segment that requires change. This packet will change one section of the segment, and the version will change to the newly specified one.

Section: The section the payload is updating.

Last section number: At the end of the transmission of all sections, the receiver has to be sure that all sections arrived. The receiver can verify it with this field and optionally also the CRC value.

Compress: There are a number of options to compress the content.

Header length: This is the length of the optional private header.

11.3.3 Digital Storage Media – Command and Control

The shorthand for this protocol is DSM-CC. This protocol was created by the Digital Audio-Visual Council (DAVIC) in 1996. The organization DAVIC no longer exists, but the standard has in part survived. The overall specification is hard to find nowadays. The protocol set covers direct user-user (U–U) connections, connections via a resource manager (U–N), file transfer, and streaming control such as RVSP. In addition, the protocol covers unicast and multicast, remote procedure calls, and data carousel multicast transfers [601]. In the context of IPTV, only the data carousel transmission is of interest and was moved to an ETSI standard [602, 603].

The protocol is based upon an Object Request Broker (ORB) defined by CORBA (Common Object Request Broker Architecture). CORBA was established in the mid-1990s as communication middleware for heterogeneous systems. CORBA programs, today, are a historical episode of a once hyped idea that was overcome by XML and SOAP (Simple Object Access Protocol) [604]. A detailed explanation of the protocol would require a lengthy introduction to ORB, which we will avoid here. So, we rather give a very high level overview.

Figure 11.13 shows the overall data structure. The User-to-User objects for the transfer are packed into a BIOP (Broadcast Inter ORB Protocol) message. The top-level objects can be just

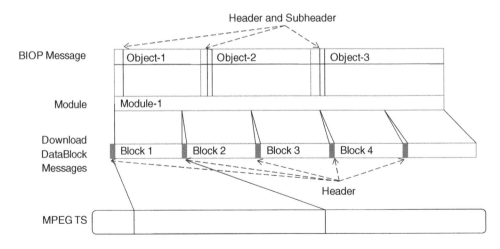

Figure 11.13 DSM-CC object carousel layers

files, complete directories, or streams. The BIOP message header contains the version of the BIOP protocol and the length. The message subheader indicates the transported object type. The modules are broadcast in downloadable data block messages in a carousel mode. Each downloadable block is packed into a Conditional Access Table (CAT) as private data in the MPEG transport stream – see Section 3.22.

11.4 Management Functions

Some parts of this section discuss little on multimedia protocols. This short section is, however, an excellent opportunity to display the management requirements for multimedia applications that serve millions of users. In many instances, these users have dedicated equipment that needs to be maintained. In addition, users expect high availability and stability when the new technology replaces an old, very reliable system such as TV or phone. To cope with the expectations, not only is the maintenance important but also the continuous search for errors and inefficiencies. The continuous improvement process during a rollout already needs automatic reporting in central databases and many analysis tools to identify the root causes of the problems – see, for example, [605]. All these back-office functions are necessary in advance to prepare for a successful product deployment.

11.4.1 Service Discovery and Selection

The data provided in this area allows the HNED to detect the services and select them. The delivery network provides an entry point for this information at a registered IP address (224.0.23.14) or under the DNS entry services.dvb.org. The information has the following overall structure (shortened):

```
Service Provider Discovery Information
    Service Provider 1
        Broadcast Discovery Information for broadcast streams
            Service Description Location 1
                Service Location
            ⋮
        Content-on-Demand (CoD) Discovery Information
            Content Description Location 1
                Content Location
            ⋮
        Services from other Service Providers
            Service Identification 1
            ⋮
        Package Discovery Information (downloadable applications)
            Package Identification 1
                Package Location
            ⋮
        Broadcast Content Guide Information
    Service Provider 2
    ⋮
```

This type of information tree is distributed to each HNED with the DBVBSTD protocol. The HNED can, in addition, send *Service Discovery Request* via HTTP for any information in the tree. This request is also answered as unicast with HTTP.

11.4.2 Broadband Content Guide

The BCG or Electronic Program Guide (EPG) is based on a former TV-Anytime specification [606–608]. This specification is well known, and there are even complete books on TV-Anytime [609]. The content can be addressed by the IETF standardized URL Content Reference Identifier (CRID) [610]. This URL has the form *crid://<DNS name>/<data>*. An additional Internet Draft for a URL of the type *dvb:dvb-path* [611] exists.

The system distributes the information in the BCG with DVBSTP by multicast or by HTTP using unicast. The initial address is available via SD&S Broadcast Content Guide. For the overall XML structure, see Ref. [607].

The base element of the data is the *Programme* for an editorial piece(s) of content. The *Programme Location* is one instance or a publication event of a *Programme*. A *Programme* might be further hierarchically substructured in *Programme Group*. Each *Programme Location* has the following overall structure:

Service Type 1 – that is, single or multiple channels
    Scheduled Event 1
        Broadcast Event consisting of time, duration, location
        ⋮
    On Demand with location
    Push Downloadable with location
Service Type 2 – that is, single or multiple channels
    ⋮

11.4.3 Remote and Firmware Management

The RMS and the FUS are the two software components to manage the software and the configuration on the HNED. The Simple Network Management Protocol (SNMP) was not considered as an RMS because it does not work over firewalls. In addition, the RMS should offer a remote procedure mechanism – that is, remote execution of programs – at the HNED.

The overall system is based on an ETSI standard [612], which underneath uses the standard T.69 of the Broadband Forum [613] – formerly DSL Forum. The T.69 standard specifies a Customer premise equipment WAN Management Protocol (CWMP). Figure 11.14 gives an overview of the architecture of the system.

A firmware update starts with the delivery of a new release plus new metadata from the vendor (**1**). The metadata is also sent to the overall RMS administrator (**2**). The FUS manager retrieves and stores data in the FUS storage (**3**) as well as the RMS administrator approves the update and forwards the metadata to the RMS manager (**4**). The RMS manager updates its configuration items and forwards the approval to the FUS manager (**5**). The FUS manager activates the firmware announcement system (**6**). The firmware announcement sends XML information with DVBSTP to all the clients (**7**). The other possible option for sending is SDP with SAP (Session Announcement Protocol) [614]. After the announcement is made, the FUS manager triggers the FUS distribution (**8**). The FUS distribution takes the package from the FUS storage and multicasts it out with FLUTE or DSM-CC (**9**). In case the HNED does not receive it with the multicast delivery, it can also retrieve it using unicast mode from the FUS distribution (**10**). For unicast, the standard allows HTTP, HTTPS, FTP, SFTP, and TFTP. With

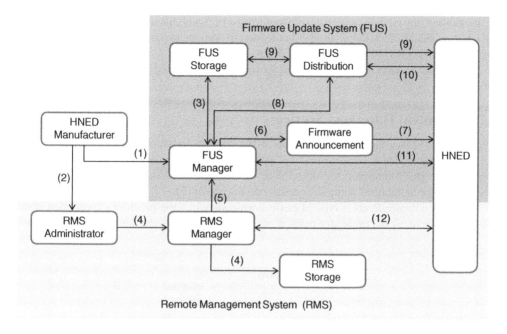

Figure 11.14   Architecture DVB for remote management and firmware update

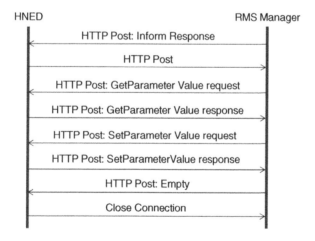

Figure 11.15   CWMP transaction session example for configuring an HNED

the Query Response Channel (QRC), the FUS manager can check if updates were successfully applied **(11)**. The RMS manager uses CWMP to configure the HNED **(12)**.

CWMP uses HTTP and for remote procedures SOAP over HTTP protocol. The change sessions are executed within the transactions to guarantee a consistent state even in case of failures during transmission or installation. Figure 11.15 shows a transaction with typical changes caused by this protocol.

In addition, the HNED can run a bootstrap from a FUS stub file server that is available from IP address 232.255.255.254 or another hardcoded IP domain address. The file can be retrieved by multicast (DVBSTP) or unicast (HTTP(S)).

For the overall RMS, interoperability tests were run in 2008 [615].

11.5 Content Download Service

Figure 11.16 shows the overall functions of the content downloading. The central content repository stores videos for downloading, and the CDS manager oversees the function of the system. The BCG maintains the description of the videos and the location. New available videos may be announced with the *Service Announcement* (**7**).

Multicast downloads are a type of push downloads when the provider expects that many users will like to see the preloaded content later. FLUTE executes the downloads by multicast,

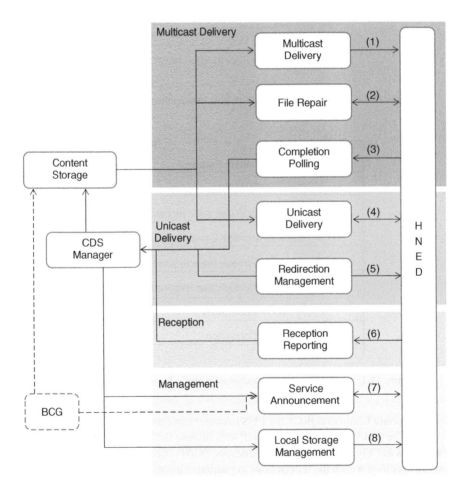

Figure 11.16   Content download – functional architecture

and either is a one-time Scheduled Multicast Download (SMD) or a Carousel Multicast Download (CMD) (**1**). In the case that a file is only partially downloaded, HNED can invoke the file repair. The HNED specifies missing parts of the file in an HTTP request for downloading (**2**).

A CMD should stop when all interested HNEDs have received the file. Therefore, the multicast delivery unit periodically includes a *Completion Poll* message within the FLUTE session. This message is added as a new header extension (65) – see Section 11.3.1. Note that this message is only seen by the end devices that still have this FLUTE connection open and wait for parts of the file. When a still waiting HNED sees this poll, it answers using a UDP message to the polling server (**3**). This behavior has the risk that the server and the network get overloaded if all HNEDs answer at once. To prevent the overload, the *Completion Poll* is accompanied by a *POLL MASK*. The HNED does a logical AND of the received mask and a random local value. Only if the result is zero, the HNED answers the message and adds his random local value plus the still expected time to receive the file. The multicast delivery unit now can adapt the POLL MASK at each send-out time using the feedback at the poll server. The delivered local random values allow "monitoring" of the progress.

In the typical case of a download requested by the user, the download runs as a unicast pull download. CDS offers two alternatives for a pull download (**4**). With the "early play-out" variant, a video can be played while it is transferred. The "early play-out" is the typical Video-on-Demand (VoD) setup. However, if a download is not marked as "early play-out," it requires a complete download before the users can view the video. The second variant is foreseen for situations when end-users are interested in offline viewing for diverse reasons. The download occurs with HTTP. In case the request goes to an overloaded server, the redirection management (**5**) redirects the HNED to a different server.

The reception reporting (**6**) not only executes the billing and charging but also verifies the availability of all CDS-related elements in the HNED. The local storage management (**5**) checks the expiry times of rented content and removes a video if required. Downloaded content is normally not transferable from the set-top box to another media.

The DVB standard [581] is the best source for CDS information. A shortened overview is available in Ref. [616].

11.6 Deployments

IPTV is typically only deployed in Telcos' own networks where they have complete access rights on all devices. This ownership is necessary for building the multicast and the RET server infrastructure. This section is based on experiences documented in Ref. [617]. We will highlight the areas of head IP-TV setup, redundancy, QoS, and testing. The head IP-TV infrastructure sits on top of the multicast tree and includes all components to generate the sent information.

The head IP-TV setup requires the utmost available performance. All network components shown in Figure 11.17 are dedicated equipment. We will not mention types and brands here, but one gets more information on the possible models by using the used terms and provider names for a search. Since the figure is a schema, each block stands for many systems of a certain type – also for failover. IP streamers are dedicated devices for sending Single or Multiple Program Transport Streams (SPTS/MPTS) over IP with a capacity up to 300 Mbit/second. Conditional Access Systems (CAS, CA) refer to DVB standards that control the access to

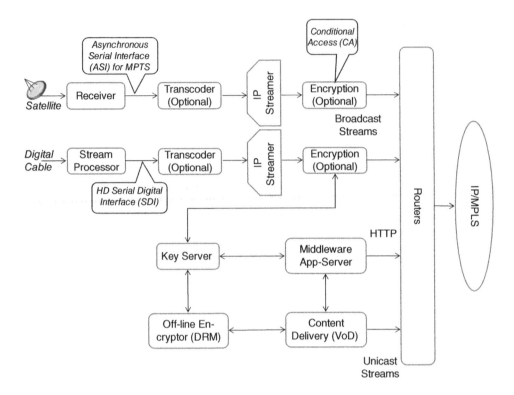

Figure 11.17   Schema of an IPTV headend setup

information by scrambling. The two asynchronous interfaces DVB ASI and SMPTE SDI can carry up to 270 Mbit/second MPTS traffic.

The complete network from the HeadEnd (HE) to the DSL subscriber is set up with redundancy and failover. Redundancy is a challenge with the multicast setup of PIM (Protocol Independent Multicast) – see Section 4.4.5. One option is to use two independent multicast trees and join duplicate packets that arrive at each redundant rendezvous point. Other options are currently under discussion in the IETF [618] to join a multicast tree via two divergent upstream paths – the so-called Multicast-only Fast ReRoute (MoFRR).

The whole path on the provider network will be set up with Diffserv. The Diffserv priority settings from high to low are Control, Voice, IPTV multicast, IPTV unicast (VoD), business data, and other data. From the service edge to the household, VLANs are recommended and in the backbone MPLS with VPLS. In the MPLS cloud, reservations are done with RSVP (Intserv).

The test equipment should include analyzers (video, protocol), MPEG generators, monitoring tools (network, impairment), and test clients in the periphery. For the networking monitoring, probes are required. The impairment monitor can simulate communication impairments for devices under test. This monitor is also able to generate diverse situations such as jitter behavior, delay, packet loss, and other faults.

11.7   Companion Screen Application

The content of this section is not solely for IPTV. Since the specification for IPTV is the most detailed of all the existing ones, this topic is treated here.

There are multiple names and diverse functionalities for this topic. Other possible names are second screen and multiscreen. In all the cases, we have two devices with two associated screens. One device with embedded screen is the smartphone or the tablet, and the other device is a TV set. Instead of a TV set, it could be any other combination of a device that produces a picture – for example, a beamer plus an Audio/Video (A/V) receiver. Currently, the functionalities can be differentiated along the following categories:

Screen movement: The screen content can be moved from one device to another one. After the move, screens can play the content in sync or one screen freezes.

Control after movement: After the move, different operations are possible for controlling the play-out. The remote control of the TV set may or may not work any longer.

Synchronized content: The second screen shows other synchronized content and not the other screen – that is, another viewing angle for the provided content or add-on information such as recipes when looking at a cookery show.

Products on the market include Apple's Airplay and Google's Chromecast. For both products, the complete protocols and implementation details are not publicized.

The Chromecast hardware or server is an HDMI stick with an integrated WiFi and a stripped-down Chrome browser. The Chromecast client software runs on many devices. Applications on the clients need special preparation to interwork with Chromecast. A prepared application can move its window to the Chromecast server that displays it over HDMI to the attached device. The video in the window on the client is no longer active after the move, but one can still control the application on the server. At this point, the inactive screen does not present any synchronized content. The remote control of the TV set can switch between the HDMI input and the TV programs but has no control over the Chromecast device.

A few elements of the Chromecast system were identified [619]. The hardware and the client detect each other using mDNS [538] – see Section 10.1. Earlier, the Netflix-developed DIAL (DIscovery And Launch) protocol was used [620]. DIAL uses the UPnP Simple Service Discovery Protocol (SSDP) and HTTP. Afterward, the client controls the Chromecast server over a WebSocket communication [473]. The media session is built directly from the server to the Internet and can execute each browser application.

UPnP approved in 2014 a specification for a multiscreen device [556] – see also Section 10.2. The documentation covers the interaction between different devices and the possibility of a remote control. The specification provides interaction primitives, but the resulting overall functionality is not clear. Consequently, if future applications on the market refer to this specification, the functionality is not evident either. The UPnP specification has not explicitly integrated the possibility of synchronized content.

DVB has recently published the detailed specification for a Companion Screen Application (CSA) [621–623] that we handle next, since it is the richest and best specified one. The first products are expected in 2015 that interwork with DVB IPTV.

A CSA offers the following functionalities in the light of the previously mentioned categories:

CSA screen movement: The same screen content is delivered in parallel to the CSA. The direction of sharing is from the TV screen to the smartphone and tablet. With this function, a viewer can leave the TV room and see the content on the CSA and later return to the TV set.

CSA screen control: The companion device does not replace the remote control for the TV set, which still stays active. However, pausing content on the CSA also pauses the playing of the TV.

CSA synchronized content: The second screen can present other viewpoints in addition to the TV screen view. Additional information about an ongoing TV program may be provided in the second screen.

Figure 11.18 provides an overview of the functional architecture of the overall system. Apart from the Material Resolution Service (MRS), the other blocks are straightforward as shown in this figure. The MRS is operated by the provider on the Internet and has all the information available to connect the IPTV content to additional resources. Note that this figure does not show, for example, the OTT streaming server of the provider in the Internet since the specification does not contain it. However, the existence of this server is given, and by specifying it this way, the current OTT servers can be embedded.

The two device types in the home network (CSA, TV device/HNED) first need to learn about each other. The Discovery and Association (DA) function (**1**) shown in Figure 11.18 is responsible for this first step. This step is accomplished by the UPnP application

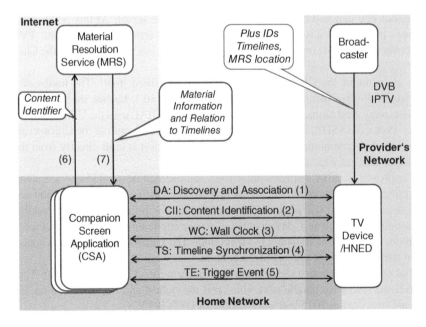

Figure 11.18 Companion screen – functional architecture

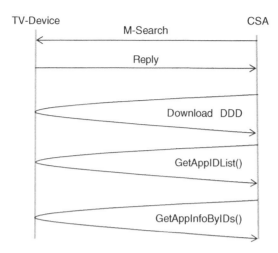

Figure 11.19   UPnP discovery and association between CSA and HNED

management Device Control Protocol (DCP). Figure 11.19 shows the steps. The Device Description Document (DDD) is retrieved after the contact. The *GetAppIDList()* for the input *Identification of TV Device* retrieves all *IDs of applications*. With the message *GetAppInfoByIDs()*, the CSA retrieves the XML documents describing all initially required information.

With the provided information from step (**1**), the device shown in Figure 11.18 initiates the "Content Identification and other Information" (CII) connection (**2**). The CII and most other services run on top of a WebSocket connection [473] that provides a two-way communication from a web page via a publicized API. With this connection, the CSA is always informed which content plays at the TV Device and exactly at which point in time. The CSA by HTTP/HTTPS queries the Material Resolution Server (MRS) with the content information (**6**) and retrieves the possible material (**7**). The material is specified in JavaScript Object Notation (JSON) [464]. The material contains URLs plus timelines. When the URLs get called, they must start the media at the specified time. Note that the start of the play-out of a video can be specified by diverse methods – see Section 4.3.1.4 – and the standard recommends possible choices. Also, the data stream for the TV device needs to be changed to carry the MRS reference information for each point in time.

The timing in the whole setup is difficult. When the TV device and the CSA want to show content in sync, they first need to have a common time that is as exact as possible. The common time is achieved with the Wall Clock (WC) (**3**). The WC uses a UDP protocol similar to NTP – see Section 4.6 – to agree and maintain a common time that is, however, not a real time. With the answer (**7**) and the earlier retrieved content identification from the server, the CSA can calculate the time offset between the Internet provided content and the content on the TV device. The Timeline Synchronization (TS) (**4**) maps all possible presentation times to the wall clock time. When the viewer requests to see the same content on both devices, it is likely that the TV device has to wait some seconds. This delay is caused by the preparation of live material for HTTP streaming, which takes more preparation time. The play-out on the two devices is afterward controlled by the wall clock time.

The situation between the CSA and the TV device is not reliable. Some typical causes of instability are: the CSA could lose connection or is overloaded, the viewer decides to switch the channel or to pause a device. This instability will result in the situation that the CSA might occasionally get unsynchronized, and pictures could freeze. We can only hope that the application on the CSA forgives many errors and still survives.

The last function is the Trigger Event (TE) **(4)** as shown in Figure 11.18. Triggers shall send information to the viewer on specific events in the stream – for example, a goal in a football match. When the CSA receives a JSON description for the trigger, it displays the information to the user. Triggers are embedded in MPEG-TS and DASH media streams.

11.8   Set-Top-Box Functions

IPTV deployments always include a set-top box or HNED in DVB IP terminology. Set-Top Boxes (STB) for TV are frequently equipped with recording facilities – Client Personal Video Recorder (CPVR). Since modern set-top boxes record during the play, they can also deliver a time-shift playing. Time-shift playing means that an ongoing live media stream is recorded and played with some delay. With this function, viewers can stop an ongoing show or sports event for some time and answer, for example, an incoming call.

Since the recorder is easily reachable from the provider, additional functions are straightforward. A typical application is a remote control for the recorder, but remote in this context means anywhere in the world. Such an application on a smartphone can access the set-top box via a proxy function of the provider. Users can start, stop, and delete recordings whenever they are connected to the Internet.

The question of whether it is required to record on a personal STB since there are recording cloud services arises. In the United States, the Supreme Court in 2014 decided that the recorder cloud provider Aereo violates the broadcasters' copyright. The same will be true with all OTT providers that distribute content without the approval of the broadcaster. For sure, broadcasters will not allow OTT distribution of high-value content without payment. At least for high-value content, an "own" recorder is required.

If one has, however, the right to receive the content, they can store the stream somewhere and view it from any location. Recording from an own set-top box can be shifted in the cloud – Network Personal Video Recorder (NPVR). This leads to the question of whether you can move additional functions of the STB in the cloud. Recent papers [624, 625] treat this topic. They show that a few elements are difficult when moving the box to the cloud. The critical areas that partly hinder the virtualization are network bandwidth, fast channel change, and QoS.

Storing videos in the cloud or at an NPVR has the problem that content rights are specific to the platform where the content was bought. A video bought by a provider of an NPVR may not be viewable on another device. The movie industry has addressed this topic and publicized the UltraViolet standards and runs a supporting system. Each video has a Common ENCryption (CENC) added to the ISOBMFF (Section 5.9), which stores the user's rights [626]. All big studios support the standard, but major distribution partners do not maintain it – for example, Disney, Apple, and Amazon.

11.9   Integration into Other Systems

We will touch on the integration of IPTV with IMS, WebRTC, and home networking. There are
also other ideas such as the integration of social media with the media presence information.
However, until now, no realistic proposals on this topic exist.

11.9.1   IPTV and IMS

The ETSI TISPAN (Telecommunications and Internet converged Services and Protocols for
Advanced Networks) has defined an integration of IPTV and IMS [627–629]. However, this
integration framework so far has neither continued with the 3GPP project nor with DVB. Only
one study paper on 3GPP in combination with WebRTC exists [487]. It seems like the market
believes more in delivery with an OTT TV on mobile phones. One of the main requirements
of IPTV was to provide a very reliable service, which is, by the nature of mobile data transmission, hard to deliver nowadays.

Since this integration in the next few years may not develop further, we only highlight
the concept. Figure 11.20 provides a high-level view of the overall system. For IMS elements, check Section 7.4.1. The Service Control Function (SCF) builds a SIP session for

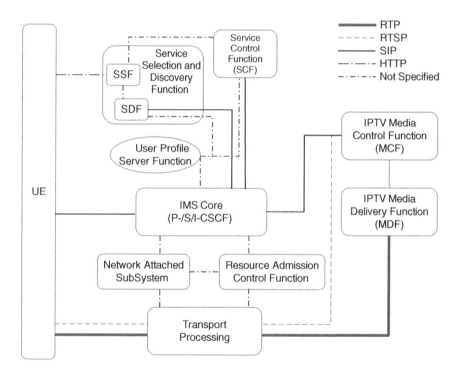

Figure 11.20   Schema of IPTV and IMS integration

IPTV services and authorizes content access with the User Profile Server Function (USPF). The Service Selection Function (SSF) interacts with the user environment for the selection. The Service Discovery Function (SDF) delivers personalized information and content identification. The Media Control Function (MCF) controls the play-out, locates the content based on the delivered identification, and does administrative tasks. The Media Delivery Function (MDF) plays the media and maintains the storage. The type of the foreseen protocols is marked in the figure. It is quite obvious that the proposal is not aligned with the DVB IPTV standard.

11.9.2   IPTV and IMS and WebRTC

Concept papers describe how the UE shown in Figure 11.20 can be replaced by a WebRTC client [487, 630]. As we have already seen in Section 8.3, this requires several codec and protocol adaptations. Many of the browser companies have allies in the HTTP-streaming industry and might not be supportive to allow connecting to the Telco's offering.

11.9.3   IPTV and Home Network

An integration in the home network occurs with the companion device – see Section 11.7.

There is a proposal by the DVB [631] to offer content within the IPTV system via UPnP for home network clients. Figure 11.21 shows the overall model. We use UPnP functionality as discussed in Section 10.2. The HNED integrates a Discovery Point (DP) and a Streaming Point (SP). The HNED-DP advertises DVB content by translating SD&S and BCG content to a UPnP Content Directory Service (CDS). In addition, any locally stored data of the HNED also gets offered by the CDS. The DP provides the standard UPnP Connection Manager (CM). The HNED-SP streams the chosen content via RTP. The DVB Renderer Point has the standard UPnP services AudioVideo Transport (AVT), Rendering Control Service (RCS), and CM.

It seems that the standard has not led to any implementation. Probably, it is easier to offer the content from central resources of the provider via the Internet than to build translation functions in set-top boxes.

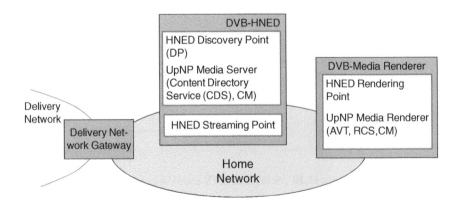

Figure 11.21   Schema of IPTV and UPnP integration

12

Solutions and Summary

The final chapter applies the gained knowledge in understanding other multimedia applications. The applications discussed in this chapter do not have complete standards as we have seen in Chapters 6 and 7 for voice and in Chapter 11 for high-end TV. Since we concentrate on standards for multimedia protocols in this book, the treatment of these applications will be, on the one hand, more compact due to missing specifications and, on the other hand, more provider-specific. The requirements for all applications in this chapter were listed already in Chapter 2.

Section 12.1 discusses what is required to create global Webcasts, which might also have its network challenges. Digital Signage is quite common today and can become challenging if globally deployed – as seen in Section 12.2. From a call center, we are only aware of the agent we are currently speaking to, but the whole system can be quite complex – see Section 12.3. Afterward, we have a closer look at one of the most challenging applications from a network perspective: High-end videoconferencing or Telepresence in Section 12.4. All sections in this chapter reuse technologies that we introduced before.

We concluded Chapter 2 on requirements with a summary in Section 2.6. In Section 12.5, we again consider the list of requirements and compare the provided solutions. This concludes this book.

12.1   Global Webcast

For the requirements of Webcasts, see also Section 2.5. As stated already, Webcasts are used for lectures, events, and meetings up to multi-thousand viewers. A typical user interface is seen in Figure 12.1. The overall application is not delay-critical, but the various shown media parts should be in sync – that is, video, audio, presentation slides, and additional information. Since thousands of users may wait for a Webcast, it is critical that a Webcast starts on time and has no problems during the transmission. After the live Webcast, the material gets published for a restricted time on a portal. Webcasts are used over the Internet and in company networks. The use within company networks is more challenging as we will see further on.

Multimedia Networks: Protocols, Design, and Applications, First Edition. Hans W. Barz and Gregory A. Bassett.
© 2016 John Wiley & Sons, Ltd. Published 2016 by John Wiley & Sons, Ltd.

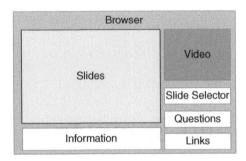

Figure 12.1   Webcast browser

A production environment for Webcasts consists of the following parts:

Production and communication coordination: The coordinator has to agree on modalities with the presenter and verifies that all technical capabilities are in place. Always required are a camera team(s), technical back-office workers, and all the technical equipments. Sometimes, additional tests are necessary. All participants need to get invites with information on how to join.

In companies, some additional steps are required. Initially, a network check is necessary – see the following discussion. Users may join with unicast or multicast and need different URLs in their invite. Some users may be out of the office and join over VPN. When VPN unicast is required, only a maximum number of parallel users is possible. For viewers in open-space offices, headsets need to be available.

Application adaptation: The customer typically requires adaptations to the web page as shown in Figure 12.1 with additional logos, links, and information.

Room support: The room support advises the presenter on good behavior during the presentation: lively but not moving from his place. If the event runs in a hall and questions are allowed, microphones need to be integrated. The lighting for the room needs to be set up together with the production team.

Audio/Video equipment: Figure 12.2 shows the typical elements for the capturing and send-out of the stream. A unique solution that does everything best in one system does not exist, and different elements must be integrated. Different audio signals require mixing before encoding. With multiple cameras, input choices are made, and pictures may be resized to adjust for the required output format. The encoders have many options to adjust best for different situations.

Question management: Users can type questions in the web browser field during the presentation. If users stay anonymous, questions can get nasty, and it is advisable to ask at least for the name or require a login to a system. Incoming questions may go directly on a screen to the presenter or to an additional person for prefiltering. Usually, this results in more questions than will happen just from the room participants. Technically, a text message system or an E-mail client can be used for this purpose.

Others: Network considerations, slide handling, and content management are treated next.

If a live Webcast is provided for the Internet, the critical elements are the number of potential users and the resources at the streaming servers inclusive of the Internet access lines. If the resources are not sufficient, one might use the line capacity in well-equipped hosting centers.

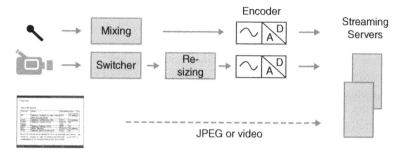

Figure 12.2   Webcast streaming

A live Webcast within a company has other challenges. Company networks are typically star-like networks with major links to regions and from regional hubs to locations. Webcast users often are densely distributed in a company. If there is a Webcast from the production head, all users of a production site want to see this Webcast. This usage creates a problem since the line capacity to this site is not built for video sessions to each user in parallel. The following options exist to solve this:

Multicast on the whole path: Typically, company networks are not configured for multicast, and this would need to be implemented. The transmission of the multicast over the WAN can also create problems with providers.

Indirect unicasts: There exists a splitter function in particular devices or within routers. A splitter in a site receives a single unicast from the streaming server. The clients at the site then connect to the local splitter function and not to the central streaming server. This setup is shown in Figure 12.3.

Caching at the sites: For this solution, the Webcast needs to be sent as an HTTP Live stream, which leads to some seconds of delay for creation, but this is not critical here. In addition, each site needs to have a caching server in place, for example, with WCCP – see Section 4.7.2. With HTTP streaming, only the first request for a particular segment of a stream is sent over the WAN. The cache serves all other requests for the same segment.

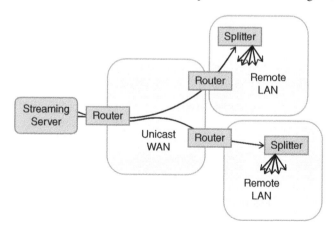

Figure 12.3   Unicast splitters

VPN users still require a unicast send-out. Here, the bandwidth on the VPN servers and their location in the network will result in a maximum allowed number of users.

The quality of the slides is essential since the content together with the audio is more interesting than the video of the presenter. One option is to send the slides as an additional video stream. Since slides only change every few minutes, this is a waste of network capacity. Another option is to create JPEG files for each slide in advance and use a cache server at each site. In this setup, the slide change needs either a trigger to the browser, or the browser must continuously poll a central server for a next slide. When a slide change occurs, the request for the JPEG file is again handled by the cache. Only the first request is sent over the WAN. Even this request can be omitted by placing the JPEG files on the cache servers in advance. This second more network–savvy solution sometimes has support problems in browsers.

After a live event, the Webcast needs preparation as an on-demand version. For the on-demand version, users want to jump directly to a certain slide as can be seen in the field *Slide Selector* in Figure 12.1. Also, different resolutions for different network speeds must be created. SMIL (Section 5.10) is an option to control the stream together with the slide changes. If the slides were sent as an additional video stream, this, however, is not required since the streams are linked in time.

The prepared on-demand version is imported into a content management application as seen in Figure 12.4. In addition to the combined video, audio, and presentations, the content management system also stores descriptions, access restrictions, and lifetime information. For effective use of the network bandwidth, the deployment function should distribute the content at low network usage times to all local cache servers. When the lifetime has ended, the caches are also cleared. In case the Webcast took place on the Internet, a replication by a content network might be useful – see Section 9.6.1.

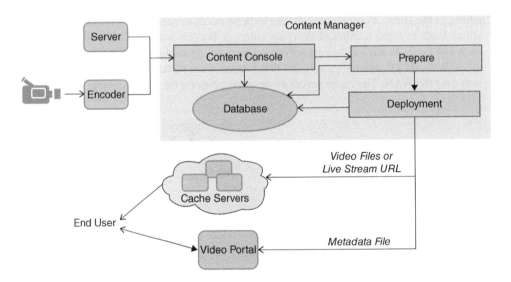

Figure 12.4   Webcast content management

12.2   Digital Signage Broadcasting

We see digital signs in many places. In a shopping mall, it shows advertisements, and in an airport, it shows departures plus arrivals. From a network perspective, this seems to be rather simple and not noteworthy to treat here. However, one can deploy a digital signage system globally within a company, and Figure 12.5 shows a typical screen. For global companies, it is a way to send the same business messages to all employees in a short and comprehensive form. Of course, this needs to be enriched by country messages also to emphasize the local view of a company. Since people regularly pass those signs and their interest in business messages might occasionally be weak, additional eye-catchers are required. Share prices and a TV part should serve this purpose. Keeping the content updated is critical to maintain the viewer's interest.

From the screen shown in Figure 12.5, it is apparent there are various information owners at different locations. The global communication group owns the company news, CNN, and the Dow Jones ticker. Local communication groups handle the country news and a local TV channel. In reality, regional messages may need to be integrated as well, but we leave this out here. Of course, global and local news is also constantly changing. Each of these screens is similar to a short presentation with some slides. Apart from the live TV channels and the share price ticker, everything else gets repeated after a few minutes.

This global content (CNN, share price ticker) can be sent as a stream from the headquarters, from regions or even locally. The decision has to take network capacity and local attachment costs into consideration. Typically, a regional setup may be the most cost-effective one. The company news and clips are not live media. Consequently, the media can be replicated at each site to a cache server. We have introduced in Section 4.7.2 WCCP for transparently accessing this information on a local cache server. With each digital signage screen, a signage player gets installed, which communicates with a central signage server. Figure 12.6 shows a typical protocol to deliver content for one window. Since there are no standard protocols for digital signage, we show an abstracted proprietary protocol.

In Figure 12.6, we see that the player asks the signage manager for a playlist. After that, the signage client accesses the content. The remote WCCP cache server checks for all HTTP calls if the URL is locally cached. For content that is not live, media is cached and the WCCP

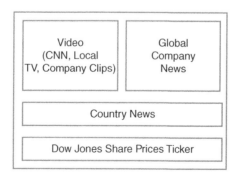

Figure 12.5   Global signage system: screen at each local affiliate

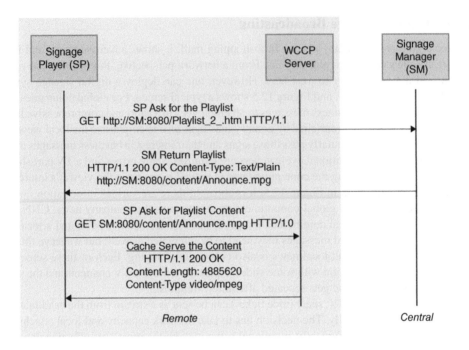

Figure 12.6   Example of a protocol between signage manager and signage player

server can answer directly to the player avoiding repeated retrieval of the information from the center.

Of course, we have to serve multiple windows, and mostly, the signage players use SMIL for coordination – see Section 5.10. The ITU article [632] states that SMIL has been deployed over 100,000 times in digital signage systems. The first call in our protocol as shown in Figure 12.6 could have been triggered by an earlier downloaded SMIL code for the signage player who starts the video for a window at a defined time.

The overall digital signage also needs a management system with the following components:

Content management: All content for displaying needs to be stored in a central system. Even if it is local content from one site and only played at the site, the central system should store it to guarantee that references from the player are correct.

Scheduling: In the signage manager, the administrator for each window defines the time and the content the system should play on each signage system. Of course, windows for multiple sites can be grouped together to also form a global window. Figure 12.7 shows a typical example of such control windows. The schedules can generate SMIL scripts that are pushed out to all signage players by publishing.

Replication: The system will replicate content to local cache servers for seamless access to content – see Figure 12.6. Replicated content will be removed if current scripts do no longer access the data.

Recently, the ITU [633] approved a service requirement statement that is not quite in alignment with practice. The ITU proposal does not cover the use of caches and external streaming

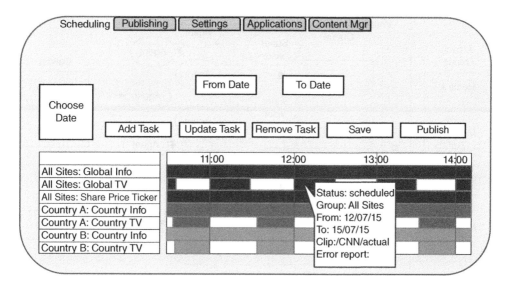

Figure 12.7 Example of a display scheduling within a signage manager

sources. FLUTE is proposed for multicast delivery to the player devices. The ITU document also does not provide any solution. Also, POPAI (Point-of-Purchase Advertising International) has published standards on Digital Signage, but they are only accessible to members. Additional literature on signage is [277, 634].

12.3 Call Center

This section first discusses the functional elements and then looks at technical components with an emphasis on the network technology.

12.3.1 Functional Components

Our first view of a call center is provided in Figure 12.8. We see that a customer may arrive at the call center using different media. This is typically per phone but can also be a video call, a text-message system, an E-mail, a voice mail, or a custom Web application.

With each arrival of a customer, this contact request needs prioritization and routing. Automated Call Distribution (ACD) performs the routing using characteristics of the client. One immediate characteristic is the client's phone number. With a phone number, Automatic Number Identification (ANI) can take place if an associated database maintains the phone number plus the required characteristics. If a customer calls again within a defined period, the system might be able to shortcut the process and add the client to the same agent queue with whom he talked before. However, the standard case is that an Interactive Voice Response (IVR) system interrogates the customer in combination with DTMF or touch tones. The routing can take into account the time of the day, the day of the week, and additional time-related information. Thereafter, the client call is initially placed in one of the queues.

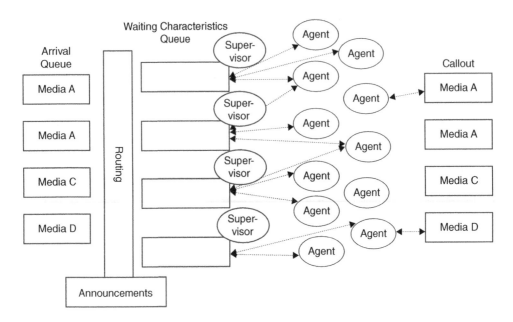

Figure 12.8   Setup of a call center

During the routing and waiting phase, announcements are made and music is played. Some announcements have information about how much more waiting time is required to communicate advertisements or propose a recall if the waiting time is too long. The queue status is constantly monitored, and clients may get requeued if their waiting time is above a defined limit.

A supervisor associates agents with queues based on their skill set. He has the Control of Agent Features (CAF) and can change skill levels of agents. Agents can get linked with multiple queues if their knowledge is usable in more than one queue. The agent's skills may then be weighted for the different areas to maintain a balance in the waiting times of multiple queues. The supervisor may join, record, or take over calls. He handles the quality of the overall service provided by a queue. His responsibility also covers required reports for different types of services. The system has to support all reporting needs.

The agents use a main application that gives them as much information on the clients as possible. When an agent begins the conversation with a customer, the main application displays numerous items. Such items are the waiting time of the customer, the results of previously asked questions, and Call Line Identity (CLI) plus customer database records. This application does not only show the individual customers' data but also presents an overview on the complete queue. Agents know if the waiting time in the queue gets too long and has to speed up conversations with customers.

Agents also handle outgoing calls or a combination of incoming and outgoing calls. Outgoing calls can be callbacks or campaign calls. Campaign calls happen for calling potential clients, for selling or marketing purposes. The application for outgoing calls dials by itself and associates an accepted call to an agent. The outgoing call application should automatically detect an answering machine, find the right dialing pace without letting the agents wait, and follow the "Do not call" rules of diverse countries.

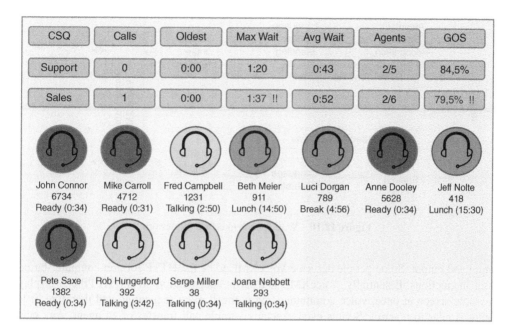

CSQ	Calls	Oldest	Max Wait	Avg Wait	Agents	GOS
Support	0	0:00	1:20	0:43	2/5	84,5%
Sales	1	0:00	1:37 !!	0:52	2/6	79,5% !!

John Connor
6734
Ready (0:34)

Mike Carroll
4712
Ready (0:31)

Fred Campbell
1231
Talking (2:50)

Beth Meier
911
Lunch (14:50)

Luci Dorgan
789
Break (4:56)

Anne Dooley
5628
Ready (0:34)

Jeff Nolte
418
Lunch (15:30)

Pete Saxe
1382
Ready (0:34)

Rob Hungerford
392
Talking (3:42)

Serge Miller
38
Talking (0:34)

Joana Nebbett
293
Talking (0:34)

Figure 12.9 Example of a call-center wallboard

Each agent has a state that shows their availability: ready, break, logged-in, logged-out, and calling. After a call, an agent may have to follow-up on certain aspects or document the call. Then, this agent is in "After call work". All times are automatically recorded and shown in the display for the agents. Typically, a maximal time for a type of call is set, and the system displays the status. The wallboard shows the complete overview of the status – see, for example, Figure 12.9. In this example, there are only two Contact Service Queues (CSQ) with currently no-one waiting and four agents in a call. The overall Grade Of Service (GOS) is also visible on the wallboard.

Agents may sit together in the same room, but often they are distributed at a site or even sit in different countries. When working time terminates at one site, a complete queue may get assigned to agents at another site around the world.

Call-center systems often have an interface to Enterprise Resource Management (ERM) and Customer Relationship Management (CRM). The integration into the customer relationship system also means that call records and any follow-ups are stored in such system. For the application system, this a Computer Telephony Integration (CTI) based on a standard API – see Section 6.6.

12.3.2 Technical Components

For the user interface in a Call Center, the W3C has developed some widely implemented specifications [635]. The overall interface framework is seen in Figure 12.10 and explains itself quite well.

In Figure 12.10, the central coordination element that interfaces with the user is the dialog manager or VoiceXML. VoiceXML specifies how the different technologies play together for

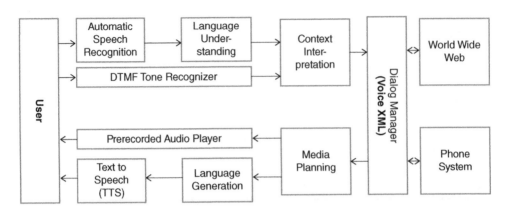

Figure 12.10   W3C speech interface framework

input and output. Some people describe VoiceXML to be the HTTP for non-computer-based user interactions. Essentially, VoiceXML is a script specification. While HTTP can display a whole screen at once, voice communication is much more sequential, and therefore, it is more similar to a script. Scripts even execute a simple call transfer of an agent. Any new connection, any playing of music, any interactions before the agent is in the call is done with scripts. VoiceXML now exists in Version 3.0 from W3C [636, 637]. For earlier versions, more literature and books are available, see Refs [638–640].

When executing VoiceXML, the components from Figure 12.10 are placed in a technical structure, which is depicted in Figure 12.11. The Media Resource Server (MRS) contains all the analysis functions used by the Voice Browser. The Voice Browser executes the VoiceXML retrieved from the web server. The functions on the MRS are Automatic Speech Recognition (ASR), Text-To-Speech (TTS), DTMF-Analysis, and audio playing and recording. The MRS might also include a Speaker Identification and Verification (SIV) function by which clients are identified by their speech.

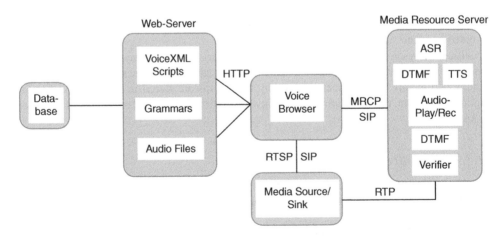

Figure 12.11   Voice XML interpretation environment

Figure 12.11 contains a new protocol Media Resource Control Protocol (MRCP), which exists in two incompatible versions MRCP and MRCPv2 [641, 642]. MRCPv2 is quite a large RFC that specifies an extensive set of capabilities including the aforementioned SIV function. MRCP protocol defines the requests, responses, and events needed to control the media processing resources. For each resource, state machines that implement all actions are specified. A control session to a resource is established by SIP and in the first version of MRCP by RTSP. A control session is run on top of TCP or TLS. The actions send in clear text in an HTTP-like manner.

To understand the overall processing, we look at a simple VoiceXML specification shown in Figure 12.12. The example uses the version 2 of VoiceXML and first asks the user on his travel plans: *San Francisco* or *Los Angeles*. For the answer, alternative names for the cities are possible *Frisco* or *LA*. Alternative names get mapped to the full names before proceeding further.

VoiceXML uses additional W3C markup languages. The block **(1)** in Figure 12.12 is surrounded by *<prompt>*, which states that it is media output. The block uses the Speech Synthesis Markup Language (SSML) [643], which describes the speed, volume, inflection, and prosody of text for TTS. Here, the city names are moderately emphasized. The box **(2)** is a Speech Recognition Grammar Specification (SRGS) [644]. Within box **(2)**, we see two additional blocks marked with **(3)**. The block **(3)** uses the Semantic Interpretation for Speech Recognition (SISR) [645], which lets a developer translate words to other internal representations – in this example, "Frisco" to "San Francisco". All grammar specifications, scripts, and audio files are stored on the web server or at the backend database – see Figure 12.11. During the execution of the VoiceXML, the voice browser retrieves the corresponding information. Input and output flow from the browser or the media resource server to the source.

```
<?xml version="1.0" encoding="ISO-8859-1"?>
<vxml version="2.0">
<field name = "to">
 <prompt>
      ┌─────────────────────────────────────────────────────────────────┐
      │ Welcome to the Californian Bus Travel. Do you want to go to   (1) │
      │ <emphasis level="moderate">San Francisco</emphasis> or           │
      │ <emphasis level="moderate">Los Angeles </emphasis>               │
      └─────────────────────────────────────────────────────────────────┘
 </prompt>
 <grammar type="application/grammar+xml" version="1.0" root="main">
      ┌─────────────────────────────────────────────────────────────────┐
      │ <rule id = "main" scope = "public"                           (2) │
      │ <one-of>                                                         │
      │ <item>San Francisco</item>                                       │
      │ <item> Frisco │ <tag> San Francisco </tag> (3) │</item>          │
      │ <item> Los Angeles </item>                                       │
      │ <item> LA │ <tag> Los Angeles </tag> (3) │ </item>               │
      │ </one-of>                                                        │
      │ </rule>                                                          │
      └─────────────────────────────────────────────────────────────────┘
 </grammar>
</field>
```

Figure 12.12   Example VoiceXML – boxes and numbers only for referencing

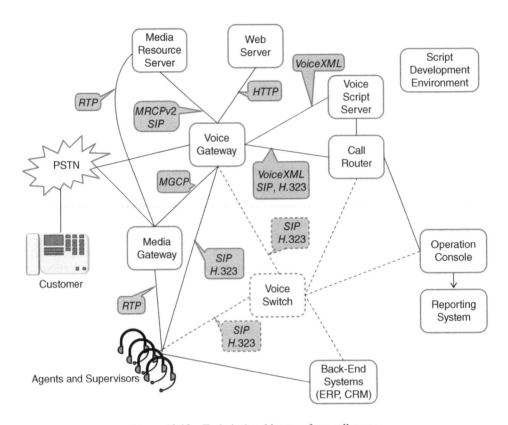

Figure 12.13 Technical architecture for a call center

With this information, we now can look at the overall still simplified, technical architecture of a Call Center as shown in Figure 12.13.

Voice gateway: This system does the PSTN signaling and acts as a media gateway controller for the media gateway – Section 7.3. In addition, the voice browser is integrated into this system. The scripts are loaded from a dedicated system. The media resource server and the web server execute elements of the scripts. The communication to the customer uses the media gateway that is the media source for the MRS. At an end of the script, the call will go on hold or already to an agent. Calls often need recording. The Session Recoding Protocol (SRP) [646, 647] provides this functionality. A Back-to-Back User Agent (B2BUA) controls the recording. B2BUA sits in the communication flow – see Section 6.2. The voice switch would be a good location to implement the B2BUA.

It is the task of the call router to decide on queues and agent associations. For call routers and their protocols, no standards exist and each implementation uses proprietary methods.

Voice script server: This component maintains all VoiceXML scripts. On request, the system sends out scripts.

Media gateway: Maps the medium between PSTN and internal network, which is controlled by the voice gateway via MGCP.

Call router: This unit routes calls based on diverse criteria and database information and also decides which agent is receiving a call. It may also use voice scripts for callout functions. These units and the used protocols are proprietary.

Voice switch: So far, we have given the impression that all proxy or gatekeeper functionality was operated by the voice gateway. However, call agents must be able to call internally within the company, check with a supervisor, and transfer calls. In such cases, the call signaling may also be handed over to the voice switch.

If the B2BUA runs in the switch, the link to the internal back-end systems would be possible instead of linking each client separately.

Other systems: Few additional systems were shown for completeness but are not detailed further: ERP, CRM, reporting, and development. The operation console may display the wallboard and supervisors may use the console also.

Figure 12.13 is still a simplified architecture since it does not show the following additional network options:

- Call agents can sit at remote sites or be connected remotely via VPN.
- Calls can come in at remote sites but get connected to a central Call Center installation. The forwarding requires that the media gateways in the sites be connected properly to the center – for example, to the MSR function.
- For load distribution, a multitude of units for one function type may be needed. Then, we also need load distribution units.
- A call center is mission-critical for many businesses and all components need to be replicated in failover mode as well as the network. Failover network setups are shown in Section 12.4.3.

In Section 6.8.3, we discussed the sizing of telephony systems. For call centers, the simple model is no longer sufficient. Different providers offer enhanced calculation tools – see also Refs [648, 649].

All major telephony system vendors offer call center with included VoiceXML and MGCP.

12.4   Videoconference and TelePresence

For the history of VideoConferencing (VC), see Section 2.4. The use of videoconferencing in companies started after the standardization of H.261 in 1992. The market only slowly accepted VC products. Of course, the devices and the MCUs have been very expensive in the beginning, but this was not the only reason for the slow market acceptance.

Figure 12.14 is based on a study from Ref. [650] and shows the criteria for a good value by the users. A system must provide a vivid, consistent, and interactive experience for a good concentration – in this order. The perceived usefulness and ease of use highly depend on the vividness. Vividness is reached by a good picture quality and sufficiently high visibility of persons. This element was a challenge at those times since the screens were small, and zooming on persons in a meeting needed manual interaction that no participant was willing to do. Picture quality also was not impressive because no one used more than two ISDN channels (128 kbit/second) for transmission. However, this was acceptable at times with low resolution in TV, cinemas, and elsewhere. Consistency was another challenge with no reliable lines and

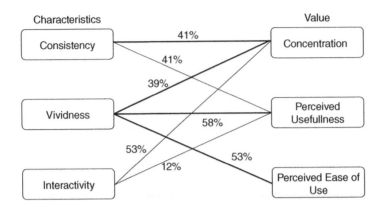

Figure 12.14 Characteristics for a videoconference and perceived value

early systems. If a videoconference gets interrupted more than once during a session, users would get bored and jump back to telephone conferences.

For a long time, the call setup of VC systems was cumbersome for the user. A user at best accepts to dial a single phone number to set up a call. However, for multisite conferences, the dialing can easily take 5 minutes. In the case of network problems during a call, the dialing procedures had to be repeated. Scheduling of global multisite meeting in addition to a telephone conference requires the reservation of VC rooms and MCU capacity.

Around the year 2000, only a handful of wealthy global companies actively used videoconferencing with more than 100 VC studios. Those successful companies had built the following elements:

Support: A specialized helpdesk was available with almost no queueing. We must imagine that the participants of those calls belong to the middle or upper management in a company. If 20 of them are in a call, one can translate this to the level of importance and the required availability of the support. The challenge was that network problems often hit multiple conferences at once and then the helpdesk got overloaded.

Fast setup: The support created dialing scripts for each videoconference in advance. For the setup of a conference, the scripts got executed on the MCU. With such a procedure, the connection can all be set up before persons enter the room at the agreed time. Also, in case of problems, fast recovery is possible.

Easy scheduling: Specialized online system for VC scheduling was available. The integration within the scheduling system of the company only became available when the market for VC got bigger.

ISDN was used for quite a while and is still an alternative for countries with very expensive line costs and rare usage.

12.4.1   Cisco's Telepresence

In 2004, the first VC system with life-sized HD picture quality was sold globally – HP's Halo. HP licensed this technology from a small company Destiny Conferencing formerly Telesuite. In the beginning, they used dedicated rented lines to guarantee the picture quality combined with an outsourced helpdesk. For the price of half a million dollar per system, the complete room setup was delivered. Rooms and tables look identical at both ends, and the tables seemed to be connected between the rooms. Shortly afterward, other companies – Cisco, Polycom – offered the same solutions, and Figure 12.15 shows such a similar system – see, for another view, Figure 2.7 in Section 2.4. These systems got the name Telepresence. Just the size of the pictures made the meeting more vivid and increased the perceived value – see Figure 12.14. The specially rented lines delivered the consistency and interactivity.

Since the Telepresence systems come with a complete room setup – tables, lighting, screens, audio – one of the major planning tasks for former VC studios was not required any longer. This setup might look trivial, but it is not at all, and it is essential how viewers hear and see each other. In companies, this setup requires strict standards for guaranteeing a valuable experience.

Telepresence systems can be used in multipoint conferences together with an MCU and the displayed screens change with the active speaker using Audio Activity Metric (AAM).

Most videoconference systems use H.264/H.265 over RTP and SRTP. Before Telepresence was introduced, session setup happened with H.323, but migrated afterward to SIP. We treated this already in Chapter 3 for compression, in Chapter 4 for network transport, and in Chapter 7 for H.323. Specific elements of the Telepresence transport are treated in Section 12.4.2.

Due to the required interactivity, the video must arrive within the 150 milliseconds delay restriction. We discussed this thoroughly in the example in Section 5.6. Creating separate network lines – as with Halo – is not a cost-effective solution since the line capacity cannot

Figure 12.15   Cisco TelePresence System. Courtesy of Cisco Systems, Inc.

otherwise be used. Consequently, a properly configured and completely redundant network needs to be built. We treat this in more detail in Section 12.4.3.

Some additional aspects are:

Media mapping: The large Telepresence systems have three screens, but there are also versions with one or two screens. When a three-screen system is connected with one-/two-screen systems, the question is where to place each screen. One-screen systems can get three screens compressed into one screen or switch between the three screens, depending on the speaker. The problem is expressed in its own RFC [651].

Each manufacturer had implemented their own mapping and could not agree on a common standard IETF screen mapping. Consequently, a nontrivial model was created where one can map the different media parts flexibly to each other. The model is named the CLUE (ControLing mUltiple streams for tElepresense) framework and exists as an unapproved RFC [652]. Due to its flexibility, SDP is no longer sufficient for the negotiation [653] and a new negotiation protocol was designed. It is momentarily unclear if this complex system can come to life. Cisco uses private extensions to RTCP for internal capability negotiations.

Slides: Slide presentations are part of each conference. In Figure 2.7, a slide is seen in an unused screen, and in Figure 12.15, we see the slide screen below the middle screen. Note that slide screens are not located on top of the telepresence screens because all persons' heads would be held high, and they would come across slighting. This is a good example for the level of required design details.

The slides need to be transported as part of the overall video transport. Section 12.4.2 shows the slide transport in an additional stream.

When additional users dial in by phone plus a video client on his computer, slides need to be shared with those users as well. Cisco has its own gateway to its WebEx infrastructure – see Section 7.2 – to accomplish this.

For Telepresence, the scheduling has been integrated into several scheduling systems. While scheduling the conference, dialing setup is stored in the conference room systems. A single click starts even multisite room conferences afterward. The earlier required support for the meetings could be reduced substantially, and it gets as easy as phoning.

The following two sections rely mostly on Ref. [654].

12.4.2   Cisco's Telepresence Transport Specifics

The Telepresence protocols are proprietary on top of existing standards. The best source for Cisco's Telepresence protocols is the Telepresence Interoperability Protocol (TIP) specification of the IMTC (International Multimedia Telecommunication Consortium) [655]. The specifications are accessible but only usable under a license agreement that even forbids the use of summaries. Consequently, we base this section on other available sources of the content [654], but the interested reader may check the TIP documentation. The TIP specification originates from Cisco but is also implemented by other leading manufacturers. The publication of those specifications has been an element of the EC approval of the takeover of Tandberg by Cisco [656].

There is just one RTP connection for all video transmissions and RTP connection for all audio transmissions per communication direction as seen in Figure 12.16. The advantage of bundling all media connection in one RTP session is the easy setup with SIP. With four separate

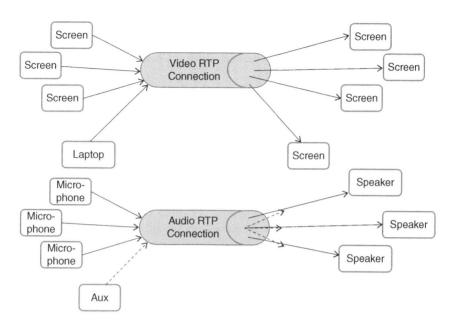

Figure 12.16 Telepresence RTP connections

RTP sessions, four different SIP sessions would need to be established – overall for both directions, it sums up to 16 connections. This setup is also easier for firewall traversal of the RTP session – Section 6.5. The optionally attached laptop screen is transmitted together with the other video information but typically transmits with only 5 frames per seconds – the other screens transmit 30 or 60 frames per second. The Binary Floor Control Protocol (BFCP) coordinates which laptop screen is shared in a meeting – see Section 6.3. In other VC systems, the analog H.239 is used.

The transmission of multiple streams within one RTP session was foreseen in the original RTP RFC [109] but was not used until the introduction of Telepresence. Since the original specification for such a case has shortcomings, an update to the original protocol exists as a non-approved draft [478].

Per screen, one microphone and one loudspeaker exist. Consequently, during a conference, one hears a person speaking from the correct screen. The fourth audio channel is used by users who dial into a conference (G.711/G.722 encoded). The audio output for dial-in users gets mixed from the other three channels. Audio cancellation is built-in to avoid that the other audio channels interfere including mobile phones that are situated close to microphones.

The encoder uses GDR to avoid peaks when sending I-frames or IDR – see Section 3.4.4. If an error occurs during the transmission, the receiver switches back to a Long-Term Reference Frame (LTRF). This frame is either the last I-frame or a GDR-frame. With an RTCP feedback [146] the receiver will tell the sender about the error and communicates the chosen LTRF frame. The encoder uses the chosen LTRF frame as a reference for its updates. All sent video frames get acknowledged every few seconds by RTCP feedback messages. This acknowledgment is required for a multisite conference since the sender can only use a proposed LTRF frame if all parties have received it correctly [657]. Complete I-frames are only sent once in several minutes.

Many more details are available in Ref. [655] but cannot be listed here because of restrictions on the use of the TIP documentation. As long as the media mapping in a teleconference is not settled, the smooth interworking between different Teleconferencing systems will be hampered.

12.4.3   Cisco's Telepresence Network Setup

Good user acceptance requires 150 milliseconds end-to-end delivery. En- and decoding plus packetization each takes roughly 15 milliseconds. The overall network path then can take 120 milliseconds. Section 5.6 details what connections can be built with these delay times.

Other specification elements are 10 milliseconds packet jitter and 50 milliseconds video frame jitter. The calculation of video frame jitter measures the first and last byte of arrival for a video frame. If the system plays 30 frames per second, a packet should arrive every 33 milliseconds. When it takes more than 33 milliseconds, it is a video jitter. Video frames are of different size – an I-frame for one screen might be 64 kB and a differential frame only 6 kB. If there is a "slow" 10 Mbit Ethernet on the path, the serialization delay for a 64 kB frame would already be at a minimum of 55 milliseconds, not including any line jitter. Consequently, there may be a big video frame jitter even if the line jitter is perfect. The video frame jitter also depends on the chosen picture quality that translates to video frame sizes. To keep the video frame delay small, big pipes help to decrease the serialization delay.

At the beginning of a Telepresence session, the jitter buffer is set to 85 milliseconds latency and can at most grow to 125 milliseconds. Both values are too high to keep the play-out timeline if the network path has 120 milliseconds as stated earlier.

Videoconferencing traffic needs QoS not only on the WAN but also on the LAN. On the LAN, we have to apply VLANs for the traffic – see Section 4.5.4. Before explaining the Cisco QoS scheme for Telepresence, we need to show additional options to prioritize traffic in Cisco devices:

Priority queuing: With PQ, there exist four queues with a strict priority as given by the following rules:

- A queue with the higher priority needs to be empty before the queue with a lower priority is served.
- Each queue has a configured maximum bandwidth. If the bandwidth capacity is reached, all additional traffic is dropped.

  These rules require that the maximum traffic for the best PQ classes need to be calculated explicitly in a traffic engineering exercise.

Custom queuing: With CQ, up to 16 queues with defined queue size exist, which are strictly served in a round-robin fashion. No queue has priority over another queue.

Weighted fair queuing: With WFQ, each flow – defined by source address, destination address, and port – is assignable to a queue that has a certain bandwidth.

The listed queuing methods can be used in various combinations.

Only on LANs, the standard DSCP settings are applied – see Section 4.5.3.

For Telepresence, Cisco proposes using DSCP CS4 (Decimal 32), but should be placed in a separate priority queue when VoIP media traffic also exists. Note that VoIP traffic is not bursty as compared to video traffic and should be served in another queue. The overall reservation for the queues shall not exceed 33% of any intermediate link to give other flows still enough capacity. For Telepresence, no shaper shall be configured because this would only delay traffic, which is counterproductive for interactive multimedia. As with all QoS settings, provider networks may treat them differently. This transition needs adaptations and discussions with the provider.

We only talked about the use of Diffserv. A few years back, Cisco recommended Intserv in the backbone for Telepresence. Cisco does not mention this recommendation any longer. Of course, service providers still use reservation over their MPLS cloud to assign the logical links for customers. In that sense, Intserv is still utilized in the backbone for Telepresence. There is also a product called Mediatrace, which allows companies and service providers to trace best their network for multimedia use. When Medianet [658] traces the network, an Intserv reservation is built and used – see also for a video [659].

The Telepresence specifications require a loss smaller than 0.05%. Due to the criticality of the user base, the network between the systems should be completely redundant. Redundant means that each network device failure or line break does not cause an interruption. Consequently, we will see a setup as shown in Figure 12.17 for a campus network. With this setup, the following elements need to be used:

Aggregation of links: Virtual Switching System (VSS) is the Cisco technology to aggregate links into one virtual segment. Therefore, the connectivity is not interrupted if one switch stops working. The corresponding IEEE standard is 802.3ad [660].

Virtual router redundancy: Hot Standby Router Protocol (HSRP) is the Cisco technology to eliminate the default router issue when one router stops working. The corresponding IETF standard is the Virtual Router Redundancy Protocol (VRRP) [661].

Gateway load balancing: This additional Cisco protocol stabilizes the default route functionality of HSRP or VRRP by distributing the load for the default route. However, each host is only served by one of the two gateways.

Event damping: This is a feature of routers that try to avoid issues when links are rapidly changing state or "flapping". By configuring this correctly, the time for routing convergence is minimized.

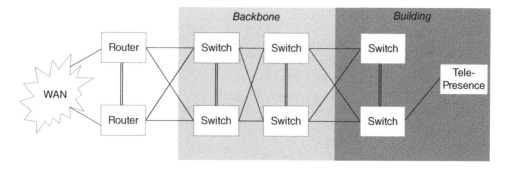

Figure 12.17   Redundant campus network also supporting Telepresence

Of course, the WAN also needs a redundant setup. Agreement with service providers on redundancy can be problematic since common dependencies are often overlooked – for example, common power supplies on the paths. If an MCU is used, a clustered MCU would also be required to cover the necessary redundancy.

12.5   Summary of Requirements versus Solutions

We first repeat the user requirement matrix from Table 2.1 (Section 2.6) as Table 12.1.

From the book, we have learned how to cope with these user requirements. This knowledge is expressed in two tables, which we show for the identical application areas:

- Table 12.2 summarizes main technical characteristics that are not end-user requirements.
- Table 12.3 summarizes required technical network implementation measures to implement the services. Note that in some cases, additional measures might be applicable but here we list solely the required ones. Also, we summarize in which type of network a given service is best deployed.

Table 12.1   Repeated requirements of typical multimedia applications – network view

	Telephony	Streaming	IP TV	Telepresence	Webcast
Availability	**Very high**	Normal	High	**Very high**	High
Stability	**Very high**	High	High	**Very high**	High
Simultaneity	**Almost**	No	Delayed	**Almost**	Delayed
Start-up delay media stream	Nationally: <1 second	>5 seconds	**<1 second**	Few seconds	<1 second
Delay within session	Not recognizable	For forward/ backward >5 seconds	Not recognizable	Not recognizable	1 second tolerable
Error	Not recognizable	Not recognizable	Not recognizable	Not recognizable	Not recognizable
Interacting devices	Yes – distrib. meetings	Yes	Yes	No	No
No. of users within same media stream	2 (seldom – 10,000)	VOD/Video-Host.: 1–5000, Live: –100,000	**100,000– 100,000,000**	2–50	20–10,000
Transported media	Audio	Audio (Stereo, mult. languages), Video – up to HD, opt. subtitle	Audio (Stereo), **HD Video**	Audio (mult. channels), Slides, **HD Video**	Audio, Video, Slides

Table 12.2 Characteristics of typical multimedia applications – network view

	IP telephony	Streaming	IP TV	Telepresence	Webcast
Treated in	Chapters 6 and 7	Chapter 9, Section 11.7*	Chapter 11	Chapter 12	Chapter 12
End-to-end delay	**<150 milliseconds**	Seconds	Seconds	**<150 milliseconds**	Seconds
Challenging for network	**Very high**	Low	**Very high**	**Very high**	Medium
Current user assessment of this technology	Works	**Great!**	Works better	**Great – but too expensive**	**Great**
Maturity of comparable "old technologies"	**Very high**	Not existent	**Very high**	Low	Not existent

* For interacting devices.

Table 12.3 Required measures taken for multimedia applications (network view) plus usability per network type

	IP telephony	Streaming	IP TV	Telepresence	Webcast
Treated in	Chapters 6 and 7	Chapter 9, Section 11.7*	Chapter 11	Chapter 12	Chapter 12
QoS on WAN	Yes	–	Yes	Yes	–
QoS on LAN	–	–	Partly	Yes	–
Redundant equipment	Yes	–	Yes – in Backbone	Yes	–
Error recovery	FEC	–	FEC++	Partly	–
Multicast	–	–	Yes	–	Rarely in company network
Caching	–	CDN	Yes	–	Yes – in company network
Operated best in which network type	Telcos & Company; Internet: weak quality	All	Telcos	Telcos & Company	All

* For interacting devices.

We see from those two Tables 12.2 and 12.3:

Challenge: When Table 12.2 states that it is a challenging application for the network, we see many more measures applied for it in Table 12.3.

Interactivity: Interactive applications – that is, end-to-end delay of less than 150 milliseconds – require many measures.

Mature technologies: When moving old, mature technologies to IP, it requires several measures. The amount of measures is caused not only by the number of features of those applications but also mainly by the user expectations on their stability.

Caching: Only noninteractive technologies can use caching.

Streaming: Streaming is a hyped application and requires considerable resources on the Internet. This technology definitively requires caching. The use of only one measure may have to change over time.

References

[1] Tanenbaum, A.S. and Wetherall, D.J. (2010) *Computer Networks*, Pearson Education Asia Pte Ltd.

[2] Fischer, W. (2010) *Digital Video and Audio Broadcasting Technology*, 3rd edn, Springer-Verlag.

[3] ITU (2010) Supplement on Global ICT fora/consortia Survey. Last retrieved March 8, 2015.

[4] Cisco (2015) https://www.youtube.com/watch?v=H30RehH-CzQ, last checked January 28, 2015.

[5] FCC (2011) Wireless E911 Location Accuracy Requirements; E911 Requirements for IP-Enabled Service Providers. http://citeseerx.ist.psu.edu/viewdoc/download?doi=10.1.1.394 .2125&rep=rep1&type=pdf, last retrieved June 2, 2015.

[6] Showler2, L. (2015) Youtube Link. https://www.youtube.com/watch?v=DQi_vXAHuTk#t=70, last retrieved June 2, 2015.

[7] Point Topic, L. (2015) IPTV statistics – market analysis. http://point-topic.com/free-analysis/ global-iptv-subscriber-numbers-q4-2014/, last retrieved June 2, 2015.

[8] ETSI (2010) Hybrid broadcast broadband TV, ETSI TS 102 796 V1.1.1. http://www.etsi.org/ standards-search, last retrieved June 2, 2015.

[9] Clark, W. (1992) Multipoint multimedia conferencing. *IEEE Communications Magazine*, **30** (5), 44–50, doi: 10.1109/35.137478.

[10] Sattlers, M. (2015) CU-SeeMe. http://sattlers.org/mickey/CU-SeeMe/index.html, last accessed on January, 28 2015.

[11] Even, R. and Ismail, N. (2006) Conferencing Scenarios, RFC 4597 (Informational).

[12] Li, Z.N., Drew, M., and Liu, J. (2014) Introduction to multimedia, in *Fundamentals of Multimedia*, Texts in Computer Science, Springer International Publishing, doi: 10.1007/978-3-319-05290-8_1.

[13] ITU (2009) Lossless compression of G.711 pulse code modulation. http://www.itu.int/rec/dologin _pub.asp?lang=e&id=T-REC-G.711.0-200909-I!!SOFT-ZST-E&type=items, last retrieved March 8, 2015.

[14] Norwich, K.H. and Wong, W. (1997) Unification of psychophysical phenomena: the complete form of Fechner's law. *Perception & Psychophysics*, **59** (6), 929–940.

[15] IMA (1992) Recommended practices for enhancing digital audio compatibility in multimedia systems, Digital Audio Focus and Technical Working Groups. http://www.cs.columbia.edu/ hgs/ audio/dvi/IMA_ADPCM.pdf, last retrieved February, 4 2015.

[16] Schulzrinne, H. and Casner, S. (2003) RTP profile for audio and video conferences with minimal control, RFC 3551 (Standard).

[17] ITU (1990) G.726: 40, 32, 24, 16 kbit/s Adaptive Differential Pulse Code Modulation (ADPCM). http://www.itu.int/rec/T-REC-G.726-199012-I/en, last retrieved March 8, 2015.

Multimedia Networks: Protocols, Design, and Applications, First Edition. Hans W. Barz and Gregory A. Bassett.
© 2016 John Wiley & Sons, Ltd. Published 2016 by John Wiley & Sons, Ltd.

[18] Spanias, A.S. (1994) Speech coding: a tutorial review. *Proceedings of the IEEE*, **82** (10), 1541–1582.

[19] Andersen, S., Duric, A., Astrom, H., Hagen, R., Kleijn, W., and Linden, J. (2004) Internet Low Bit Rate Codec (iLBC). RFC 3951 (Experimental).

[20] ITU (2006) Dual rate speech coder for multimedia communications transmitting at 5.3 and 6.3 kbit/s G.723.1. http://www.itu.int/rec/T-REC-G.711-198811-I/en, last retrieved March 8, 2015.

[21] Sun, L., Wade, G., Lines, B., and Ifeachor, E. (2001) Impact of packet loss location on perceived speech quality, in 2nd IP-Telephony Workshop, pp. 114–122.

[22] ITU (2007) G.729: Coding of speech at 8 kbit/s using conjugate-structure algebraic-code-excited linear prediction (CS-ACELP). http://www.itu.int/rec/T-REC-G.729/e, last retrieved March 8, 2015.

[23] ISO/IEC (1993) MPEG: 11172-2 information technology – coding of moving pictures and associated audio for digital storage media at up to about 1,5 Mbit. http://www.iso.org/iso/home/store/catalogue_ics.htm, last retrieved January, 18 2015.

[24] ISO/IEC (1998) 13818-3: Information technology – generic coding of moving pictures and associated audio information. http://www.iso.org/iso/home/store/catalogue_ics.htm, last retrieved June 2, 2015.

[25] ISO/IEC (2006) 13818-7: Information technology -generic coding of moving pictures and associated audio information – part 7: Advanced audio coding (AAC). http://www.iso.org/iso/home/store/catalogue_ics.htm, last retrieved June 2, 2015.

[26] ISO/EC (2009) 14496-3: Information technology – coding of audio-visual objects – Part 3: Audio. http://www.iso.org/iso/home/store/catalogue_ics.htm, last retrieved June 2, 2015.

[27] ISO/IEC (2002) 15938-4: Information technology – multimedia content description interface – Part 4: Audio. http://www.iso.org/iso/home/store/catalogue_ics.htm, last retrieved June 2, 2015.

[28] ISO/IEC (2004) 21000-1: Information technology – multimedia framework (mpeg-21) – part 1: Vision, technologies and strategy. http://www.iso.org/iso/home/store/catalogue_ics.htm, last accessed June 14, 2015.

[29] Pohlmann, K.C. (2011) *Principles of Digital Audio*, vol. **6**, McGraw-Hill, New York.

[30] Pan, D. (1995) A tutorial on mpeg/audio compression. *IEEE Multimedia*, **2** (2), 60–74, doi: 10.1109/93.388209.

[31] ITU (2007) BT.601: Studio encoding parameters of digital television for standard 4:3 and wide screen 16:9 aspect ratios. http://www.itu.int/rec/R-REC-BT.601/en, last retrieved March 8, 2015.

[32] Compuserve (1989) Graphics interchange format. http://www.w3.org/Graphics/GIF/spec-gif89a .txt, last retrieved June 14, 2015.

[33] Shannon, C. (1948, 1963) *A Mathematical Theory of Communication*, vol. **27**, University of Illinois Press.

[34] Huffman, D. (1952) A method for the construction of minimum-redundancy codes. *Proceedings of the I.R.E.*, **40** (9), 1098–1101. https://www.ic.tu-berlin.de/fileadmin/fg121/Source-Coding _WS12/selected-readings/10_04051119.pdf, last retrieved June 2, 2015.

[35] Langdon, G.G. Jr. (1984) An introduction to arithmetic coding. *IBM Journal of Research and Development*, **28** (2), 135–149.

[36] Moffat, A., Neal, R.M., and Witten, I.H. (1998) Arithmetic coding revisited. *ACM Transactions on Information Systems (TOIS)*, **16** (3), 256–294.

[37] Pratt, W., Chen, W.H., and Welch, L. (1974) Slant transform image coding. *IEEE Transactions on Communications*, **22** (8), 1075–1093.

[38] Clarke, R. (1981) Relation between the Karhunen Loève and cosine transforms. *Communications, Radar and Signal Processing, IEE Proceedings F*, **128** (6), 359–360, doi: 10.1049/ip-f-1.1981.0061.

[39] Rao, K.R. and Yip, P. (2014) *Discrete Cosine Transform: Algorithms, Advantages, Applications*, Academic Press.

[40] ITU (1992) Recommendation T.81 – digital compression and coding of continuous-tone still images – requirements and guidelines. http://www.itu.int/rec/T-REC-T.81-199209-I/en, last retrieved March 8, 2015.

[41] Miano, J. (1999) *Compressed Image File Formats: JPEG, PNG, GIF, XBM, BMP*, Addison-Wesley Professional.

[42] Marcellin, M.W., Gormish, M.J., Bilgin, A., and Boliek, M.P. (2000) An overview of jpeg-2000, in Data Compression Conference, 2000. Proceedings. DCC 2000, IEEE, pp. 523–541.

[43] ISO/IEC (2013) MPEG: 13818-1: Information technology – generic coding of moving pictures and associated audio information – Part 1: Systems. http://www.iso.org/iso/home/store/catalogue_ics.htm, last retrieved March 8, 2015.

[44] ISO/IEC (2015) MPEG: 13818-2 information technology- generic coding of moving pictures and associated audio information – Part 2: Video. http://www.iso.org/iso/home/store/catalogue_ics.htm, last retrieved March 8, 2015.

[45] ISO/IEC (2004) MPEG: 14496-2 information technology – coding of audio-visual objects – part 2: Visual. http://www.iso.org/iso/home/store/catalogue_ics.htm, last retrieved June 2, 2015.

[46] ITU (2014) H.264: Advanced video coding for generic audiovisual services. http://www.itu.int/rec/T-REC-H.264-201402-I/en, last retrieved March 8, 2015.

[47] ITU (2014) High efficiency video coding. http://www.itu.int/rec/T-REC-H.265-201410-I/en, last retrieved March 8, 2015.

[48] ITU (1993) H.261 video codec for audiovisual services at p x 64 kbits. http://www.itu.int/rec/T-REC-H.261-199303-I/en, last retrieved March 8, 2015.

[49] ITU (2009) H.221: Frame structure for a 64 to 1920 kbit/s channel in audiovisual teleservices. http://www.itu.int/rec/T-REC-H.221-200903-I/en, last retrieved March 8, 2015.

[50] ITU (2009) H.230: Frame-synchronous control and indication signals for audiovisual systems. http://www.itu.int/rec/T-REC-H.230-200903-I/en, last retrieved March 8, 2015.

[51] ITU (2009) H.242: System for establishing communication between audiovisual terminals using digital channels up to 2 Mbit/s. http://www.itu.int/rec/T-REC-H.242-200903-I/en, last retrieved March 8, 2015.

[52] ITU (2011) H.245: Control protocol for multimedia communication. http://www.itu.int/rec/T-REC-H.245-201105-I/en, last retrieved March 8, 2015.

[53] ITU (2005) H.263: Video coding for low bit rate communication. http://www.itu.int/rec/T-REC-H.263-200501-I/en, last retrieved March 8, 2015.

[54] ITU (2001) H.223: Multiplexing protocol for low bit rate multimedia communication. http://www.itu.int/rec/T-REC-H.223-200107-I/en, last retrieved March 8, 2015.

[55] ITU (2010) Implementors Guide for H.264: "Advanced video coding for generic audiovisual services". http://www.itu.int/rec/T-REC-H.Imp264-201007-S/en, last retrieved March 8, 2015.

[56] Lim, Y., Park, K., Lee, J.Y., Aoki, S., and Fernando, G. (2013) MMT: an emerging MPEG standard for multimedia delivery over the internet. *IEEE Multimedia*, **20** (1), 80–85, doi: 10.1109/MMUL.2013.7.

[57] Pang, K. and Tan, T. (1994) Optimum loop filter in hybrid coders. *IEEE Transactions on Circuits and Systems for Video Technology*, **4** (2), 158–167, doi: 10.1109/76.285622.

[58] Stephen Gould, X.H. (2014) Scene understanding by labeling pixels. *Communications of the ACM*, **57** (11), 68–77.

[59] Wiegand, T., Sullivan, G., Bjontegaard, G., and Luthra, A. (2003) Overview of the H.264/AVC video coding standard. *IEEE Transactions on Circuits and Systems for Video Technology*, **13** (7), 560–576, doi: 10.1109/TCSVT.2003.815165.

[60] Richardson, I.E. (2010) *The H.264 Advanced Video Compression Standard*, 2nd edn, John Wiley & Sons, Inc.

[61] Zhou, Z., Sun, M.T., and Hsu, Y.F. (2004) Fast variable block-size motion estimation algorithm based on merge and split procedures for H.264/MPEG-4 AVC, in Circuits and Systems, 2004.

ISCAS '04. Proceedings of the 2004 International Symposium on, vol. **3**, pp. III–725–8 doi: 10.1109/ISCAS.2004.1328849.

[62] Zhu, C., Lin, X., Chau, L.P., Lim, K.P., Ang, H.A., and Ong, C.Y. (2001) A novel hexagon-based search algorithm for fast block motion estimation, in Acoustics, Speech, and Signal Processing, 2001. Proceedings. (ICASSP '01). 2001 IEEE International Conference on, vol. **3**, pp. 1593–1596, doi: 10.1109/ICASSP.2001.941239.

[63] Yu, L., Li, J., and Shen, Y. (2006) Fast frame/field coding for H.264/AVC, in Digital Telecommunications, 2006. ICDT '06. International Conference on, pp. 18–18, doi: 10.1109/ICDT.2006.36.

[64] Marpe, D., Schwarz, H., and Wiegand, T. (2003) Context-based adaptive binary arithmetic coding in the H. 264/AVC video compression standard. *IEEE Transactions on Circuits and Systems for Video Technology*, **13** (7), 620–636.

[65] Ndjiki-Nya, P., Makai, B., Blattermann, G., Smolic, A., Schwarz, H., and Wiegand, T. (2003) Improved H. 264/AVC coding using texture analysis and synthesis, in Image Processing, 2003. ICIP 2003. Proceedings 2003 International Conference on, vols 2 and 3, IEEE, pp. III–849–52.

[66] Kim, Y.H., Yoo, Y.J., Shin, J., Choi, B., and Paik, J. (2006) Memory-efficient H.264/AVC CAVLC for fast decoding. *IEEE Transactions on Consumer Electronics*, **52** (3), 943–952, doi: 10.1109/TCE.2006.1706492.

[67] Wiegand, T., Noblet, L., and Rovati, F. (2009) Scalable video coding for IPTV services. *IEEE Transactions on Broadcasting*, **55** (2), 527–538, doi: 10.1109/TBC.2009.2020954.

[68] Schwarz, H., Marpe, D., and Wiegand, T. (2007) Overview of the Scalable Video Coding Extension of the H. 264/AVC Standard. *IEEE Transactions on Circuits and Systems for Video Technology*, **17** (9), 1103–1120.

[69] Ohm, J. and Sullivan, G. (2013) High efficiency video coding: the next frontier in video compression [standards in a nutshell]. *IEEE Signal Processing Magazine*, **30** (1), 152–158, doi: 10.1109/MSP.2012.2219672.

[70] Sullivan, G., Ohm, J., Han, W.J., and Wiegand, T. (2012) Overview of the high efficiency video coding (HEVC) standard. *IEEE Transactions on Circuits and Systems for Video Technology*, **22** (12), 1649–1668, doi: 10.1109/TCSVT.2012.2221191.

[71] Pourazad, M., Doutre, C., Azimi, M., and Nasiopoulos, P. (2012) HEVC: the new gold standard for video compression: how does HEVC compare with H.264/AVC? *IEEE Consumer Electronics Magazine*, **1** (3), 36–46, doi: 10.1109/MCE.2012.2192754.

[72] Nguyen, T., Helle, P., Winken, M., Bross, B., Marpe, D., Schwarz, H., and Wiegand, T. (2013) Transform coding techniques in HEVC. *IEEE Journal on Selected Topics in Signal Processing*, **7** (6), 978–989, doi: 10.1109/JSTSP.2013.2278071.

[73] Wien, M. (2014) *High Efficiency Video Coding: Coding Tools and Specification*, Springer-Verlag.

[74] Winkler, S. and Mohandas, P. (2008) The evolution of video quality measurement: from PSNR to hybrid metrics. *IEEE Transactions on Broadcasting*, **54** (3), 660–668.

[75] Ohm, J., Sullivan, G., Schwarz, H., Tan, T.K., and Wiegand, T. (2012) Comparison of the coding efficiency of video coding standards-including high efficiency video coding (HEVC). *IEEE Transactions on Circuits and Systems for Video Technology*, **22** (12), 1669–1684, doi: 10.1109/TCSVT.2012.2221192.

[76] Sullivan, G.J. (2008), Video coding: H264/MPEG-4 AVC and beyond. ftp://ftp.imtc-files .org/Previous Forum Presentations/2008%20Fall/Presentations/2008_11_13_IMTCFallForum -Sullivan_r0.pdf, last retrieved February, 23 2015.

[77] Kalva, H. and Lee, J. (2008) *The VC-1 and H. 264 Video Compression Standards for Broadband Video Services*, Springer-Verlag, New York.

[78] Srinivasan, S., Hsu, P., Holcomb, T., Mukerjee, K., Regunathan, S., Lin, B., Liang, J., Lee, M., and Ribas-Corbera, J. (2004) Windows Media Video 9: overview and applications. *Signal Processing: Image Communication*, **19** (9), 851–875.

[79] Srinivasan, S. and Regunathan, S. (2005) An overview of VC-1. *Proceedings of SPIE*, **5960**, 596025.

[80] Borer, T. and Davies, T. (2005) Dirac – video compression using open technology, Tech. Rep. WHP 117, EBU Technical Review. http://tech.ebu.ch/docs/techreview/trev_303-borer.pdf, last retrieved June 2, 2015.

[81] Bankoski, J., Koleszar, J., Quillio, L., Salonen, J., Wilkins, P., and Xu, Y. (2011) VP8 data format and decoding guide, RFC 6386 (Informational).

[82] Grange, A. and Alvestrand, H. (2013) A VP9 Bitstream Overview – Internet Draft. https://tools .ietf.org/html/draft-grange-vp9-bitstream-00, last retrieved February 24, 2015.

[83] Yu, L., Yi, F., Dong, J., and Zhang, C. (2005) Overview of AVS-Video: tools, performance and complexity. *Proceedings of SPIE*, **5960**, 679–690.

[84] Ma, S., Huang, T., Reader, C., and Gao, W. (2015) AVS2 ? Making video coding smarter [standards in a nutshell]. *IEEE Signal Processing Magazine*, **32** (2), 172–183, doi: 10.1109/MSP. 2014.2371951.

[85] Rao, K. and Kim, D.N. (2010) Current video coding standards: H.264/AVC, DIRAC, AVS China and VC-1, in 42nd Southeastern Symposium on System Theory (SSST), 2010, pp. 1–8, doi: 10.1109/SSST.2010.5442824.

[86] Grois, D., Marpe, D., Mulayoff, A., Itzhaky, B., and Hadar, O. (2013) Performance comparison of H. 265/MPEG-HEVC, VP9, and H. 264/MPEG-AVC encoders, in PCS, pp. 394–397.

[87] Kim, I.K., Lee, S., Piao, Y., and Chen, J. (2014) Coding efficiency comparison of new video coding standards: HEVC VS VP9 VS AVS2 video, in Multimedia and Expo Workshops (ICMEW), 2014 IEEE International Conference on, pp. 1–6, doi: 10.1109/ICMEW.2014.6890700.

[88] MPEGLA (2015) HEVC patent portfolio license briefing. http://www.mpegla.com/main/ programs/HEVC/Documents/HEVCweb.pdf, last retrieved June 2, 2015.

[89] Smolic, A., Mueller, K., Merkle, P., Fehn, C., Kauff, P., Eisert, P., and Wiegand, T. (2006) 3D video and free viewpoint video – technologies, applications and MPEG standards, in 2006 IEEE International Conference on Multimedia and Expo, pp. 2161–2164, doi: 10.1109/ICME.2006.262683.

[90] Puri, A., Kollarits, R., and Haskell, B. (1997) Basics of stereoscopic video, new compression results with MPEG-2 and a proposal for MPEG-4. *Signal Processing: Image Communication*, **10** (1–3), 201–234.

[91] ETSI (2012) DVB: GEM profile for plano-stereoscopic 3DTV – ETSI TS 101 600 V1.1.1. http:// www.etsi.org/standards-search, last retrieved June 2, 2015.

[92] ETSI (2011) DVB: Plano-stereoscopic 3DTV (DVB-3DTV) – ETSI TS 101 547-2. http://www .etsi.org/standards-search, last retrieved February, 25 2015.

[93] Postley, H. (2012) Sports: 3-D TV's toughest challenge. *IEEE Spectrum*, **49** (11), 40–66.

[94] Aylsworth, W.L., Postley, H.J., Sandrew, B.B., Dobrin, B.E., and Husak, W. (2013) Stereographic digital cinema: production and exhibition techniques in 2012. *Proceedings of the IEEE*, **101** (1), 169–189.

[95] Ozaktas, H. and Onural, L. (eds) (2008) in *Three-Dimensional Television: Capture, Transmission, Display*, Signals and Communication Technology, Springer-Verlag.

[96] Levine, D., Lynch, W., and Le-Ngoc, T. (2007) Observations on error detection in H.264, in MWS-CAS 2007. 50th Midwest Symposium on Circuits and Systems, pp. 815–818, doi: 10.1109/MWS-CAS.2007.4488698.

[97] Takishima, Y., Wada, M., and Murakami, H. (1995) Reversible variable length codes. *IEEE Transactions on Communications*, **43** (234), 158–162, doi: 10.1109/26.380026.

[98] Carle, G. and Biersack, E. (1997) Survey of error recovery techniques for IP-based audio-visual multicast applications. *IEEE Network*, **11** (6), 24–36.

[99] Zhou, X. and Kuo, C. (2008) Robust streaming of offline coded H. 264/AVC video via alternative macroblock coding. *IEEE Transactions on Circuits and Systems for Video Technology*, **18** (4), 425–438.

[100] Purandare, R., Kshirsagar, S., Koli, S., and Gohokar, V. (2011) Impact of bit error on video transmission over wireless networks and error resiliency, in Image Information Processing (ICIIP), 2011 International Conference on, pp. 1–6, doi: 10.1109/ICIIP.2011.6108871.

[101] Burmeister, C., Hakenberg, R., Miyazaki, A., Ott, J., Sato, N., and Fukunaga, S. (2006) Extended RTP Profile for Real-time Transport Control Protocol (RTCP)-Based Feedback: Results of the Timing Rule Simulations, RFC 4586 (Informational).

[102] Stockhammer, T. and Zia, W. (2007) Error-resilient coding and decoding strategies for video communication, *Multimedia Over IP and Wireless Networks: Compression, Networking, and Systems*, Academic Press, pp. 13–58.

[103] Perkins, C., Hodson, O., and Hardman, V. (1998) A survey of packet loss recovery techniques for streaming audio. *IEEE Network*, **12** (5), 40–48.

[104] Ahmad, I., Wei, X., Sun, Y., and Zhang, Y.Q. (2005) Video transcoding: an overview of various techniques and research issues. *IEEE Transactions on Multimedia*, **7** (5), 793–804, doi: 10.1109/TMM.2005.854472.

[105] Vetro, A., Christopoulos, C., and Sun, H. (2003) Video transcoding architectures and techniques: an overview. *IEEE Signal Processing Magazine*, **20** (2), 18–29, doi: 10.1109/MSP.2003.1184336.

[106] Xin, J., Lin, C., and Sun, M. (2005) Digital video transcoding. *Proceedings of the IEEE*, **93** (1), 84–97.

[107] Elarabi, T., Abdelgawad, A., and Bayoumi, M. (2014) *Real-Time Heterogenous Video Transcoding for Low-Power Applications*, Springer International Publishing, doi: 10.1007/978-3-319-06071-2.

[108] Schulzrinne, H., Casner, S., Frederick, R., and Jacobson, V. (1996) RTP: A Transport Protocol for Real-Time Applications, RFC 1889 (Proposed Standard).

[109] Schulzrinne, H., Casner, S., Frederick, R., and Jacobson, V. (2003) RTP: A transport protocol for real-time applications, RFC 3550 (Standard).

[110] Perkins, C. (2003) *RTP: Audio and Video for the Internet*, Addison-Wesley.

[111] Wang, Y.K., Even, R., Kristensen, T., and Jesup, R. (2011) RTP Payload Format for H.264 Video, RFC 6184 (Proposed Standard).

[112] Wenger, S. (2003) H. 264/AVC over IP. *IEEE Transactions on Circuits and Systems for Video Technology*, **13** (7), 645–656.

[113] Petit-Huguenin, M. and Zorn, G. (2014) Support for Multiple Clock Rates in an RTP Session, RFC 7160 (Proposed Standard).

[114] Wenger, S., Wang, Y., and Schierl, T. (2007) Transport and signaling of SVC in IP networks. *IEEE Transactions on Circuits and Systems for Video Technology*, **17** (9), 1164–1173.

[115] Wenger, S., Wang, Y.K., Schierl, T., and Eleftheriadis, A. (2011) RTP Payload Format for Scalable Video Coding, RFC 6190 (Proposed Standard).

[116] Hoffman, D., Fernando, G., Goyal, V., and Civanlar, M. (1998) RTP payload format for MPEG1/MPEG2 video, RFC 2250 (Proposed Standard).

[117] Schierl, T., Hellge, C., Mirta, S., Grueneberg, K., and Wiegand, T. (2006) Using H.264/AVC-based scalable video coding (SVC) for real time streaming in wireless IP networks,

[118] Kikuchi, Y., Nomura, T., Fukunaga, S., Matsui, Y., and Kimata, H. (2000) RTP payload format for MPEG-4 audio/visual streams.

[119] Goor, S., Murphy, S., and Murphy, J. (2004) Experimental performance analysis of RTP-based transmission techniques for MPEG-4, in Proceedings of IEEE Packet Video (PV 04), Irvine, CA, USA.

[120] Wang, Y.-K., Sanchez, Y., Schierl, T., Wenger, S., and Hannuksela, M.M. (2014) RTP payload format for high efficiency video coding – draft-ietf-payload-rtp-h265-10.txt. https://datatracker .ietf.org/doc/draft-ietf-payload-rtp-h265/, last retrieved June 2, 2015.

[121] Baugher, M., McGrew, D., Naslund, M., Carrara, E., and Norrman, K. (2004) The Secure Real-time Transport Protocol (SRTP), RFC 3711 (Proposed Standard).

[122] Kristensen, T. and Luthi, P. (2011) RTP Payload Format for H.264 Reduced-Complexity Decoding Operation (RCDO) Video, RFC 6185 (Standard).

[123] Johansson, I. and Westerlund, M. (2009) Support for Reduced-Size Real-Time Transport Control Protocol (RTCP): Opportunities and Consequences, RFC 5506 (Proposed Standard).

[124] Wing, D. (2007) Symmetric RTP / RTP Control Protocol (RTCP), RFC 4961 (Best Current Practice).

[125] Ott, J. and Perkins, C. (2010) Guidelines for Extending the RTP Control Protocol (RTCP), RFC 5968 (Informational).

[126] Bormann, C. (ed.), Burmeister, C. et al. (2001) RObust Header Compression (ROHC): Framework and four profiles: RTP, UDP, ESP, and uncompressed, RFC 3095 (Proposed Standard).

[127] Jonsson, L.E., Sandlund, K., Pelletier, G., and Kremer, P. (2007) RObust Header Compression (ROHC): Corrections and Clarifications to RFC 3095, RFC 4815 (Proposed Standard).

[128] Casner, S. and Jacobson, V. (1999) Compressing IP/UDP/RTP Headers for Low-Speed Serial Links, RFC 2508 (Proposed Standard).

[129] Koren, T., Casner, S., and Bormann, C. (2003) IP Header Compression over PPP, RFC 3544 (Proposed Standard).

[130] Jacobson, V. (1990) Compressing TCP/IP Headers for Low-Speed Serial Links, RFC 1144 (Proposed Standard).

[131] Koren, T., Casner, S., Geevarghese, J., Thompson, B., and Ruddy, P. (2003) Enhanced Compressed RTP (CRTP) for Links with High Delay, Packet Loss and Reordering, RFC 3545 (Proposed Standard).

[132] Dimitriadis, G., Karapantazis, S., and Pavlidou, F.N. (2008) Performance evaluation of CRTP and enhanced CRTP within the DVB/rcs context, in *IP Networking over Next-Generation Satellite Systems* (eds L. Fan, H. Cruickshank, and Z. Sun), Springer-Verlag, New York, pp. 317–327.

[133] Handley, M. and Jacobson, V. (1998) SDP: Session description protocol.

[134] Handley, M., Jacobson, V., and Perkins, C. (2006) SDP: Session Description Protocol, RFC 4566 (Proposed Standard).

[135] Ott, J., Bormann, C., Sullivan, G., Wenger, S., and Even, R. (2007) RTP Payload Format for ITU-T Rec. H.263 Video, RFC 4629 (Proposed Standard).

[136] Rosenberg, J. and Schulzrinne, H. (2002) An offer/answer model with session description protocol (SDP), RFC 3264 (Proposed Standard).

[137] Johnston, A. and Sparks, R. (2005) Session description protocol (SDP) offer/answer examples, RFC 4317 (Informational).

[138] Andreasen, F. (2010) Session Description Protocol (SDP) Capability Negotiation, RFC 5939 (Proposed Standard).

[139] Gilman, R., Even, R., and Andreasen, F. (2013) Session Description Protocol (SDP) Media Capabilities Negotiation, RFC 6871 (Proposed Standard).

[140] Schierl, T. and Wenger, S. (2009) Signaling Media Decoding Dependency in the Session Description Protocol (SDP), RFC 5583 (Proposed Standard).

[141] Camarillo, G., Eriksson, G., Holler, J., and Schulzrinne, H. (2002) Grouping of media lines in the session description protocol (SDP),

[142] Casner, S. (2003) Session Description Protocol (SDP) Bandwidth Modifiers for RTP Control Protocol (RTCP) Bandwidth, RFC 3556 (Proposed Standard).

[143] Huitema, C. (2003) Real Time Control Protocol (RTCP) attribute in Session Description Protocol (SDP), RFC 3605 (Proposed Standard).

[144] Garcia-Martin, M., Bormann, C., Ott, J., Price, R., and Roach, A.B. (2003) The Session Initiation Protocol (SIP) and Session Description Protocol (SDP) Static Dictionary for Signaling Compression (SigComp), RFC 3485 (Proposed Standard).

[145] Price, R., Bormann, C., Christoffersson, J., Hannu, H., Liu, Z., and Rosenberg, J. (2003) Signaling compression (SigComp),

[146] Ott, J., Wenger, S., Sato, N., Burmeister, C., and Rey, J. (2006) Extended RTP Profile for Real-time Transport Control Protocol (RTCP)-Based Feedback (RTP/AVPF), RFC 4585 (Proposed Standard). Updated by RFC 5506.

[147] Uberti, J., Jennings, C., and Rescorla, E. (2015) Javascript session establishment protocol – draft-ietf-rtcweb-jsep-11. https://tools.ietf.org/html/draft-ietf-rtcweb-jsep-11, last retrieved July 15, 2015.

[148] Demetriades, G. (2003) *Streaming Media: Building and Implementing a Complete Streaming System*, Wiley Publishing.

[149] Van Deursen, D., Van Lancker, W., and Van de Walle, R. (2010) On media delivery protocols in the web, in Multimedia and Expo (ICME), 2010 IEEE International Conference on, pp. 1028–1033, doi: 10.1109/ICME.2010.5582620.

[150] Schulzrinne, H., Rao, A., and Lanphier, R. (1998) Real time streaming protocol (RTSP), RFC 2326 (Proposed Standard).

[151] Fielding, R., Gettys, J., Mogul, J., Frystyk, H., Masinter, L., Leach, P., and Berners-Lee, T. (1999) Hypertext Transfer Protocol – HTTP/1.1, RFC 2616 (Draft Standard).

[152] ISO/IEC (2004) Data elements and interchange formats – information exchange – representation of dates and times – 23001-1. http://www.iso.org/iso/home/store/catalogue_ics.htm, last retrieved March 8, 2015.

[153] Faith, R. and Martin, B. (1997) A Dictionary Server Protocol, RFC 2229 (Informational).

[154] Conklin, G., Greenbaum, G., Lillevold, K., Lippman, A., and Reznik, Y. (2001) Video coding for streaming media delivery on the Internet. *IEEE Transactions on Circuits and Systems for Video Technology*, **11** (3), 269–281.

[155] Schulzrinne, H., Rao, A., Lanphier, R., and Westerlund, M. (2014) Real time streaming protocol 2.0 (RTSP) (draft). https://tools.ietf.org/html/draft-ietf-mmusic-rfc2326bis-40, last retrieved March 15, 2015.

[156] W3C (2012) Media fragments URI 1.0 (basic) recommendation. http://www.w3.org/TR/media-frags/, last retrieved June 2, 2015.

[157] Van Deursen, D., Troncy, R., Mannens, E., Pfeiffer, S., Lafon, Y., and Van de Walle, R. (2010) Implementing the media fragments URI specification, in Proceedings of the 19th international conference on World wide web, ACM, pp. 1361–1364.

[158] Deering, S. (1989) Host extensions for IP multicasting, RFC 1112 (INTERNET STANDARD). Updated by RFC 2236.

[159] Holbrook, H., Cain, B., and Haberman, B. (2006) Using Internet Group Management Protocol Version 3 (IGMPv3) and Multicast Listener Discovery Protocol Version 2 (MLDv2) for Source-Specific Multicast, RFC 4604 (Proposed Standard).

[160] Cotton, M., Vegoda, L., and Meyer, D. (2010) IANA Guidelines for IPv4 Multicast Address Assignments, RFC 5771 (Best Current Practice).

[161] Bhattacharyya, S. (2003) An Overview of Source-Specific Multicast (SSM), RFC 3569 (Informational).

[162] Fenner, W. (1997) Internet Group Management Protocol, Version 2, RFC 2236 (Proposed Standard). Updated by RFC 3376.

[163] Cain, B., Deering, S., Kouvelas, I., Fenner, B., and Thyagarajan, A. (2002) Internet Group Management Protocol, Version 3, RFC 3376 (Proposed Standard).

[164] Liu, H., Cao, W., and Asaeda, H. (2010) Lightweight Internet Group Management Protocol Version 3 (IGMPv3) and Multicast Listener Discovery Version 2 (MLDv2) Protocols, RFC 5790 (Proposed Standard).

[165] Deering, S., Fenner, W., and Haberman, B. (1999) Multicast Listener Discovery (MLD) for IPv6, RFC 2710 (Proposed Standard). Updated by RFCs 3590, 3810.

[166] Vida, R. and Costa, L. (2004) Multicast Listener Discovery Version 2 (MLDv2) for IPv6, RFC 3810 (Proposed Standard).

[167] Waitzman, D., Partridge, C., and Deering, S. (1988) Distance Vector Multicast Routing Protocol, RFC 1075 (Experimental).

[168] Moy, J. (1994) Multicast Extensions to OSPF, RFC 1584 (Historic).

[169] Adams, A., Nicholas, J., and Siadak, W. (2005) Protocol Independent Multicast – Dense Mode (PIM-DM): Protocol Specification (Revised), RFC 3973 (Experimental).

[170] Deering, S., Estrin, D., Farinacci, D., Jacobson, V., Liu, C.G., and Wei, L. (1996) The PIM architecture for wide-area multicast routing. *IEEE/ACM Transactions on Networking*, **4** (2), 153–162, doi: 10.1109/90.490743.

[171] Fenner, B., Handley, M., Holbrook, H., and Kouvelas, I. (2006) Protocol Independent Multicast – Sparse Mode (PIM-SM): Protocol Specification (Revised), RFC 4601 (Proposed Standard).

[172] Ballardie, A. (1997) Core Based Trees (CBT) Multicast Routing Architecture, RFC 2201 (Historic).

[173] Thaler, D. (2004) Border Gateway Multicast Protocol (BGMP): Protocol Specification, RFC 3913 (Historic).

[174] Bhaskar, N., Gall, A., Lingard, J., and Venaas, S. (2008) Bootstrap Router (BSR) Mechanism for Protocol Independent Multicast (PIM), RFC 5059 (Proposed Standard).

[175] ETSI (2011) Digital Video Broadcasting (DVB); Guidelines for the implementation of DVB-IPTV Phase specifications – TS 102 542-1. http://www.etsi.org/standards-search, last retrieved September 22, 2015.

[176] Speakman, T., Crowcroft, J., Gemmell, J., Farinacci, D., Lin, S., Leshchiner, D., Luby, M., Montgomery, T., Rizzo, L., Tweedly, A., Bhaskar, N., Edmonstone, R., Sumanasekera, R., and Vicisano, L. (2001) PGM Reliable Transport Protocol Specification, RFC 3208 (Experimental).

[177] Boivie, R., Feldman, N., Imai, Y., Livens, W., and Ooms, D. (2007) Explicit Multicast (Xcast) Concepts and Options, RFC 5058 (Experimental).

[178] Hosseini, M., Ahmed, D., Shirmohammadi, S., and Georganas, N. (2007) A survey of application-layer multicast protocols. *IEEE Communications Surveys & Tutorials*, **9** (3), 58–74.

[179] Banerjee, S., Bhattacharjee, B., and Kommareddy, C. (2002) Scalable application layer multicast, in Proceedings of the 2002 conference on Applications, technologies, architectures, and protocols for computer communications, ACM, pp. 205–217.

[180] Tran, D., Hua, K., and Do, T. (2003) ZIGZAG: an efficient peer-to-peer scheme for media streaming, in INFOCOM 2003. Twenty-Second Annual Joint Conference of the IEEE Computer and Communications. IEEE Societies, vol. **2**, IEEE, pp. 1283–1292.

[181] Banerjee, S., Kommareddy, C., Kar, K., Bhattacharjee, B., and Khuller, S. (2003) Construction of an efficient overlay multicast infrastructure for real-time applications, in INFOCOM 2003. Twenty-Second Annual Joint Conference of the IEEE Computer and Communications. IEEE Societies, vol. **2**, IEEE, pp. 1521–1531.

[182] Oikarinen, J. and Reed, D. (1993) Internet Relay Chat Protocol, RFC 1459 (Experimental).

[183] Kalt, C. (2000) Internet Relay Chat: Architecture, RFC 2810 (Informational).

[184] Camarillo, G. and IAB (2009) Peer-to-Peer (P2P) Architecture: Definition, Taxonomies, Examples, and Applicability, RFC 5694 (Informational).

[185] BitTorrent (2008) The bittorrent protocol specification. http://www.bittorrent.org/beps/bep_0003 .html, last retrieved June 2, 2015.

[186] Ghodsi, A. (2006) Distributed K-Ary System: Algorithms for Distributed Hash Tables, Ph.D. thesis, The Royal Institute of Technology (KTH), School of Information and Communication Technology, Department of Electronic, Computer, and Software Systems. http://www.sics.se/~ali/ thesis/dks.pdf, last retrieved June 2, 2015.

[187] Lau, G., Jasseemuddin, M., and Ravindran, G. (2005) Proactive neighbor replacement for RAON, in Wireless And Mobile Computing, Networking And Communications, 2005. (WiMob'2005), IEEE International Conference on, vol. **3**, pp. 260–267, doi: 10.1109/WIMOB.2005.1512912.

[188] Hei, X., Liang, C., Liang, J., Liu, Y., and Ross, K. (2007) A measurement study of a large-scale P2P IPTV system. *IEEE Transactions on Multimedia*, **9** (8), 1672–1687.

[189] Braden, R., Zhang, L., Berson, S., Herzog, S., and Jamin, S. (1997) Resource ReSerVation Protocol (RSVP) – Version 1 Functional Specification, RFC 2205 (Proposed Standard).

[190] Shenker, S., Partridge, C., and Guerin, R. (1997) Specification of Guaranteed Quality of Service, RFC 2212 (Proposed Standard).

[191] Wroclawski, J. (1997) Specification of the Controlled-Load Network Element Service, RFC 2211 (Proposed Standard).

[192] Zhang, L., Deering, S., Estrin, D., Shenker, S., and Zappala, D. (1993) RSVP: a new resource reservation protocol. *IEEE Network*, **7** (5), 8–18, doi: 10.1109/65.238150.

[193] Mankin, A., Baker, F., Braden, B., Bradner, S., O'Dell, M., Romanow, A., Weinrib, A., and Zhang, L. (1997) Resource ReSerVation Protocol (RSVP) – Version 1 Applicability Statement Some Guidelines on Deployment, RFC 2208 (Informational).

[194] Braden, R. and Zhang, L. (1997) Resource ReSerVation Protocol (RSVP) – Version 1 Message Processing Rules, RFC 2209 (Informational).

[195] Wroclawski, J. (1997) The Use of RSVP with IETF Integrated Services, RFC 2210 (Proposed Standard).

[196] Herzog, S. (2000) RSVP Extensions for Policy Control, RFC 2750 (Proposed Standard).

[197] Awduche, D., Berger, L., Gan, D., Li, T., Srinivasan, V., and Swallow, G. (2001) RSVP-TE: Extensions to RSVP for LSP Tunnels, RFC 3209 (Proposed Standard).

[198] Shenker, S. and Wroclawski, J. (1997) General Characterization Parameters for Integrated Service Network Elements, RFC 2215 (Proposed Standard).

[199] Nichols, K., Blake, S., Baker, F., and Black, D. (1998) Definition of the Differentiated Services Field (DS Field) in the IPv4 and IPv6 Headers, RFC 2474 (Proposed Standard).

[200] Grossman, D. (2002) New Terminology and Clarifications for Diffserv, RFC 3260 (Informational).

[201] Babiarz, J., Chan, K., and Baker, F. (2006) Configuration Guidelines for DiffServ Service Classes, RFC 4594 (Informational). Updated by RFC 5865.

[202] Davie, B., Charny, A., Bennet, J., Benson, K., Boudec, J.L., Courtney, W., Davari, S., Firoiu, V., and Stiliadis, D. (2002) An Expedited Forwarding PHB (Per-Hop Behavior), RFC 3246 (Proposed Standard).

[203] Baker, F., Polk, J., and Dolly, M. (2010) A Differentiated Services Code Point (DSCP) for Capacity-Admitted Traffic, RFC 5865 (Proposed Standard).

[204] Heinanen, J., Baker, F., Weiss, W., and Wroclawski, J. (1999) Assured Forwarding PHB Group, RFC 2597 (Proposed Standard).

[205] Ramakrishnan, K., Floyd, S., and Black, D. (2001) The Addition of Explicit Congestion Notification (ECN) to IP, RFC 3168 (Proposed Standard).

[206] Floyd, S. (2000) Congestion Control Principles, RFC 2914 (Best Current Practice). Updated by RFC 7141.

[207] Braden, B., Clark, D., Crowcroft, J., Davie, B., Deering, S., Estrin, D., Floyd, S., Jacobson, V., and Minshal, G. (1998) Recommendations on Queue Management and Congestion Avoidance in the Internet, RFC 2309 (Informational).

[208] Bless, R., Nichols, K., and Wehrle, K. (2003) A Lower Effort Per-Domain Behavior (PDB) for Differentiated Services, RFC 3662 (Informational).

[209] Eardley, P. (2009) Pre-Congestion Notification (PCN) Architecture, RFC 5559 (Informational).

[210] Menth, M., Briscoe, B., and Tsou, T. (2012) Precongestion notification: new QoS support for differentiated services IP networks. *IEEE Communications Magazine*, **50** (3), 94–103, doi: 10.1109/MCOM.2012.6163587.

[211] IEEE (2005) Virtual bridged local area networks. http://standards.ieee.org/getieee802/download/802.1Q-2005.pdf, last retrieved March 8, 2015.

[212] Lasserre, M. and Kompella, V. (2007) Virtual Private LAN Service (VPLS) Using Label Distribution Protocol (LDP) Signaling, RFC 4762 (Proposed Standard).

[213] IEEE (2006) Multiple registration protocol 802.1ak. http://www.ieee802.org/11/, last accessed February 5, 2015.

[214] Yu, M., Rexford, J., Sun, X., Rao, S., and Feamster, N. (2011) A survey of virtual lan usage in campus networks. *IEEE Communications Magazine*, **49** (7), 98–103, doi: 10.1109/MCOM.2011.5936161.

[215] Xiao, X. (2008) *Technical, Commercial and Regulatory Challenges of QoS: An Internet Service Model Perspective*, The Morgan Kaufmann Series in Networking, Elsevier.

[216] Marsden, C.T. (2010) *Net Neutrality: Towards a Co-Regulatory Solution*, Bloomsbury, USA.

[217] But, J., Armitage, G., and Stewart, L. (2008) Outsourcing automated QoS control of home routers for a better online game experience. *IEEE Communications Magazine*, **46** (12), 64–70.

[218] Mills, D., Martin, J., Burbank, J., and Kasch, W. (2010) Network Time Protocol Version 4: Protocol and Algorithms Specification, RFC 5905 (Proposed Standard).

[219] Mills, D. (1991) Internet time synchronization: the network time protocol. *IEEE Transactions on Communications*, **39**, 1482–1493, doi: 10.1109/26.103043.

[220] Garner, G. and Ryu, H. (2011) Synchronization of audio/video bridging networks using IEEE 802.1AS. *IEEE Communications Magazine*, **49** (2), 140–147, doi: 10.1109/MCOM.2011.5706322.

[221] Ratzel, R. and Greenstreet, R. (2012) Toward higher precision. *Communications of the ACM*, **55** (10), 38–47, doi: 10.1145/2347736.2347750.

[222] Pathan, M., Sitaraman, R.K., and Robinson, D. (eds) (2014) *Advanced Content Delivery, Streaming, and Cloud Services*, John Wiley & Sons, Inc., ISBN: 978-1-118-90964-5.

[223] Chattopadhyay, S., Ramaswamy, L., and Bhandarkar, S. (2007) A framework for encoding and caching of video for quality adaptive progressive download, in Proceedings of the 15th international conference on Multimedia, ACM, pp. 778.

[224] Liu, J. (2005) Streaming Media Caching, in *Web Content Delivery*, vol. **II**, Springer-Verlag, pp. 197–214.

[225] Wang, J. (1999) A survey of web caching schemes for the internet. *ACM Computer Communication Review*, **29** (5), 36–46.

[226] Barish, G. and Obraczke, K. (2002) World wide web caching: trends and techniques. *IEEE Communications Magazine*, **38** (5), 178–184.

[227] McLaggan, D. (2012), Web cache communication protocol V2, revision 1. https://tools.ietf.org/id/draft-mclaggan-wccp-v2rev1-00.txt, last accessed March 14, 2015.

[228] Wessels, D. and Claffy, K. (1997) Internet Cache Protocol (ICP), version 2, RFC 2186 (Informational).

[229] Valloppillil, V. and Ross, K.W. (1998) Cache array routing protocol V1.0 – <draft-vinod-carp-v1-03.txt>. http://tools.ietf.org/html/draft-vinod-carp-v1-03, last accessed February 5, 2015.

[230] Cieslak, M., Forster, D., Tiwana, G., and Wilson, R. (2001) Web cache communication protocol V2.0 – <draft-wilson-wrec-wccp-v2-01.txt>. http://tools.ietf.org/id/draft-wilson-wrec-wccp-v2-01.txt, last retrieved June 2, 2015.

[231] Morris, S. and Clarkson, R. (2007) US Patent 7,305,479: Methods and apparatus for delivery of content requests within a content delivery network. http://www.google.de/patents/US7305479?hl=de&dq=WCCP+patent, last retrieved March 8, 2015.

[232] Seo, K.D., Jung, T.J., Yoo, J., Kim, C.K., and Hong, J. (2012) A new timing model design for MPEG media transport (MMT), in Broadband Multimedia Systems and Broadcasting (BMSB), 2012 IEEE International Symposium on, IEEE, pp. 1–5.

[233] ITU-T (2000) B-ISDN ATM layer cell transfer performance – I.356. https://www.itu.int/rec/dologin_pub.asp?lang=e&id=T-REC-I.356-200003-I!!PDF-E&type=items, last retrieved May 6, 2015.

[234] Morton, A. and Claise, B. (2009) Packet Delay Variation Applicability Statement, RFC 5481 (Informational).

[235] Yajnik, M., Kurose, J., and Towsley, D. (1996) Packet loss correlation in the MBone multicast network, in Global Telecommunications Conference, 1996. GLOBECOM'96.'Communications: The Key to Global Prosperity, IEEE, pp. 94–99.

[236] Ribas-Corbera, J., Chou, P., and Regunathan, S. (2003) A generalized hypothetical reference decoder for H. 264/AVC. *IEEE Transactions on Circuits and Systems for Video Technology*, **13** (7), 674–687.

[237] Wagner, M., Kopilovic, I., Dohla, S., Farber, N., and Fuchs, H. (2009) Towards an RTP profile for IPTV. *IEEE Transactions on Broadcasting*, **55** (2), 343–352, doi: 10.1109/TBC.2009.2019422.

[238] ETSI (2009) IP datacast over DVB-H: Content delivery protocols – TS 102 472 V1.3.1. http://www.etsi.org/standards-search, last retrieved June 2, 2015.

[239] Blakowski, G. and Steinmetz, R. (1996) A media synchronization survey: reference model, specification, and case studies. *IEEE Journal on Selected Areas in Communications*, **14** (1), 5–35, doi: 10.1109/49.481691.

[240] Laoutaris, N. and Stavrakakis, I. (2002) Intrastream synchronization for continuous media streams: a survey of playout schedulers. *IEEE Network*, **16** (3), 30–40.

[241] van Brandenburg, R., Stokking, H., van Deventer, O., Boronat, F., Montagud, M., and Gross, K. (2014) Inter-Destination Media Synchronization (IDMS) Using the RTP Control Protocol (RTCP), RFC 7272 (Proposed Standard).

[242] Boronat, F., Lloret, J., and García, M. (2009) Multimedia group and inter-stream synchronization techniques: a comparative study. *Information Systems*, **34** (1), 108–131.

[243] Steinbach, E., Liang, Y., Kalman, M., and Girod, B. (2007) Adaptive media playout, in *Multimedia Over IP amd Wireless Networks*, Chapter 16 (eds P. Chou and M. van derSchaar), Elsevier, pp. 527–556.

[244] Allman, M., Paxson, V., and Blanton, E. (2009) TCP Congestion Control, RFC 5681 (Draft Standard).

[245] Perkins, C. and Singh, V. (2013) Multimedia congestion control: circuit breakers for unicast RTP sessions draft-ietf-avtcore-rtp-circuit-breakers-02. https://tools.ietf.org/html/draft-ietf-avtcore-rtp-circuit-breakers-02, last retrieved March 20, 2015.

[246] Zanaty, M., V. Singh, Nandakumar, S., and Sarker, Z. (2014) RTP application interaction with congestion control – draft-zanaty-rmcat-app-interaction-01. http://www.ietf.org/archive/id/draft-zanaty-rmcat-app-interaction-01.txt, last retrieved March 21, 2015.

[247] Zhu, X., Ramalho, M.A., de la Cruz, S.M., and Ganzhorn, C. (2014), NADA: A unified congestion control scheme for real-time-media draft-zhu-rmcat-nada-04. https://tools.ietf.org/id/draft-zhu-rmcat-nada-04.txt, last retrieved on March 20, 2015.

[248] Welzl, M. and Eddy, W. (2010) Congestion Control in the RFC Series, RFC 5783 (Informational).

[249] Szigeti, T. and Hattingh, C. (2004) *End-to-End QoS Network Design*, Cisco Press.

[250] Claise, B. (2004) Cisco Systems NetFlow Services Export Version 9, RFC 3954 (Informational).

[251] Welch, J. and Clark, J. (2006) A Proposed Media Delivery Index (MDI), RFC 4445 (Informational).

[252] Makofske, D.B. and Almeroth, K.C. (2000) Real-time multicast tree visualization and monitoring. *Software-Practice and Experience*, **30** (9), 1047–1065.

[253] BISmark, P. (2015) http://networkdashboard.org/, last accessed on March 15, 2015 – navigate to Router in Munich.

[254] Jacobson, V. (2006) A rant on queues. http://www.pollere.net/Pdfdocs/QrantJul06.pdf, last retrieved March 20, 2015.

[255] Gettys, J. and Nicols, K. (2011) Bufferbloat: dark buffers in the internet. *IEEE Internet Computing*, **15**, 57–65.

[256] Floyd, S. and Jacobson, V. (1993) Random early detection gateways for congestion avoidance. *IEEE/ACM Transactions on Networking*, **1** (4), 397–413.

[257] Nichols, K. and Jacobson, V. (2013) Controlled delay active queue management. https://tools.ietf.org/html/draft-nichols-tsvwg-codel-02, last retrieved March 20, 2015.

[258] Pan, R., Natarajan, P., Piglione, C., Prabhu, M., Subramanian, V., Baker, F., and Steeg, B.V. (2013), PIE: a lightweight control scheme to address the bufferbloat problem – draft-pan-aqm-pie-00. https://tools.ietf.org/html/draft-pan-aqm-pie-00, last retrieved March 20, 1015.

[259] Schwardmann, J., Wagner, D., and Kühlewind, M. (2014) Evaluation of ARED, CoDeL and PIE, in *Advances in Communication Networking*, Springer-Verlag, pp. 185–191.

[260] Huang, T.Y., Handigol, N., Heller, B., McKeown, N., and Johari, R. (2012) Confused, timid, and unstable: picking a video streaming rate is hard, in Proceedings of the 2012 ACM conference on Internet measurement conference, ACM, pp. 225–238.

[261] Mansy, A., Ver Steeg, B., and Ammar, M. (2013) SABRE: a client based technique for mitigating the buffer bloat effect of adaptive video flows, in Proceedings of the 4th ACM Multimedia Systems Conference, ACM, pp. 214–225.

[262] Brown, P. (2010) *Silverlight 4 in Action*, Manning Publications Co.

[263] W3C (2015) Media source extensions – W3C candidate recommendation 31 march 2015. http://www.w3.org/TR/media-source/, last retrieved May 12, 2015.

[264] Park, A. and Watson, M. (2013) HTML5 video at Netflix, Netflix Tech Blog. http://techblog.netflix.com/2013/04/html5-video-at-netflix.html, last Referenced May 5, 2015.

[265] ISO/IEC (2012) ISO base media file format – 14496-12. http://www.iso.org/iso/home/store/catalogue_ics.htm, last retrieved May 5, 2015.

[266] ISO/IEC (2003) Information technology – coding of audio-visual objects – Part 14: MP4 file format 14496-14. http://www.iso.org/iso/home/store/catalogue_ics.htm, last retrieved May 12, 2015.

[267] ISO/IEC (2005) Information technology – coding of audio-visual objects – Part 11: Scene description and application engine – 14496-11. http://www.iso.org/iso/home/store/catalogue_ics.htm, last retrieved May 11, 2015.

[268] ISO/IEC (2004) Information technology – coding of audio-visual objects – Part 15: Advanced video coding (AVC) file format -14496-15. http://www.iso.org/iso/home/store/catalogue_ics.htm, last retrieved May 15, 2015.

[269] W3C (2014) ISO BMFF byte stream format. http://www.w3.org/2013/12/byte-stream-format-registry/isobmff-byte-stream-format.html, last retrieved May 22, 2015.

[270] MP4REG (2015) Mp 4 registry codes. http://www.mp4ra.org/atoms.html, last retrieved May 30, 2015.

[271] ETSI (2014) 3GPP: Transparent end-to-end packet switched streaming service (PSS); 3GPP file format (3GP) – 3GPP TS 26.244 version 12.3.0 release. http://www.etsi.org/standards-search, last retrieved May 30, 2015.

[272] Adobe (2010) Adobe flash video file format specification – version 10.1. http://download.macromedia.com/f4v/video_file_format_spec_v10_1.pdf, last retrieved May 27, 2015.

[273] Apple (2015) Introduction to QuickTime File Format specification. https://developer.apple.com/library/mac/documentation/QuickTime/QTFF/QTFFPreface/qtffPreface.html, last retrieved May 28, 2015.

[274] W3C (2008) Synchronized multimedia integration language. http://www.w3.org/TR/2008/REC-SMIL3-20081201/, last retrieved April 13, 2015.

[275] Bodendorf, F., Schertler, M., and Cohen, E. (2005) Producing reusable web-based multimedia presentations. *Interdisciplinary Journal of E-Learning and Learning Objects*, **1** (1), 127–142.

[276] Eidenberger, H. (2003) SMIL and SVG in teaching, IN Electronic Imaging 2004, International Society for Optics and Photonics, pp. 69–80.

[277] Jung, H.T., Lee, J.S., Jeong, Y.J., and Yoon, K.S. (2012) Digital signage system for supporting high quality resolution, in Computing and Convergence Technology (ICCCT), 2012 7th International Conference on, pp. 1277–1280.

[278] Microsoft (2015) IIS smooth streaming manifest structure. https://msdn.microsoft.com/en-us/library/ff436055(v=vs.90).aspx, last retrieved March 20, 2015.

[279] W3C (2009) Scalable vector graphics 1.1. http://www.w3.org/TR/SVG/, last retrieved May 12, 2015.

[280] Rizzo, L. (2012) Revisiting network I/O APIs: the netmap framework. *Communications of the ACM*, **55** (3), 45–51, doi: 10.1145/2093548.2093565.

[281] Everts, T. (2013) Rules for mobile performance optimization. *Communications of the ACM*, **56** (8), 52–59, doi: 10.1145/2492007.2492024.

[282] Dean, J. and Barroso, L.A. (2013) The tail at scale. *Communications of the ACM*, **56** (2), 74–80, doi: 10.1145/2408776.2408794.

[283] Belshe, M. and Peon, R. (2012), SPDY protocol draft-mbelshe-httpbis-spdy-00. https://tools.ietf .org/html/draft-mbelshe-httpbis-spdy-00, last retrieved March 20, 2015.

[284] Thomas, B., Jurdak, R., and Atkinson, I. (2012) SPDYing up the Web. *Communications of the ACM*, **55** (12), 64–73, doi: 10.1145/2380656.2380673.

[285] Zhang, Y. and Ansari, N. (2014) On protocol-independent data redundancy elimination. *IEEE Communications Surveys & Tutorials*, **16** (1), 455–472.

[286] Rosenberg, J., Schulzrinne, H., Camarillo, G., Johnston, A., Peterson, J., Sparks, R., Handley, M., and Schooler, E. (2002) SIP: Session Initiation Protocol, RFC 3261 (Proposed Standard).

[287] Johnston, A., Soroushnejad, M., and Venkataramanan, V. (2015) Shared Appearances of a Session Initiation Protocol (SIP) Address of Record (AOR), RFC 7463 (Proposed Standard).

[288] Rosenberg, J. and Schulzrinne, H. (2006) Guidelines for Authors of Extensions to the Session Initiation Protocol (SIP), RFC 4485 (Informational).

[289] Peterson, J., Jennings, C., and Sparks, R. (2010) Change Process for the Session Initiation Protocol (SIP) and the Real-time Applications and Infrastructure Area, RFC 5727 (Best Current Practice).

[290] Rosenberg, J. (2009) A Hitchhiker's Guide to the Session Initiation Protocol (SIP), RFC 5411 (Informational).

[291] Sinnreich, H. and Johnston, A. (2006) *Internet Communications Using SIP: Delivering VoIP and Multimedia Services with Session Initiation Protocol (Networking = Council)*, John Wiley & Sons, Inc., New York.

[292] Johnston, A. (2014) *SIP: Understanding the Session Initiation Protocol*, 3rd edn, Artech House Publishers.

[293] Swale, R. and Collins, D. (2013) *Carrier Grade Voice Over IP*, McGraw-Hill Professional.

[294] Yergeau, F. (2003) UTF-8, a transformation format of ISO 10646, RFC 3629 (INTERNET STANDARD).

[295] Schulzrinne, H. (2004) The tel URI for Telephone Numbers, RFC 3966 (Proposed Standard).

[296] Freed, N., Klensin, J., and Hansen, T. (2013) Media Type Specifications and Registration Procedures, RFC 6838 (Best Current Practice).

[297] Okumura, S., Sawada, T., and Kyzivat, P. (2011) Session Initiation Protocol (SIP) Usage of the Offer/Answer Model, RFC 6337 (Informational).

[298] Rosenberg, J., Schulzrinne, H., and Kyzivat, P. (2004) Indicating User Agent Capabilities in the Session Initiation Protocol (SIP), RFC 3840 (Proposed Standard).

[299] Sollins, K. (1992) The TFTP Protocol (Revision 2), RFC 1350 (INTERNET STANDARD). Updated by RFCs 1782, 1783, 1784, 1785, 2347, 2348, 2349.

[300] Gulbrandsen, A., Vixie, P., and Esibov, L. (2000) A DNS RR for specifying the location of services (DNS SRV), RFC 2782 (Proposed Standard).

[301] Mealling, M. (2002) Dynamic Delegation Discovery System (DDDS) Part Three: The Domain Name System (DNS) Database, RFC 3403 (Proposed Standard).

[302] Rosenberg, J. and Schulzrinne, H. (2002) Session initiation protocol (SIP): Locating sip servers, RFC 3263 (Proposed Standard).

[303] Rosenberg, J. (2002) The Session Initiation Protocol (SIP) UPDATE Method, RFC 3311 (Proposed Standard).

[304] Rosenberg, J. and Schulzrinne, H. (2002) Reliability of Provisional Responses in Session Initiation Protocol (SIP), RFC 3262 (Proposed Standard).

[305] Holmberg, C., Burger, E., and Kaplan, H. (2011) Session Initiation Protocol (SIP) INFO Method and Package Framework, RFC 6086 (Proposed Standard).

[306] ITU-T (1988) Q.23: Technical features of push-button telephone sets. http://www.itu.int/rec/ T-REC-Q.23-198811-I/en, last retrieved April 13, 2015.

[307] Campbell, B., Rosenberg, J., Schulzrinne, H., Huitema, C., and Gurle, D. (2002) Session Initiation Protocol (SIP) Extension for Instant Messaging, RFC 3428 (Proposed Standard).

[308] Roach, A. (2012) SIP-Specific Event Notification, RFC 6665 (Proposed Standard).

[309] Rosenberg, J., Schulzrinne, H., and Levin, O. (2006) A Session Initiation Protocol (SIP) Event Package for Conference State, RFC 4575 (Proposed Standard).

[310] Rosenberg, J. (2004) A Presence Event Package for the Session Initiation Protocol (SIP), RFC 3856 (Proposed Standard).

[311] Sparks, R. (2003) The Session Initiation Protocol (SIP) Refer Method, RFC 3515 (Proposed Standard).

[312] Sparks, R. (2004) The Session Initiation Protocol (SIP) Referred-By Mechanism, RFC 3892 (Proposed Standard).

[313] ITU (2005) The international public telecommunication numbering plan. http://www.itu.int/rec/dologin_pub.asp?lang=e&id=T-REC-E.164-200502-I!!PDF-E&type=items, last retrieved March 8, 2015.

[314] Bradner, S., Conroy, L., and Fujiwara, K. (2011) The E.164 to Uniform Resource Identifiers (URI) Dynamic Delegation Discovery System (DDDS) Application (ENUM), RFC 6116 (Proposed Standard).

[315] Hoeneisen, B., Mayrhofer, A., and Livingood, J. (2011) IANA Registration of Enumservices: Guide, Template, and IANA Considerations, RFC 6117 (Proposed Standard).

[316] Livingood, J., Pfautz, P., and Stastny, R. (2009) The E.164 to Uniform Resource Identifiers (URI) Dynamic Delegation Discovery System (DDDS) Application for Infrastructure ENUM, RFC 5526 (Informational).

[317] Haberler, M., Lendl, O., and Stastny, R. (2009) Combined User and Infrastructure ENUM in the e164.arpa Tree, RFC 5527 (Informational).

[318] ETSI (2009) Telecommunications and internet converged services and protocols for advanced networking (TISPAN); infrastructure ENUM options for a TISPAN IPX; TR 184 008 version 2.1.1. http://www.etsi.org/standards-search, last retrieved June 2, 2015.

[319] Malas, D. and Livingood, J. (2011) Session PEERing for Multimedia INTerconnect (SPEERMINT) Architecture, RFC 6406 (Informational).

[320] Uzelac, A. and Lee, Y. (2011) Voice over IP (VoIP) SIP Peering Use Cases, RFC 6405 (Informational).

[321] Strand, L. and Leister, W. (2010), A survey of SIP peering, NATO ASI-Architects of secure Networks (ASIGE10). http://www.nr.no/ strand/papers/conf/asige10/Strand-ASurveyOfSIPPeering-ASIGE10.pdf, last retrieved April 13, 2015.

[322] Seedorf, J., Niccolini, S., Chen, E., and Scholz, H. (2011) Session PEERing for Multimedia INTerconnect (SPEERMINT) Security Threats and Suggested Countermeasures, RFC 6404 (Informational).

[323] Rosenberg, J. (2006) A framework for conferencing with the session initiation protocol (SIP), RFC 4353 (Informational).

[324] Barnes, M., Boulton, C., and Levin, O. (2008) A Framework for Centralized Conferencing, RFC 5239 (Proposed Standard).

[325] Dyke, J.V., Burger, E., and Spitzer, A. (2007) Media Server Control Markup Language (MSCML) and Protocol, RFC 5022 (Informational).

[326] Melanchuk, T. (2009) An Architectural Framework for Media Server Control, RFC 5567 (Informational).

[327] Johnston, A. and Levin, O. (2006) Session Initiation Protocol (SIP) Call Control – Conferencing for User Agents, RFC 4579 (Best Current Practice).

[328] Camarillo, G., Ott, J., and Drage, K. (2006) The Binary Floor Control Protocol (BFCP), RFC 4582 (Proposed Standard).

[329] Camarillo, G. (2006) Session Description Protocol (SDP) Format for Binary Floor Control Protocol (BFCP) Streams, RFC 4583 (Proposed Standard).

[330] Camarillo, G., Drage, K., Kristensen, T., Ott, J., and Eckel, C. (2015) The binary floor control protocol (BFCP) – draft-ietf-bfcpbis-rfc4582bis-13. https://tools.ietf.org/html/draft-ietf-bfcpbis-rfc4582bis-13, last retrieved April 13, 2015.

[331] Camarillo, G. and Kristensen, T. (2015) Session description protocol (SDP) format for Binary Floor Control Protocol (BFCP) streams – draft-ietf-bfcpbis-rfc4583bis-11. https://tools.ietf.org/html/draft-ietf-bfcpbis-rfc4583bis-11, last retrieved April 13, 2015.

[332] ITU-T (2014) Role management and additional media channels for ITU-T H.300-series terminals – H.239. http://handle.itu.int/11.1002/1000/12235-en?locatt=format:pdf&auth, last retrieved May 22, 2015.

[333] Barnes, M., Boulton, C., Romano, S., and Schulzrinne, H. (2012) Centralized Conferencing Manipulation Protocol, RFC 6503 (Proposed Standard).

[334] Barnes, M., Miniero, L., Presta, R., Romano, S.P., and Schulzrinne, H. (2010) CCMP: a novel standard protocol for conference management in the XCON framework, in Principles, Systems and Applications of IP Telecommunications, ACM, pp. 91–100.

[335] Novo, O., Camarillo, G., Morgan, D., and Urpalainen, J. (2012) Conference Information Data Model for Centralized Conferencing (XCON), RFC 6501 (Proposed Standard).

[336] Barnes, M., Miniero, L., Presta, R., and Romano, S.P. (2012) Centralized Conferencing Manipulation Protocol (CCMP) Call Flow Examples, RFC 6504 (Informational).

[337] Shekh-Yusef, R. and Barnes, M. (2013) Indication of Conference Focus Support for the Centralized Conferencing Manipulation Protocol (CCMP), RFC 7082 (Informational).

[338] Boulton, C., Melanchuk, T., and McGlashan, S. (2011) Media Control Channel Framework, RFC 6230 (Proposed Standard).

[339] Boulton, C. and Barnes, N. (2012) An XCON client conference control package for the media control channel framework – draft-boulton-xcon-conference-control-package-07. ftp://ftp.bme.hu/documents/internet-drafts/draft-boulton-xcon-conference-control-package-07.txt, last retrieved April 13, 2015.

[340] Rosenberg, J. (2006) A Data Model for Presence, RFC 4479 (Proposed Standard).

[341] Niemi, A. (2004) Session Initiation Protocol (SIP) Extension for Event State Publication, RFC 3903 (Proposed Standard).

[342] Rosenberg, J. (2013) SIMPLE Made Simple: An Overview of the IETF Specifications for Instant Messaging and Presence Using the Session Initiation Protocol (SIP), RFC 6914 (Informational).

[343] Saint-Andre, P. (2011) Extensible Messaging and Presence Protocol (XMPP): Core, RFC 6120 (Proposed Standard).

[344] Saint-Andre, P., Houri, A., and Hildebrand, J. (2014) Interworking between the Session Initiation Protocol (SIP) and the Extensible Messaging and Presence Protocol (XMPP): Architecture, Addresses, and Error Handling, RFC 7247 (Proposed Standard).

[345] ETSI (2007) Universal mobile telecommunications system (UMTS); presence service; architecture and functional description; stage 2; TS 123 141; version 6.10.0. http://www.etsi.org/standards-search, last retrieved June 2, 2015.

[346] ATIS (2009) ATIS service oriented networks (SON) assessment and work plan. http://www.itu.int/dms_pub/itu-t/oth/21/05/T21050000010061PDFE.pdf, last retrieved April 13, 2015.

[347] Rosenberg, J. (2007) Presence Authorization Rules, RFC 5025 (Proposed Standard).

[348] Schulzrinne, H. (2006) CIPID: Contact Information for the Presence Information Data Format, RFC 4482 (Proposed Standard).

[349] Schulzrinne, H. (2006) Timed Presence Extensions to the Presence Information Data Format (PIDF) to Indicate Status Information for Past and Future Time Intervals, RFC 4481 (Proposed Standard).

[350] Lonnfors, M., Leppanen, E., Khartabil, H., and Urpalainen, J. (2008) Presence Information Data Format (PIDF) Extension for Partial Presence, RFC 5262 (Proposed Standard).

[351] Lonnfors, M., Costa-Requena, J., Leppanen, E., and Khartabil, H. (2008) Session Initiation Protocol (SIP) Extension for Partial Notification of Presence Information, RFC 5263 (Proposed Standard).

[352] Thomson, M. and Winterbottom, J. (2015) Representation of Uncertainty and Confidence in the Presence Information Data Format Location Object (PIDF-LO), RFC 7459 (Proposed Standard).

[353] Winterbottom, J., Thomson, M., Barnes, R., Rosen, B., and George, R. (2013) Specifying Civic Address Extensions in the Presence Information Data Format Location Object (PIDF-LO), RFC 6848 (Proposed Standard).

[354] Schulzrinne, H., Gurbani, V., Kyzivat, P., and Rosenberg, J. (2006) RPID: Rich Presence Extensions to the Presence Information Data Format (PIDF), RFC 4480 (Proposed Standard).

[355] Microsoft (2010) Presence protocol specification. http://msdn.microsoft.com/en-us/library/cc431501.aspx, last retrieved April 13, 2015.

[356] Boulton, C., Rosenberg, J., Camarillo, G., and Audet, F. (2011) NAT Traversal Practices for Client-Server SIP, RFC 6314 (Informational).

[357] Audet, F. and Jennings, C. (2007) Network Address Translation (NAT) Behavioral Requirements for Unicast UDP, RFC 4787 (Best Current Practice). Updated by RFC 6888.

[358] Rosenberg, J., Weinberger, J., Huitema, C., and Mahy, R. (2003) STUN – Simple Traversal of User Datagram Protocol (UDP) Through Network Address Translators (NATs), RFC 3489 (Proposed Standard). Obsoleted by RFC 5389.

[359] Rosenberg, J., Mahy, R., Matthews, P., and Wing, D. (2008) Session Traversal Utilities for NAT (STUN), RFC 5389 (Proposed Standard).

[360] MacDonald, D. and Lowekamp, B. (2010) NAT Behavior Discovery Using Session Traversal Utilities for NAT (STUN), RFC 5780 (Experimental).

[361] Perreault, S. and Rosenberg, J. (2010) Traversal Using Relays around NAT (TURN) Extensions for TCP Allocations, RFC 6062 (Proposed Standard).

[362] Rosenberg, J. and Schulzrinne, H. (2003) An Extension to the Session Initiation Protocol (SIP) for Symmetric Response Routing, RFC 3581 (Proposed Standard).

[363] Gurbani, V., Mahy, R., and Tate, B. (2010) Connection Reuse in the Session Initiation Protocol (SIP), RFC 5923 (Proposed Standard).

[364] Cheshire, S. and Krochmal, M. (2013) NAT Port Mapping Protocol (NAT-PMP), RFC 6886 (Informational).

[365] Wing, D., Cheshire, S., Boucadair, M., Penno, R., and Selkirk, P. (2013) Port Control Protocol (PCP), RFC 6887 (Proposed Standard).

[366] Jennings, C., Mahy, R., and Audet, F. (2009) Managing Client-Initiated Connections in the Session Initiation Protocol (SIP), RFC 5626 (Proposed Standard).

[367] Rosenberg, J., Keranen, A., Lowekamp, B.B., and Roach, A.B. (2012) TCP Candidates with Interactive Connectivity Establishment (ICE), RFC 6544 (Proposed Standard).

[368] Rosenberg, J. (2010) Indicating Support for Interactive Connectivity Establishment (ICE) in the Session Initiation Protocol (SIP), RFC 5768 (Proposed Standard).

[369] Rosenberg, J. (2010) Interactive Connectivity Establishment (ICE): A Protocol for Network Address Translator (NAT) Traversal for Offer/Answer Protocols, RFC 5245 (Proposed Standard). Updated by RFC 6336.

[370] Mahy, R., Matthews, P., and Rosenberg, J. (2010) Traversal Using Relays around NAT (TURN): Relay Extensions to Session Traversal Utilities for NAT (STUN), RFC 5766 (Proposed Standard).

[371] Lennox, J. and Schulzrinne, H. (2000) Call Processing Language Framework and Requirements, RFC 2824 (Informational).

[372] Lennox, J., Wu, X., and Schulzrinne, H. (2004) Call Processing Language (CPL): A Language for User Control of Internet Telephony Services, RFC 3880 (Proposed Standard).

[373] W3C (2011) Voice browser call control: CCXML version 1.0. http://www.w3.org/TR/ccxml/, last retrieved June 14, 2015.

[374] Java Community Process (2008) JSR 289: SIP Servlet v1.1. https://jcp.org/en/jsr/detail?id=289, last retrieved May 22, 2015.

[375] Java Community Process (2002) JSR 21: Jaintm JCC specification. https://jcp.org/en/jsr/detail?id=21, last retrieved May 22, 2015.

[376] Microsoft (2015) TAPI 3.0 connection and media services. https://technet.microsoft.com/en-us/library/bb742543.aspx, last retrieved May 22, 2015.

[377] Sisalem, D., Floroiu, J., Kuthan, J., Abend, U., and Schulzrinne, H. (2009) *SIP Security*, John Wiley & Sons, Inc.

[378] Keromytis, A. (2012) A comprehensive survey of voice over ip security research. *IEEE Communications Surveys Tutorials*, **14** (2), 514–537, doi: 10.1109/SURV.2011.031611.00112.

[379] Franks, J., Hallam-Baker, P., Hostetler, J., Lawrence, S., Leach, P., Luotonen, A., and Stewart, L. (1999) HTTP Authentication: Basic and Digest Access Authentication, RFC 2617 (Draft Standard).

[380] Dierks, T. and Rescorla, E. (2008) The Transport Layer Security (TLS) Protocol Version 1.2, RFC 5246 (Proposed Standard).

[381] Audet, F. (2009) The Use of the SIPS URI Scheme in the Session Initiation Protocol (SIP), RFC 5630 (Proposed Standard).

[382] Zimmermann, P., Johnston, A., and Callas, J. (2011) ZRTP: Media Path Key Agreement for Unicast Secure RTP, RFC 6189 (Informational).

[383] Rescorla, E. and Modadugu, N. (2012) Datagram Transport Layer Security Version 1.2, RFC 6347 (Proposed Standard).

[384] Fischl, J., Tschofenig, H., and Rescorla, E. (2010) Framework for Establishing a Secure Real-time Transport Protocol (SRTP) Security Context Using Datagram Transport Layer Security (DTLS), RFC 5763 (Proposed Standard).

[385] Grimes, R.A. (2006) *Professional Windows Desktop & Server Hardening*, John Wiley & Sons, Inc.

[386] Kim, J.Y., Song, W., and Schulzrinne, H. (2006) An enhanced VOIP emergency services prototype, in ISCRAM, Newark, NJ, pp. 102–109.

[387] IEEE (2009) Standard for local and metropolitan area networks – station and media access control connectivity discovery, doi: 10.1109/IEEESTD.2009.5251812.

[388] TIA (2011) ANSI/TIA: Telecommunications ip telephony infrastructure link layer discovery protocol for media endpoint devices, TIA-1057. https://global.ihs.com/tia_free.cfm?rid=TIA, last accessed April 14, 2015.

[389] Schulzrinne, H. (2008) A Uniform Resource Name (URN) for Emergency and Other Well-Known Services, RFC 5031 (Proposed Standard). Updated by RFC 7163.

[390] Holmberg, C. and Sedlacek, I. (2014) URN for Country-Specific Emergency Services, RFC 7163 (Proposed Standard).

[391] Rosen, B., Schulzrinne, H., Polk, J., and Newton, A. (2011) Framework for Emergency Calling Using Internet Multimedia, RFC 6443 (Informational).

[392] Thomson, M. and Winterbottom, J. (2008) Revised Civic Location Format for Presence Information Data Format Location Object (PIDF-LO), RFC 5139 (Proposed Standard).

[393] Camarillo, G. and García-Martín, M. (2008) *The 3G IP Multimedia Subsystem (IMS)*, 3rd edn, John Wiley & Sons, Ltd.

[394] Poikselkä, M., Holma, H., Hongisto, J., Kallio, J., and Toskala, A. (2012) *Voice over LTE (VoLTE)*, John Wiley & Sons, Ltd.

[395] Firestone, S., Ramalingam, T., and Fry, S. (2007) *Voice and Video Conferencing Fundamentals*, Cisco Press.

[396] ITU (2009) Packet-based multimedia communications systems H.323. http://www.itu.int/rec/dologin_pub.asp?lang=e&id=T-REC-H.323-200912-I!!PDF-E&type=items, last retrieved March 8, 2015.

[397] Rose, M. and Cass, D. (1987) ISO Transport Service on top of the TCP Version: 3, RFC 1006 (Standard).

[398] Pouffary, Y. and Young, A. (1997) ISO Transport Service on top of TCP (ITOT), RFC 2126 (Proposed Standard).

[399] ITU-T (2010) Procedures for real-time group 3 facsimile communication over IP networks – T.38. http://www.itu.int/rec/T-REC-T.38-201009-I, last retrieved April 18, 2015.

[400] ITU-T (2008) Information technology – Abstract Syntax Notation One (ASN.1): Specification of basic notation – X.680. http://www.itu.int/rec/T-REC-X.680-200811-I, last retrieved April 17, 2015.

[401] ITU-T (2008) Information technology – Abstract Syntax Notation One (ASN.1): Information object specification – X.681. http://www.itu.int/rec/T-REC-X.681-200811-I, last retrieved April 15, 2015.

[402] ITU-T (2008) Information technology – Abstract Syntax Notation One (ASN.1): Constraint specification – X.682. http://www.itu.int/rec/T-REC-X.682-200811-I, last retrieved April 16, 2015.

[403] ITU-T (2008) Information technology – Abstract Syntax Notation One (ASN.1): Parameterization of ASN.1 specifications – X.683. http://www.itu.int/rec/T-REC-X.683-200811-I, last retrieved April 16, 2015.

[404] ITU (2008) Advanced multimedia system (AMS). http://www.itu.int/ITU-T/studygroups/com16/ams/index.html, last retrieved March 8, 2015.

[405] ITU (1998) ISDN user-network interface layer 3 specification for basic call control Q.931. http://www.itu.int/rec/dologin_pub.asp?lang=e&id=T-REC-Q.931-199805-I!!PDF-E&type=items, last retrieved March 8, 2015.

[406] ITU-T (2011) Generic functional protocol for the support of supplementary services in ITU-T H.323 systems – H.450.1. http://www.itu.int/rec/T-REC-H.450.1-201105-I/en, last retrieved April 17, 2015.

[407] ITU (2009) Call signalling protocols and media stream packetization for packet-based multimedia communication systems H.225.0. http://www.itu.int/rec/dologin_pub.asp?lang=e&id=T-REC-H.225.0-200912-I!!PDF-E&type=items, last retrieved March 8, 2015.

[408] Nortel Networks (2000) A Comparison of H.323v4 and SIP, Tech. Rep. Tdoc S2-000505, Nortel, 3GPP S2, Tokyo, Japan. http://www.cs.columbia.edu/sip/drafts/sip_h323v4.doc, last retrieved April 18, 2015.

[409] Papageorgiou, P. (2001) A Comparison of H. 323 vs SIP, Tech. Rep., University of Maryland at College Park, USA. http://www.cs.umd.edu/pavlos/papers/unpublished/papageorgiou01-comparison.pdf, last retrieved June 2, 2015.

[410] ITU (2007) Data protocols for multimedia conferencing T.120. http://www.itu.int/rec/dologin_pub.asp?lang=e&id=T-REC-T.120-200701-I!!PDF-E&type=items, last retrieved March 8, 2015.

[411] ITU (2008) Multipoint application sharing. http://www.itu.int/rec/dologin_pub.asp?lang=e&id=T-REC-T.128-200806-I!!PDF-E&type=items, last retrieved March 8, 2015.

[412] ITU (1998) Multipoint communication service – service definition T.122. http://www.itu.int/rec/dologin_pub.asp?lang=e&id=T-REC-T.122-199802-I!!PDF-E&type=items, last retrieved March 8, 2015.

[413] ITU (1998) Multipoint communication service protocol specification T.125. http://www.itu.int/rec/dologin_pub.asp?lang=e&id=T-REC-T.125-199802-I!!PDF-E&type=items, last retrieved March 8, 2015.

[414] Chandra, S., Boreczky, J., and Rowe, L.A. (2014) High performance many-to-many intranet screen sharing with DisplayCast. *ACM Transactions on Multimedia Computing, Communications, and Applications*, **10** (2), 19:1–19:22. doi: 10.1145/2534328.

[415] Miao, D., Fu, J., Lu, Y., Li, S., and Chen, C.W. (2014) High frame rate screen video coding for screen sharing applications, in Circuits and Systems (ISCAS), 2014 IEEE International Symposium on, pp. 2157–2160, doi: 10.1109/ISCAS.2014.6865595.

[416] Musa, N., Abiddin, S.Z.Z., and Omar, N. (2013) Shared services in collaborative system configurations. *International Journal of Engineering Research*, **4** (5), 124–129.

[417] Andreasen, F. and Foster, B. (2003) Media Gateway Control Protocol (MGCP) Version 1.0, RFC 3435 (Informational).

[418] Foster, B. and Sivachelvan, C. (2003) Media Gateway Control Protocol (MGCP) Return Code Usage, RFC 3661 (Informational).

[419] Groves, C., Pantaleo, M., Anderson, T., and Taylor, T. (2003) Gateway Control Protocol Version 1, RFC 3525 (Historic).

[420] ITU (2013) Gateway control protocol: H.248.1. https://www.itu.int/rec/dologin_pub.asp?lang=e&id=T-REC-H.248.1-201303-I!!PDF-E&type=items, last retrieved October 30, 2015.

[421] ITU (2008) H.248.1 version 2 implementors guide and its corrigendum 1 (03/2004). http://www.itu.int/rec/T-REC-H.Imp248.1-200805-I/dologin.asp?lang=e&id=T-REC-H.Imp248.1-200805-I!!PDF-E&type=items, last retrieved March 8, 2015.

[422] Crocker, D. and Overell, P. (2008) Augmented BNF for Syntax Specifications: ABNF, RFC 5234 (Standard).

[423] Morneault, K., Rengasami, S., Kalla, M., and Sidebottom, G. (2006) Integrated Services Digital Network (ISDN) Q.921-User Adaptation Layer, RFC 4233 (Proposed Standard).

[424] Rosenberg, J., Schulzrinne, H., and Camarillo, G. (2005) The Stream Control Transmission Protocol (SCTP) as a Transport for the Session Initiation Protocol (SIP), RFC 4168 (Proposed Standard).

[425] Stewart, R. (2007) Stream Control Transmission Protocol, RFC 4960 (Proposed Standard).

[426] Tuexen, M., Ruengeler, I., and Stewart, R. (2013) SACK-IMMEDIATELY Extension for the Stream Control Transmission Protocol, RFC 7053 (Proposed Standard).

[427] Paasch, C. and Bonaventure, O. (2014) Multipath TCP. *Communications of the ACM*, **57** (4), 51–57, doi: 10.1145/2578901.

[428] Wen-Yuan, T., De-Wu, X., and Wei, C. (2009) Research on the transport performance of SCTP, in Computer Network and Multimedia Technology, 2009. CNMT 2009. International Symposium on, pp. 1–3, doi: 10.1109/CNMT.2009.5374598.

[429] Rosenbrock, K., Sanmugam, R., Bradner, S., and Klensin, J. (2001) 3GPP-IETF Standardization Collaboration, RFC 3113 (Informational).

[430] Bradner, S., Calhoun, P., Cuschieri, H., Dennett, S., Flynn, G., Lipford, M., and McPheters, M. (2001) 3GPP2-IETF Standardization Collaboration, RFC 3131 (Informational).

[431] ITU (2006) Ims for next generation networks Y.2021. http://www.itu.int/rec/T-REC-Y.2021/recommendation.asp?lang=en&parent=T-REC-Y.2021-200609-I, last retrieved March 8, 2015.

[432] ETSI (2015) Digital cellular telecommunications system (Phase 2+); IP multimedia call control protocol based on session initiation protocol (SIP) and session description protocol (SDP); Stage 3 (3GPP TS 24.229 version 12.7.0 release 12). http://www.etsi.org/standards-search, last retrieved April 25, 2015.

[433] Fajardo, V., Arkko, J., Loughney, J., and Zorn, G. (2012) Diameter Base Protocol, RFC 6733 (Proposed Standard). Updated by RFC 7075.

[434] Garcia-Martin, M., Belinchon, M., Pallares-Lopez, M., Canales-Valenzuela, C., and Tammi, K. (2006) Diameter Session Initiation Protocol (SIP) Application, RFC 4740 (Proposed Standard).

[435] Khlifi, H. and Gregoire, J.C. (2008) Ims application servers: roles, requirements, and implementation technologies. *IEEE Internet Computing*, **12** (3), 40–51, doi: 10.1109/MIC.2008.57.

[436] Niemi, A., Arkko, J., and Torvinen, V. (2002) Hypertext Transfer Protocol (HTTP) Digest Authentication Using Authentication and Key Agreement (AKA), RFC 3310 (Informational).

[437] Arkko, J., Torvinen, V., Camarillo, G., Niemi, A., and Haukka, T. (2003) Security Mechanism Agreement for the Session Initiation Protocol (SIP), RFC 3329 (Proposed Standard).

[438] Kent, S. and Seo, K. (2005) Security Architecture for the Internet Protocol, RFC 4301 (Proposed Standard). Updated by RFC 6040.

[439] Rosenberg, J. (2009) Obtaining and Using Globally Routable User Agent URIs (GRUUs) in the Session Initiation Protocol (SIP), RFC 5627 (Proposed Standard).

[440] Camarillo, G., Marshall, W., and Rosenberg, J. (2002) Integration of resource management and session initiation protocol (SIP), RFC 3312 (Proposed Standard).

[441] ITU-T (2003) Wideband coding of speech at around 16 kbit/s using adaptive multi-rate wideband (AMR-WB) – G.722.2. http://www.itu.int/rec/T-REC-G.722.2-200307-I/en, last retrieved April 5, 2015.

[442] ETSI (2015) Signalling flows for the session setup in the IP multimedia core network subsystem (IMS) based on session initiation protocol (SIP) and session description protocol (SDP); Stage 3; ETSI TR 124 930, V.12.1.0. http://www.etsi.org/standards-search, last retrieved April, 25 2015.

[443] GSM-Association (2013) IMS profile for voice and SMS, version 7.0. http://www.gsma.com/newsroom/wp-content/uploads/2013/04/IR.92-v7.0.pdf, last retrieved April 28, 2015.

[444] ETSI (2015) Circuit switched fallback in evolved packet system (EPS) ETSI 123 272 V12.5.0, TS23.272. http://www.etsi.org/standards-search, last retrieved April 27, 2015.

[445] ETSI (2015) Single radio voice call continuity (SRVCC), 3GPP, TS 123 216 V12.2.0. http://www.etsi.org/standards-search.

[446] Paisal, V. (2010) Seamless voice over LTE, in Internet Multimedia Services Architecture and Application(IMSAA), 2010 IEEE 4th International Conference on, pp. 1–5, doi: 10.1109/IMSAA.2010.5729423.

[447] Desclaux, P.B.F. (2006) Silver needle in the Skype, BlackHat Europe. http://www.oklabs.net/wp-content/uploads/2012/06/bh-eu-06-Biondi.pdf, last retrieved April 28, 2015.

[448] Azab, A., Watters, P., and Layton, R. (2012) Characterising network traffic for Skype forensics, in Cybercrime and Trustworthy Computing Workshop (CTC), 2012 Third, pp. 19–27, doi: 10.1109/CTC.2012.14.

[449] Valin, J., Vos, K., and Terriberry, T. (2012) Definition of the Opus Audio Codec, RFC 6716 (Proposed Standard).

[450] Rämö, A. and Toukomaa, H. (2011) Voice quality characterization of IETF Opus Codec, in INTERSPEECH, pp. 2541–2544.

[451] Xu, Y., Yu, C., Li, J., and Liu, Y. (2012) Video telephony for end-consumers: measurement study of Google+, iChat, and Skype, in Proceedings of the 2012 ACM conference on Internet measurement conference, ACM, pp. 371–384.

[452] Baset, S.A. and Schulzrinne, H. (2004), An analysis of the Skype peer-to-peer internet telephony protocol. http://arxiv.org/pdf/cs.NI/0412017&ei=IEUnQ53uHLT2YI-ptNAN, last retrieved June 2, 2015.

[453] Yuan, Z., Du, C., Chen, X., Wang, D., and Xue, Y. (2014) SkyTracer: Towards fine-grained identification for Skype traffic via sequence signatures, in Computing, Networking and Communications (ICNC), 2014 International Conference on, IEEE, pp. 1–5.

[454] Sinam, T., Singh, I., Lamabam, P., Devi, N., and Nandi, S. (2014) A technique for classification of VoIP flows in UDP media streams using VoIP signalling traffic, in Advance Computing Conference (IACC), 2014 IEEE International, pp. 354–359, doi: 10.1109/IAdCC.2014.6779348.

[455] W3C (2014) HTML elements video. https://www.w3.org/wiki/HTML/Elements/video, last retrieved May 28, 2015.

[456] Google (2011) Google release of WebRTC source code. http://lists.w3.org/Archives/Public/public-webrtc/2011May/0022.html, last accessed May 23, 2015.

[457] W3C (2015) WebRTC 1.0: Real-time communication between browsers – working draft. http://www.w3.org/TR/webrtc/, last retrieved May 22, 2015.

[458] Loreto, S. and Romano, S.P. (2014) *Real-Time Communication with WebRTC: Peer-to-Peer in the Browser*, O'Reilly Media, Inc.

[459] Valin, J. and Bran, C. (2015), WebRTC audio codec and processing requirements – draft-ietf-rtcweb-audio-08. https://tools.ietf.org/html/draft-ietf-rtcweb-audio-08, last retrieved June 2, 2015.

[460] Roach, A. (2015) WebRTC video processing and codec requirements – draft-ietf-rtcweb-video-05. https://tools.ietf.org/html/draft-ietf-rtcweb-video-05, last retrieved June 2, 2015.

[461] Castillo, I.B., Villegas, J.M., and Pascual, V. (2014) The WebSocket Protocol as a Transport for the Session Initiation Protocol (SIP), RFC 7118 (Proposed Standard).

[462] Salgueiro, G., Pascual, V., Roman, A., and Garcia, S. (2014) Indicating WebSocket Protocol as a Transport in the Session Initiation Protocol (SIP) Common Log Format (CLF), RFC 7355 (Informational).

[463] Bertin, E., Cubaud, S., Tuffin, S., Cazeaux, S., Crespi, N., and Beltran, V. (2013) WebRTC, the day after. http://servicearchitecture.wp.tem-tsp.eu/files/2014/04/1569766727.pdf, last retrieved May 27, 2015.

[464] Bray, T. (2014) The JavaScript Object Notation (JSON) Data Interchange Format, RFC 7159 (Proposed Standard).

[465] Marjou, X., Proust, S., Bogineni, K., et al. (2013) WebRTC audio codecs for interoperability with legacy networks – draft-marjou-rtcweb-audio-codecs-for-interop-01. https://tools.ietf.org/html/draft-marjou-rtcweb-audio-codecs-for-interop-01, last retrieved May 22, 2015.

[466] W3C (2015) ORTC (Object Real-Time Communications) Community Group. https://www.w3.org/community/ortc/#, last retrieved May 22, 2015.

[467] Microsoft (2015) Customizable, ubiquitous real-time communication over the web. http://lists.w3.org/Archives/Public/public-webrtc/2012Oct/att-0076/realtime-media.html, last retrieved May 26, 2015.

[468] McGrew, D. and Rescorla, E. (2010) Datagram Transport Layer Security (DTLS) Extension to Establish Keys for the Secure Real-time Transport Protocol (SRTP), RFC 5764 (Proposed Standard).

[469] Petit-Huguenin, M. and Salgueiro, G. (2015) Multiplexing scheme updates for secure real-time transport protocol (SRTP) extension for datagram transport layer security (DTLS) – draft-ietf-avtcore-rfc5764-mux-fixes-02. https://www.ietf.org/id/draft-ietf-avtcore-rfc5764-mux-fixes-02.txt, last retrieved May 23, 2015.

[470] Stewart, R., Loreto, S., Tuexen, M., and Seggelmann, R. (2013) A new data chunk for stream control transmission protocol – draft-stewart-tsvwg-sctp-ndata-03.txt. https://tools.ietf.org/html/draft-stewart-tsvwg-sctp-ndata-03, last retrieved May 23, 2015.

[471] Jesup, R., Loreto, S., and Tuexen, M. (2015), Webrtc data channel establishment protocol – draft-ietf-rtcweb-data-protocol-09.txt. https://www.ietf.org/id/draft-ietf-rtcweb-data-protocol-09.txt, last retrieved May 23, 2015.

[472] Jesup, R., Loreto, S., and Tuexen, M. (2015), WebRTC data channels – draft-ietf-rtcweb-data-channel-13.txt. https://www.ietf.org/id/draft-ietf-rtcweb-data-channel-13.txt, last retrieved May 23, 2015.

[473] Fette, I. and Melnikov, A. (2011) The WebSocket Protocol, RFC 6455 (Proposed Standard).

[474] Becke, M., Rathgeb, E.P., Werner, S., Rungeler, I., Tüxen, M., and Stewart, R. (2013) Data channel considerations for RTCWeb. *IEEE Communications Magazine*, **51** (4), 34–41.

[475] XMPP Standard Association (2009) Xep-0166: Jingle. http://xmpp.org/extensions/xep-0166.html, last retrieved May 26, 2015.

[476] Ivov, E., Rescorla, E., and Uberti, J. (2015) Trickle ICE: Incremental provisioning of candidates for the interactive – connectivity establishment (ICE) protocol – draft-ietf-mmusic-trickle-ice-02. https://tools.ietf.org/html/draft-ietf-mmusic-trickle-ice-02, last retrieved May 25, 2015.

[477] Nandakumar, S. and Jennings, C. (2015) SDP for the WebRTC – draft-nandakumar-rtcweb-sdp-07. https://tools.ietf.org/html/draft-nandakumar-rtcweb-sdp-07, last retrieved May 27, 2015.

[478] Lennox, J., Westerlund, M., Wu, Q., and Perkins, C. (2015) Sending multiple media streams in a single RTP session – draft-ietf-avtcore-rtp-multi-stream-07. https://tools.ietf.org/html/draft-ietf-avtcore-rtp-multi-stream-07, last retrieved May 18, 2015.

[479] Perkin, C.S., Westerlund, M., and Ott, J. (2015), Web real-time communication (WebRTC): Media transport and use of RTP – draft-ietf-rtcweb-rtp-usage-23. https://www.ietf.org/id/draft-ietf-rtcweb-rtp-usage-23.txt, last retrieved May 26, 2015.

[480] Wenger, S., Chandra, U., Westerlund, M., and Burman, B. (2008) Codec Control Messages in the RTP Audio-Visual Profile with Feedback (AVPF), RFC 5104 (Proposed Standard).

[481] Westerlund, M. and Wenger, S. (2015) RTP topologies – draft-ietf-avtcore-rtp-topologies-update -07. https://tools.ietf.org/html/draft-ietf-avtcore-rtp-topologies-update-07, last rerieved May 27, 2015.

[482] Kaplan, H. and Pascual, V. (2013) A Taxonomy of Session Initiation Protocol (SIP) Back-to-Back User Agents, RFC 7092 (Informational).

[483] Miniero, L., Murillo, S.G., and Pascual, V. (2015), Guidelines to support RTCP end-to-end in back-to-back user agents (B2BUAs) – draft-ietf-straw-b2bua-rtcp-07. https://tools.ietf.org/html/draft-ietf-straw-b2bua-rtcp-07, last retrieved June 2, 2015.

[484] Amirante, A., Castaldi, T., Miniero, L., and Romano, S.P. (2014) Janus: a general purpose WebRTC gateway, in *Proceedings of the Conference on Principles, Systems and Applications of IP Telecommunications*, ACM, New York, inIPTComm'14, pp. 7:1–7:8, doi: 10.1145/2670386.2670389.

[485] Amirante, A., Castaldi, T., Miniero, L., and Romano, S.P. (2013) On the seamless interaction between WebRTC browsers and SIP-based conferencing systems. *IEEE Communications Magazine*, **51** (4), 42–47.

[486] Ericsson (2012) WebRTC interworking with traditional telephony services. http://www.ericsson.com/research-blog/context-aware-communication/webrtc-interworking-traditional-telephony-services/, last retrieved May 25, 2015.

[487] 3GPP (2013) Study on web real time communication (WebRTC) access to IP multimedia subsystem (IMS) – TR 23.701 V123GPP. http://www.3gpp.org/ftp/Specs/archive/23_series/23.701/23701-c00.zip, last retrieved May 23, 2015.

[488] Johnston, A., Yoakum, J., and Singh, K. (2013) Taking on WebRTC in an enterprise. *IEEE Communications Magazine*, **51** (4), 48–54.

[489] Pantos, R.E. (2015) HTTP live streaming, Draft RFC. http://www.ietf.org/id/draft-pantos-http-live-streaming-07.txt, last retrieved April 16, 2015.

[490] Zambelli, A. (2009) IIS smooth streaming technical overview. http://www.microsoft.com/en-us/download/details.aspx?id=17678, last retrieved June 2, 2015.

[491] Various (2015) Comparison of streaming media systems. http://en.wikipedia.org/wiki/Comparison_of_streaming_media_systems, last referenced March 12, 2015.

[492] ISO/IEC (2014) Dynamic adaptive streaming over HTTP (DASH) – 23009-1. http://www.iso.org/iso/home/store/catalogue_ics.htm, last retrieved March 8, 2015.

[493] Hartung, F., Kesici, S., and Catrein, D. (2011) DRM protected dynamic adaptive HTTP streaming. ACM Computer Communication Review, pp. 277–282.

[494] OIPF (2014) Volume 2A – HTTP adaptive streaming. http://www.oipf.tv/docs/OIPF-T1-R2-Specification-Volume-2a-HTTP-Adaptive-Streaming-v2_3-2014-01-24.pdf, last accessed April 16, 2015.

[495] Begen, A., Akgul, T., and Baugher, M. (2011) Watching video over the web: Part 1: Streaming protocols. *IEEE Internet Computing*, **15** (2), 54–63, doi: 10.1109/MIC.2010.155.

[496] Sodagar, I. (2011) The MPEG-DASH standard for multimedia streaming over the internet. *IEEE MultiMedia*, **18** (4), 62–67, doi: 10.1109/MMUL.2011.71.

[497] Lohmar, T., Einarsson, T., Frojdh, P., Gabin, F., and Kampmann, M. (2011) Dynamic adaptive HTTP streaming of live content, in World of Wireless, Mobile and Multimedia Networks (WoWMoM), 2011 IEEE International Symposium on a, pp. 1–8, doi: 10.1109/WoWMoM.2011.5986186.

[498] Adhikari, V.K., Guo, Y., Hao, F., Varvello, M., Hilt, V., Steiner, M., and Zhang, Z.L. (2012) Unreeling netflix: understanding and improving multi-CDN movie delivery, in INFOCOM, 2012 Proceedings IEEE, IEEE, pp. 1620–1628.

[499] Yang, H., Chen, X., Yang, Z., Zhu, X., and Chen, Y. (2014) Opportunities and challenges of HTTP adaptive streaming. *International Journal of Future Generation Communication & Networking*, **7** (6), 165–180.

[500] Akhshabi, S., Narayanaswamy, S., Begen, A.C., and Dovrolis, C. (2012) An experimental evaluation of rate-adaptive video players over HTTP. *Signal Processing: Image Communication*, **27** (4), 271–287.

[501] Jiang, J., Sekar, V., and Zhang, H. (2012) Improving fairness, efficiency, and stability in HTTP-based adaptive video streaming with festive, in Proceedings of the 8th international conference on Emerging networking experiments and technologies, ACM, pp. 97–108.

[502] Akhshabi, S., Anantakrishnan, L., Begen, A.C., and Dovrolis, C. (2012) What happens when HTTP adaptive streaming players compete for bandwidth? in Proceedings of the 22nd International Workshop on Network and Operating System Support for Digital Audio and Video, ACM, pp. 9–14.

[503] Chen, P., Lim, J., Lee, B. et al. (2007) A network-adaptive SVC streaming architecture. 9th International Conference on Advanced Communication Technology, vol. **2**, pp. 955–960.

[504] Leighton, T. (2009) Improving performance on the internet. *Communications for the ACM*, **52** (2), 44–51.

[505] Roverso, R., El-Ansary, S., and Högqvist, M. (2013) On HTTP live streaming in large enterprises, in Proceedings of the ACM SIGCOMM 2013 conference on SIGCOMM, ACM, pp. 489–490.

[506] ETSI (2011) DVB-IPTV Phase 1 specifications; Part 5: Content download service (CDS) – ETSI TS 102 542-5 V1.3.1. http://www.etsi.org/standards-search, last retrieved June 2, 2015.

[507] Liu, R., Yin, H., Cai, X., Zhu, G., Yu, L., Fan, Q., and Xu, J. (2013) Cooperative caching scheme for content oriented networking. *IEEE Communications Letters*, **17** (4), 781–784, doi: 10.1109/LCOMM.2013.020513.121680.

[508] Wilkinson, G. (2014) CDN architecture and sizing, Cisco Live Conference – Published Presentation. http://d2zmdbbm9feqrf.cloudfront.net/2014/eur/pdf/BRKSPV-2160.pdf, last Accessed May 1, 2015.

[509] Dawkins, S. (2013) Content delivery networks interconnection – working group charter, ITEF Approved Document. https://datatracker.ietf.org/doc/charter-ietf-cdni/, last retrieved June 2, 2015.

[510] Zhuang, Z. and Guo, C. (2011) Optimizing CDN infrastructure for live streaming with constrained server chaining, in Parallel and Distributed Processing with Applications (ISPA), 2011 IEEE 9th International Symposium on, pp. 183–188, doi: 10.1109/ISPA.2011.44.

[511] Su, A.J., Choffnes, D.R., Kuzmanovic, A., and Bustamante, F.E. (2006) Drafting behind Akamai (travelocity-based detouring), in *SIGCOMM '06: Proceedings of the 2006 Conference on Applications, Technologies, Architectures, and Protocols for Computer Communications*, ACM, New York, pp. 435–446, doi: 10.1145/1159913.1159962.

[512] Nygren, E., Sitaraman, R.K., and Sun, J. (2010) The Akamai network: a platform for high-performance internet applications. *SIGOPS – Operating Systems Review*, **44** (3), 2–19, doi: 10.1145/1842733.1842736.

[513] Sitaraman, R.K., Kasbekar, M., Lichtenstein, W., and Jain, M. (2014) Overlay networks: an Akamai perspective, in *Advanced Content Delivery, Streaming, and Cloud Services* (eds M. Pathan, R.K. Sitaraman, and D. Robinson, John Wiley & Sons, Inc.

[514] Lederer, S., Mueller, C., Timmerer, C., and Hellwagner, H. (2014) Adaptive multimedia streaming in information-centric networks. *IEEE Network*, **28** (6), 91–96.

[515] Liu, X., Dobrian, F., Milner, H., Jiang, J., Sekar, V., Stoica, I., and Zhang, H. (2012) A case for a coordinated internet video control plane, in Proceedings of the ACM SIGCOMM 2012 Conference

on Applications, Technologies, Architectures, and Protocols for Computer Communication, ACM, pp. 359–370.

[516] Hoßfeld, T., Schatz, R., Biersack, E., and Plissonneau, L. (2013) Internet video delivery in Youtube: from traffic measurements to quality of experience, in *Data Traffic Monitoring and Analysis*, Lecture Notes in Computer Science, vol. **7754** (eds E. Biersack, C. Callegari, and M. Matijasevic), Springer-Verlag, Berlin Heidelberg, pp. 264–301, ISBN: 978-3-642-36783-0, doi: 10.1007/978-3-642-36784-7_11.

[517] Lardinois, F. (2015) Youtube now streams HTML5 video by default, TechCrunch News Website. http://techcrunch.com/2015/01/27/youtube-goes-html5-flash-is-now-deader/, last retrieved June 2, 2015.

[518] Rao, A., Legout, A., Lim, Y.S., Towsley, D., Barakat, C., and Dabbous, W. (2011) Network characteristics of video streaming traffic, in Proceedings of the Seventh Conference on emerging Networking Experiments and Technologies, ACM, p. 25.

[519] Adhikari, V., Jain, S., Chen, Y., and Zhang, Z.L. (2012) Vivisecting Youtube: an active measurement study, in INFOCOM, 2012 Proceedings IEEE, IEEE, pp. 2521–2525.

[520] Netflix (2015) Netflix Q1 2015 earning report, Netflix Investor Newsletter. http://ir.netflix.com/common/download/download.cfm?companyid=NFLX&fileid=821407&filekey=db785b50-90fe-44da-9f5b-37dbf0dcd0e1&filename=Q1_15_Earnings_Letter_final_tables.pdf, last Referenced May 5, 2015.

[521] Sandvine Incorporated ULC (2014) Global internet phenomena report 2H 2014, White Paper. https://www.sandvine.com/downloads/general/global-internet-phenomena/2014/1h-2014-global-internet-phenomena-report.pdf, last Referenced May 5, 2015.

[522] Cockcroft, A. (2011) Netflix cloud architecture, Conference Presentation. http://www.slideshare.net/adrianco/netflix-velocity-conference-2011, last retrieved June 2, 2015.

[523] Antonov, L. (2011) Simulating HTTP LiveStreaming (HLS) – A way to ensure video playback works great, Hulu Tech Blog. http://tech.hulu.com/blog/2011/07/25/simulating-http-livestreaming/, last accessed May 6, 2015.

[524] Dreier, T. (2014) Hulu: 'DASH is definitely the future for US', Article – streamingmedia.com. http://www.streamingmedia.com/Articles/Editorial/Featured-Articles/Hulu-DASH-Is-Definitely-the-Future-for-Us-97468.aspx, last accessed May 6, 2015.

[525] Adhikari, V.K., Guo, Y., Hao, F., Hilt, V., and Zhang, Z.L. (2012) A tale of three CDNs: an active measurement study of Hulu and its CDNs, in Computer Communications Workshops (INFOCOM WKSHPS), 2012 IEEE Conference on, IEEE, pp. 7–12.

[526] Conviva (2015) Viewer experience report, Company Report. http://www.conviva.com/vxr-home/vxr2015/, last accessed May 6, 2015.

[527] OpenSignal (2014) Android Framgentation Visualized, Tech Blog. http://opensignal.com/reports/2014/android-fragmentation/, last accessed May 7, 2015.

[528] ISO/IEC (2013) Picture Transfer Protocol (PTP) for digital still photography devices – ISO 15740. http://www.iso.org/iso/home/store/catalogue_ics.htm, last retrieved June 9, 2015.

[529] Bigioi, P., Susanu, G., Steinberg, E., and Corcoran, P. (2005) PTP/IP-a new transport specification for wireless photography. *IEEE Transactions on Consumer Electronics*, **51** (1), 240–244.

[530] CIPA (2015) PTP-IP standard. http://www.cipa.jp/ptp-ip/contents_e/01guide_e.html, last retrieved June 10, 2015.

[531] I3C (1997) Internet imaging protocol – internet imaging protocol. http://iipimage.sourceforge.net/IIPv105.pdf, last retrieved June 8, 2015.

[532] Microsoft (2005) Media transfer protocol implementation details. http://download.microsoft.com/download/9/8/f/98f3fe47-dfc3-4e74-92a3-088782200fe7/TWMD05003_WinHEC05.ppt, last retrieved June 8, 2015.

[533] Microsoft (2015) Overview windows protocols. https://msdn.microsoft.com/en-us/library/hh128055.aspx, last retrieved June 8, 2015.

[534] Cheshire, S. and Steinberg, D.H. (2006) *Zero Configuration Networking: The Definitive Guide*, O'Reilly Media.

[535] Guttman, E. (2002) Autoconfiguration for IP networking: enabling local communication. *IEEE Internet Computing*, **5** (3), 81–86.

[536] Cheshire, S., Aboba, B., and Guttman, E. (2005) Dynamic Configuration of IPv4 Link-Local Addresses, RFC 3927 (Proposed Standard).

[537] Thomson, S., Narten, T., and Jinmei, T. (2007) IPv6 Stateless Address Autoconfiguration, RFC 4862 (Draft Standard).

[538] Cheshire, S. and Krochmal, M. (2013) Multicast DNS, RFC 6762 (Proposed Standard).

[539] Aboba, B., Thaler, D., and Esibov, L. (2007) Link-local Multicast Name Resolution (LLMNR), RFC 4795 (Informational).

[540] Cheshire, S. and Krochmal, M. (2013) DNS-Based Service Discovery, RFC 6763 (Proposed Standard).

[541] Guttman, E., Perkins, C., Veizades, J., and Day, M. (1999) Service Location Protocol, Version 2, RFC 2608 (Proposed Standard).

[542] Guttman, E. (2002) Vendor Extensions for Service Location Protocol, Version 2, RFC 3224 (Proposed Standard).

[543] Guttman, E., Perkins, C., and Kempf, J. (1999) Service Templates and Service: Schemes, RFC 2609 (Proposed Standard).

[544] Smith, M. and Howes, T. (2006) Lightweight Directory Access Protocol (LDAP): String Representation of Search Filters, RFC 4515 (Proposed Standard).

[545] Zhao, W., Schulzrinne, H., Guttman, E., Bisdikian, C., and Jerome, W. (2004) Remote Service Discovery in the Service Location Protocol (SLP) via DNS SRV, RFC 3832 (Experimental).

[546] Bishaj, B. (2007) Comparison of service discovery protocols, Tech. Rep., Helsinki University of Technology. http://www.tml.tkk.fi/Publications/C/23/papers/Bishaj_final.pdf, last retrieved June 6, 2015.

[547] Wisner, P., Reynolds, F., Källstrom, L., Suoranta, S., Mikkonen, T., and Saarinen, J. (2007) Device and service discovery in home networks, in *Technologies for Home Networking*, Chapter 9 (eds S. Dixit and R. Prasad), John Wiley & Sons, Inc., pp. 153–182.

[548] W3C (2007) Soap version 1.2. http://www.w3.org/TR/soap/, last retrieved June 10, 2015.

[549] UPnP (2008) UPnP device architecture – version 1.1. http://upnp.org/specs/arch/UPnP-arch-DeviceArchitecture-v1.1.pdf, last retrieved June 8, 2015.

[550] UPnP (2015) UPnP AV architecture:2. http://upnp.org/specs/av/UPnP-av-AVArchitecture-v2.pdf, last retrieved June 6, 2015.

[551] UPnP (2015) Contentdirectory:4 service. http://upnp.org/specs/av/UPnP-av-ContentDirectory-v4-Service.pdf, last retrieved June 8, 2015.

[552] UPnP (2015) Standards: device control protocols. http://upnp.org/sdcps-and-certification/standards/sdcps/, last retrieved June 8, 2015.

[553] W3C (2004) XHTML-print. http://www.w3.org/TR/2004/CR-xhtml-print-20040120/, last retrieved June 8, 2015.

[554] UPnP (2012) Telephonyarchitecture:2. http://upnp.org/specs/phone/UPnP-phone-TelephonyArchitecture-v2.pdf, last retrieved June 8, 2015.

[555] Sugano, H., Fujimoto, S., Klyne, G., Bateman, A., Carr, W., and Peterson, J. (2004) Presence Information Data Format (PIDF), RFC 3863 (Proposed Standard).

[556] UPnP (2014) Multiscreenarchitecture:1. http://upnp.org/specs/ms/UPnP-ms-MultiScreenArchitecture-v1.pdf, last retrieved May 11, 2015.

[557] Goland, Y.Y., Chai, T., Leach, P., Gu, Y., and Albright, S. (1999) Simple service discovery protocol/1.0 operating without an arbiter – <draft-cai-ssdp-v1-03.txt>. https://tools.ietf.org/html/draft-cai-ssdp-v1-03, last retrieved June 10, 2015.

[558] Goland, Y.Y. (1999) Multicast and Unicast UDP HTTP messages – <draft-goland-http-udp-01.txt>. https://tools.ietf.org/html/draft-goland-http-udp-01, last retrieved June 10, 2015.

[559] UPnP (2013) Avtransport:3 service. http://upnp.org/specs/av/UPnP-av-AVTransport-v3-Service.pdf, last retrieved June 8, 2015.

[560] ISO (2005) ISO/IEC 21000-2:2005 information technology – multimedia framework (MPEG-21) – part 2: Digital item declaration. http://www.iso.org/iso/home/store/catalogue_tc/catalogue_detail.htm?csnumber=41112, last retrieved June 8, 2015.

[561] UPnP (2013) Mediarenderer:3 device template. http://upnp.org/specs/av/UPnP-av-MediaRenderer-v3-Device.pdf, last retrieved June 6, 2015.

[562] UPnP (2013) Mediaserver:4 device. http://upnp.org/specs/av/UPnP-av-MediaServer-v4-Device.pdf, last retrieved June 6, 2015.

[563] UPnP (2013) ConnectionManager3 service. http://upnp.org/specs/av/UPnP-av-ConnectionManager-v3-Service.pdf, last retrieved June 6, 2015.

[564] Curbera, F., Duftler, M., Khalaf, R., Nagy, W., Mukhi, N., and Weerawarana, S. (2002) Unraveling the web services web: an introduction to SOAP, WSDL, and UDDI. *IEEE Internet Computing*, **6** (2), 86–93.

[565] UPnP (2013) Renderingcontrol:3 service. http://upnp.org/specs/av/UPnP-av-RenderingControl-v3-Service.pdf, last retrieved June 8, 2015.

[566] Diehl, E. (2012) *Securing Digital Video: Techniques for DRM and Content Protection*, Springer Science & Business Media.

[567] Kalofonos, D.N. and Saaranen, M. (2007) A network connectivity power-saving mechanism for mobile devices in DLNA home networks, in 4th IEEE Consumer Communications and Networking Conference, CCNC, pp. 645–649.

[568] Kum, S.W., Kim, K.W., Lee, Y.J., Lim, T.B., and Lee, S.P. (2011) Design and implementation of aggregation framework for UPNP/DLNA media server, in Computer Communications and Networks (ICCCN), 2011 Proceedings of 20th International Conference on, pp. 1–4, doi: 10.1109/ICCCN.2011.6005817.

[569] Heredia, E., Sturm, O., Durzi, G., Greenlee, M., Davim, J., Rainer, R., Turban, E., Splettstoesser-Hogeterp, I., Sanchez-Rodriguez, C., Horstmann, C. et al. (2011) *An Introduction to the DLNA Architecture: Network Technologies for Media Devices*, John Wiley & Sons, Inc.

[570] IEC (2013) Digital living network alliance (DLNA) home networked device interoperability guidelines – Part 1: Archtecture and protocols; IEC 62481-1:2013. https://webstore.iec.ch/publication/7083, last retrieved June 7, 2015.

[571] Whiteaker, J., Schneider, F., Teixeira, R., Diot, C., Soule, A., Picconi, F., and May, M. (2012) Expanding home services with advanced gateways. *ACM SIGCOMM Computer Communication Review*, **42** (5), 37–43.

[572] ETSI (2009) Customer network gateway (CNG) – architecture and reference points; TS 185 003 V2.3.1. http://www.etsi.org/standards-search, last retrieved June 5, 2015.

[573] ETSI (2009) Protocols for advanced networking (TISPAN); customer devices architecture and reference points – 185 006 V2.3.1. http://www.etsi.org/standards-search, last accessed June 6, 2015.

[574] Home-Gateway-Initiative (2009) IMS enabled HG HGI-GD006-R2. http://www.homegatewayinitiative.org/publis/HGI-GD006-R2.pdf, last retrieved June 10, 2015.

[575] 3GPP (2013) Study on wireless local area network (WLAN) – 3GPP radio interworking TR 37.834 V12.0.0. http://www.3gpp.org/ftp/Specs/archive/37_series/37.834/37834-c00.zip, last retrieved June 5, 2015.

[576] Home-Gateway-Initiative (2009) Requirement for the HG interworking with an external NT – HGI-RD007-R2. http://www.homegatewayinitiative.org/publis/HGI-RD007-R2.pdf, last retrieved June 10, 2015.

[577] Yiakoumis, Y., Yap, K.K., Katti, S., Parulkar, G., and McKeown, N. (2011) Slicing home networks, in Proceedings of the 2nd ACM SIGCOMM Workshop on Home Networks, ACM, pp. 1–6.

[578] Krishnan, S.S. and Sitaraman, R.K. (2013) Video stream quality impacts viewer behavior: inferring causality using quasi-experimental designs. *IEEE/ACM Transactions on Networking*, **21** (6), 2001–2014.

[579] Maisonneuve, J., Deschanel, M., Heiles, J., Li, W., Liu, H., Sharpe, R., and Wu, Y. (2009) An overview of iptv standards development. *IEEE Transactions on Broadcasting*, **55** (2), 315–328, doi: 10.1109/TBC.2009.2020451.

[580] Reimers, U. (2011) *DVB: The Family of International Standards for Digital Video Broadcasting*, Springer-Verlag.

[581] ETSI (2014) Digital Video Broadcasting (DVB); transport of MPEG-2 TS based DVB services over IP based networks – ETSI TS 102 034 V1.5.2. http://www.etsi.org/standards-search.

[582] SMPTE (2007) Forward error correction for real-time video/audio transport over IP networks – SMPTE 2022-1-2007. http://standards.smpte.org/content/978-1-61482-615-6/st-2022-1-2007/SEC1.body.pdf+html, last accessed May 7, 2015.

[583] Begen, A. (2010) RTP payload format for 1-D interleaved parity forward error correction (FEC), RFC 6015.

[584] Begen, A. and Stockhammer, T. (2012) Guidelines for Implementing Digital Video Broadcasting – IPTV (DVB-IPTV) Application-Layer Hybrid Forward Error Correction (FEC) Protection, RFC 6683 (Informational).

[585] ETSI (2015) Multimedia broadcast/multicast service: Protocols and codecs – TS 126 346 vers. V.12.5. http://www.etsi.org/standards-search, last accessed September 22, 2015.

[586] Watson, M., Luby, M., and Vicisano, L. (2007) Forward Error Correction (FEC) Building Block, RFC 5052 (Proposed Standard).

[587] Rey, J., Leon, D., Miyazaki, A., Varsa, V., and Hakenberg, R. (2006) RTP Retransmission Payload Format, RFC 4588 (Proposed Standard).

[588] ETSI (2010) DVB-IPTV Phase 1 specifications; Part 3: Error recovery; sub-part 3: Retransmission (RET) – TS 102 542-3-3 V1.3.1. http://www.etsi.org/standards-search, last retrieved May 2, 12015.

[589] Ott, J., Chesterfield, J., and Schooler, E. (2010) RTP Control Protocol (RTCP) Extensions for Single-Source Multicast Sessions with Unicast Feedback, RFC 5760 (Proposed Standard).

[590] Begen, A. (2011) RTP Control Protocol (RTCP) Port for Source-Specific Multicast (SSM) Sessions, RFC 6128 (Proposed Standard).

[591] Manzato, D. and da Fonseca, N. (2013) A survey of channel switching schemes for IPTV. *IEEE Communications Magazine*, **51** (8), doi: 10.1109/MCOM.2013.6576349.

[592] Ramos, F. (2013) Mitigating IPTV zapping delay. *IEEE Communications Magazine*, **51** (8), doi: 10.1109/MCOM.2013.6576350.

[593] Steeg, B.V., Begen, A., Caenegem, T.V., and Vax, Z. (2011) Unicast-Based Rapid Acquisition of Multicast RTP Sessions, RFC 6285 (Proposed Standard).

[594] Begen, A., Wing, D., and Caenegem, T.V. (2011) Port Mapping between Unicast and Multicast RTP Sessions, RFC 6284 (Proposed Standard).

[595] ETSI (2009) IP Datacast over DVB-H: Content delivery protocols – TS 102 472 V1.3.1. http://www.etsi.org/standards-search, last retrieved June 2, 2015.

[596] Paila, T., Luby, M., Lehtonen, R., Roca, V., and Walsh, R. (2004) FLUTE – File Delivery over Unidirectional Transport, RFC 3926 (Experimental). Obsoleted by RFC 6726.

[597] Luby, M., Gemmell, J., Vicisano, L., Rizzo, L., and Crowcroft, J. (2002) Asynchronous Layered Coding (ALC) Protocol Instantiation, RFC 3450 (Experimental). Obsoleted by RFC 5775.

[598] Luby, M., Watson, M., and Vicisano, L. (2010) Asynchronous Layered Coding (ALC) Protocol Instantiation, RFC 5775 (Proposed Standard).

[599] Paila, T., Walsh, R., Luby, M., Roca, V., and Lehtonen, R. (2012) FLUTE – File Delivery over Unidirectional Transport, RFC 6726 (Proposed Standard).

[600] ETSI (2014) IP Multimedia Subsystem (IMS) based Packet Switch Streaming (PSS) and multimedia broadcast/multicast service (MBMS) user service protocols – TS 126 237 V12.1.0. http://www.etsi.org/standards-search, last retrieved May 4, 2015.

[601] Balabanian, V., Casey, L., Greene, N., and Adams, C. (1996) An introduction to digital storage media-command and control. *IEEE Communications Magazine*, **34** (11), 122–127, doi: 10.1109/35.544202.

[602] ETSI (2003) Digital Video Broadcasting (DVB); implementation guidelines for data broadcasting – TR 101 202 V1.2.1. http://www.etsi.org/standards-search, last retrieved June 2, 2015.

[603] ETSI (2008) Digital Video Broadcasting (DVB); specification for system software update in DVB systems – TS 102 006 V1.3.2. http://www.etsi.org/standards-search, last retrieved June 2, 2015.

[604] Henning, M. (2006) The rise and fall of CORBA. *Queue*, **4** (5), 28–34.

[605] Song, H.H., Ge, Z., Mahimkar, A., Wang, J., Yates, J., and Zhang, Y. (2011) Analyzing IPTV set-top box crashes, in Proceedings of the 2nd ACM SIGCOMM Workshop on Home Networks, ACM, pp. 31–36.

[606] ETSI (2010) Carriage of Broadband Content Guide (BCG) information over Internet Protocol (IP); TS 102 539; version 1.3.1. http://www.etsi.org/standards-search, last retrieved June 2, 2015.

[607] ETSI (2012) Broadcast and On-Line Services: Search, select, and rightful use of content on personal storage systems ("TV-Anytime"); Part 3: Metadata; Sub-part 1: Phase 1 – metadata schemas; TS 102 822-3-1; version 1.8.1. http://www.etsi.org/standards-search, last retrieved June 2, 2015.

[608] ETSI (2012) Digital Video Broadcasting (DVB); carriage and signalling of TV-Anytime information in DVB transport streams; TS 102 323; version 1.5.1. http://www.etsi.org/standards-search, last retrieved June 2, 2015.

[609] Gil Solla, A. and Sotelo Bovino, R.G. (2013) *TV-Anytime*, Springer-Verlag.

[610] Earnshaw, N., Aoki, S., Ashley, A., and Kameyama, W. (2005) The TV-Anytime Content Reference Identifier (CRID), RFC 4078 (Informational).

[611] McRoberts, M. and Adolf, A. (2013), Uniform Resource Identifier (URI) scheme for Digital Video Broadcasting (DVB) programme resources – draft-mcroberts-uri-dvb-10. https://tools.ietf.org/html/draft-mcroberts-uri-dvb-10, last retrieved May 5, 2015.

[612] ETSI (2010) Remote management and firmware update system for DVB IPTV services (phase 2) – etsi ts 102 824. http://www.etsi.org/standards-search, last retrieved June 2, 2015.

[613] Broadband-Forum (2007) CPE wan management protocol V1.1 TR-69. http://www.broadband-forum.org/technical/download/TR-069_Amendment-2.pdf, last retrieved June 2, 2015.

[614] Handley, M., Perkins, C., and Whelan, E. (2000) Session Announcement Protocol, RFC 2974 (Experimental).

[615] Wey, J., Luken, J., and Heiles, J. (2009) Standardization Activities for IPTV Set-Top Box Remote Management. *IEEE Internet Computing*, **13** (3), 32–39, doi: 10.1109/MIC.2009.69.

[616] Stockhammer, T. and Heiles, J. (2009) DVB-IPTV content download services – IPTV services anytime and anywhere, in Telecommunications, 2009. ConTEL 2009. 10th International Conference on, pp. 413–420.

[617] Chaudhuri, R. (2008) End to end IPTV design and implementation, how to avoid pitfalls, in Telecommunications Network Strategy and Planning Symposium, 2008. Networks 2008. The 13th International, pp. 1–119, doi: 10.1109/NETWKS.2008.4763765.

[618] Wijnands, I.J., De Ghein, L., Enyedi, G., Csaszar, A., and Tantsura, J. (2014) Tree notification to improve multicast fast reroute – draft-wijnands-rtgwg-mcast-frr-tn-02. https://tools.ietf.org/html/draft-wijnands-rtgwg-mcast-frr-tn-02, last retrieved May 8, 2015.

[619] Lavrell, N. (2014) Integrating the Google Cast Technology in a Second-screen Solution, Master's thesis, Linköpings universiteit. http://www.diva-portal.org/smash/get/diva2:728193/FULLTEXT01.pdf, last retrieved May 10, 2015.

[620] Netflix (2015) Dial discovery and launch protocol specification, version 1.7.1. http://www.dial-multiscreen.org/dial-protocol-specification/DIAL-2ndScreenProtocol-1.7.1.pdf?attredirects=0&d=1, last retrieved May 10, 2015.

[621] DVB (2014) Companion screens and streams. https://www.dvb.org/resources/public/standards/a167-1_dvb-css_spec.pdf, last retrieved March 12, 2015.

[622] DVB (2014) Companion screens and streams; Part 2: Content identification and media synchronisation. https://www.dvb.org/resources/public/standards/a167-2_dvb-css_part_2-content-id-and-media-sync.zip, last retrieved May 22, 2015.

[623] DVB (2014) Companion screens and streams; Part 3: Discovery. https://www.dvb.org/resources/public/standards/a167-3_dvb-css_discovery.pdf, last retrieved May 22, 2015.

[624] Mikityuk, A., Seifert, J.P., and Friedrich, O. (2013) The virtual Set-Top Box: On the shift of IPTV service execution, service amp; UI composition into the cloud, in Intelligence in Next Generation Networks (ICIN), 2013 17th International Conference on, pp. 1–8, doi: 10.1109/ICIN.2013.6670887.

[625] Mikityuk, A., Seifert, J.P., and Friedrich, O. (2014) Paradigm shift in IPTV service generation: Comparison between locally- and cloud-rendered IPTV UI, in Consumer Communications and Networking Conference (CCNC), 2014 IEEE 11th, pp. 205–212, doi: 10.1109/CCNC.2014.6866572.

[626] Kalker, T., Samtani, R., and Wang, X. (2012) Ultraviolet: redefining the movie industry? *IEEE Multimedia*, **19** (1), 7, doi: 10.1109/MMUL.2012.14.

[627] ETSI (2011) Telecommunications and internet converged services and protocols for advanced networking (TISPAN); IPTV architecture; IPTV functions supported by the IMS subsystem; TS 182 027; version 3.5.1. http://www.etsi.org/standards-search, last retrieved April 21, 2015.

[628] ETSI (2011) Protocols for advanced networking (TISPAN); NGN integrated IPTV subsystem architecture – TS 182 028 V3.5.1. http://www.etsi.org/standards-search, last retrieved May 8, 2015.

[629] Mikoczy, E., Sivchenko, D., Xu, B., and Moreno, J. (2008) IPTV systems, standards and architectures: Part II – IPTV services over IMS: Architecture and standardization. *IEEE Communications Magazine*, **46** (5), 128–135, doi: 10.1109/MCOM.2008.4511659.

[630] Bach, T., Zimmermann, J., Maruschke, M., Hänsge, K., and Baumgart, M. (2014) Combination of IMS-based IPTV services with WebRTC, in ICCGI 2014, The Ninth International Multi-Conference on Computing in the Global Information Technology, pp. 140–145.

[631] ETSI (2010) Digital Video Broadcasting (DVB); technical specification for DVB services in the home network phase 1 – TS 102 905 V1.1.1. http://www.etsi.org/standards-search, last retrieved June 2, 2015.

[632] ITU-T (2011) Digital Signage: the right information in all the right places. http://www.itu.int/oth/T2301000015/en, last retrieved May 16, 2015.

[633] ITU-T (2012) Digital Signage: Service requirements and IPTV-based architecture – recommendation H.780. https://www.itu.int/rec/dologin_pub.asp?lang=e&id=T-REC-H.780-201206-I!!PDF-E&type=items, last retrieved May 16, 2015.

[634] Lundström, L.I. (2008) *Digital Signage Broadcasting – Content Management and Distribution Techniques*, Focal Press, Elsevier.

[635] W3C (2000) Introduction and overview of W3C speech interface framework. http://www.w3.org/TR/voice-intro/, last retrieved May 14, 2015.

[636] W3C (2010) Voice extensible markup language (voicexml) 3.0. http://www.w3.org/TR/voicexml30/, last retrieved May 13, 2015.

[637] Froumentin, M. (2005) The W3C Speech Interface Framework Media Types: application/voicexml+xml, application/ssml+xml, application/srgs, application/srgs+xml, application/ccxml+xml, and application/pls+xml, RFC 4267 (Informational).

[638] Larson, J.A. (2002) *Voicexml: Introduction to Developing Speech Applications: Version*, Prentice Hall PTR, Upper Saddle River, NJ.

[639] Lucas, B. (2000) Voicexml. *Communications of the ACM*, **43** (9), 53–57.

[640] Larson, J. (2003) Voicexml and the W3C speech interface framework. *IEEE MultiMedia*, **10** (4), 91–93, doi: 10.1109/MMUL.2003.1237554.

[641] Shanmugham, S., Monaco, P., and Eberman, B. (2006) A Media Resource Control Protocol (MRCP) Developed by Cisco, Nuance, and Speechworks, RFC 4463 (Informational).

[642] Burnett, D. and Shanmugham, S. (2012) Media Resource Control Protocol Version 2 (MRCPv2), RFC 6787 (Proposed Standard).

[643] W3C (2010) Speech Synthesis Markup Language (SSML) version 1.1. http://www.w3.org/TR/speech-synthesis11/, last retrieved May 15, 2015.

[644] W3C (2004) Speech recognition grammar specification version 1.0. http://www.w3.org/TR/speech-grammar/#S2.7, last retrieved May 15, 2015.

[645] W3C (2007) Semantic Interpretation for Speech Recognition (SISR) version 1.0. http://www.w3.org/TR/semantic-interpretation/, last retrieved May 15, 2015.

[646] Portman, L., Lum, H., Eckel, C., Johnston, A., and Hutton, A. (2015), Session recording protocol – draft-ietf-siprec-protocol-16. https://www.ietf.org/id/draft-ietf-siprec-protocol-16.txt, last retrieved May 15, 2015.

[647] Hutton, A., Portman, L., Jain, R., and Rehor, K. (2014) An Architecture for Media Recording Using the Session Initiation Protocol, RFC 7245 (Informational).

[648] Koole, G. and Mandelbaum, A. (2002) Queueing models of call centers: an introduction. *Annals of Operations Research*, **113** (1–4), 41–59.

[649] Koole, G. (2013) *Call Center Optimizations*, MG Books, Amsterdam.

[650] Wu, W., Arefin, A., Rivas, R., Nahrstedt, K., Sheppard, R., and Yang, Z. (2009) Quality of experience in distributed interactive multimedia environments: toward a theoretical framework, in *Proceedings of the Seventeen ACM International Conference on Multimedia*, ACM, New York, MM'09, pp. 481–490.

[651] Romanow, A., Botzko, S., Duckworth, M., and Even, R. (2014) Use Cases for Telepresence Multistreams, RFC 7205 (Informational).

[652] Duckworth, M., Pepperell, A., and Wenger, S. (2015), Framework for telepresence multi-streams – draft-ietf-clue-framework-22.txt. https://www.ietf.org/id/draft-ietf-clue-framework-22.txt, last retrieved June 2, 2015.

[653] Romanow, A., Andreasen, F., and Krishna, A. (2012) Investigation of Session Description Protocol (SDP) usage for controlling multiple streams for telepresence (CLUE) – draft-romanow-clue-sdp-usage-02. https://tools.ietf.org/html/draft-romanow-clue-sdp-usage-02, last retrieved June 2, 2015.

[654] Szigeti, T., McMenamy, K., Saville, R., and Glowacki, A. (2009) *Cisco TelePresence Fundamentals*, Cisco Press.

[655] IMTC (2010) Telepresence Interoperability Protocol ("TIP"). http://portal.imtc.org/DesktopModules/Inventures_Document/FileDownload.aspx?ContentID=21240, last retrieved May 22, 2015.

[656] Graef, I. (2014) How can software interoperability be achieved under European competition law and related regimes? *Journal of European Competition Law & Practice*, **5** (1), 6–19.

[657] Cisco (2012) Basic video concepts. http://www.cisco.com/c/en/us/td/docs/voice_ip_comm/uc_system/design/guides/videodg/vidguide/basics.pdf, last retrieved May 17, 2015.

[658] Cisco (2011) Configuring mediatrace. http://www.cisco.com/c/en/us/td/docs/ios/media_monitoring/configuration/guide/15_1m_and_t/mm_15_1m_and_t/mm_mediatrace.html, last retrieved May 22, 2015.

[659] Parson, K. (2013) Liveaction and Cisco medianet. https://www.youtube.com/watch?v=UT0McI4Jooc, last retrieved May 21, 2015.

[660] IEEE (2000) Amendment to carrier sense multiple access with collision detection (CSMA/CD) access method and physical layer specifications-aggregation of multiple link segments. *IEEE Std 802.3ad-2000*, pp. i–173, doi: 10.1109/IEEESTD.2000.91610.

[661] Nadas, S. (2010) Virtual Router Redundancy Protocol (VRRP) Version 3 for IPv4 and IPv6, RFC 5798 (Proposed Standard).

Index

Multimedia Networks: Protocols, Design, and Applications, First Edition. Hans W. Barz and Gregory A. Bassett.
© 2016 John Wiley & Sons, Ltd. Published 2016 by John Wiley & Sons, Ltd.

Printed and bound by CPI Group (UK) Ltd, Croydon, CR0 4YY

27/10/2024

14580359-0004